ENGINEERING GEOLOGY

ENGINEERING GEOLOGY

Editor

Jabril Hakim

scitus
academics

Engineering Geology

Edited by **Jabril Hakim**

This Book is Printed on Acid-Free Paper. ♾

ISBN: 9781681178950

Library of Congress Control Number: 2017948923

616, Corporate Way, Suite 2, 4766,

Valley Cottage, NY 10989

www.scitusacademics.com

Printed in United States of America.

Preface

Engineering geology is the application of the geology to engineering study for the purpose of assuring that the geological factors regarding the location, design, construction, operation and maintenance of engineering works are recognized and accounted for. In recent years, advances in technology and knowledge of geology have resulted in the accessibility of civil engineering project and high resolution to identify. Producing the earthquake in sea triggered the tsunami. An earthquake may serve as a warning that a tsunami is coming, and so may a rapid fall or rise in coastal waters. Earthquakes are an unavoidable natural hazard, but proper planning and a well-informed and well prepared public can reduce their impact. Construction of dam, tunnel, reservoir and other civil engineering projects under the consideration of safety precaution, and using material are more stable, durable and economical due to the knowledge of geology. The field of Geological Engineering includes a wide range of activities including the geological characterization of complex foundations of major buildings and structures, development of natural resources (mining, water, hydroelectricity, forestry, oil and gas), investigation and assessment of groundwater movement and quality, the engineering safety of major infrastructure (dams, reservoirs, offshore drilling platforms, pipelines, roads and railways), and the assessment of geohazard risk (landslides, earthquakes, volcanoes, and the stability of natural dams). It also contains aspects of such fields as project finance and insurance, land-use planning, forensic geological engineering, and the application of geological knowledge to the repair and preservation of cultural heritage sites.

This volume 'Engineering Geology' aims to understand the structure of the Earth and its surface features, what causes earthquakes and tsunamis, and why volcanoes form and erupt. It discusses about minerals, which form the building blocks of rocks, and how rocks are made and destroyed. It presents emphasis on applied geomorphology and structural geology, applied geophysics and geochemistry, environmental geology and hydrogeology, land use planning, natural hazards, remote sensing techniques, soil and rock mechanics and applied geotechnical engineering. This book will appeal to academic scientists, industry and applied researchers, and policy and decision makers.

Table of Contents

CHAPTER 1

A General Analytical Solution for the Evolution of Cliffs Accounting for Strength Degradation, Seismic Action, Formation of Tension Cracks and Seepage

C. Voulgari, S. Utili

School of Engineering, University of Warwick, CV4 7AL, Coventry, UK

ABSTRACT

The evolution of natural slopes over time is ruled by several concurrent physical phenomena, namely the strength of its component geomaterials and their weakening over time due to weathering processes, the occurrence of seismic events, seepage and the formation of tension cracks. The paper presents analytical solutions obtained considering a succession of discrete failure events (landslides) progressively altering the slope morphology over time. The model, derived in the framework of limit analysis assuming plane strain conditions, provides a tool for the assessment of whether manufacts and/or infrastructures located on a slope subject to various natural degradation phenomena will be affected by the occurrence of failures.

Unlike current empirical and semi-empirical models of slope evolution, the analytical solution that is here presented is derived by applying principles of soil and rock mechanics, therefore it is of general validity, so that no ad-hoc calibration against past observations of the evolving slope is needed. This analytical technique only requires knowledge of the (geotechnical) parameters characterising the geomaterials comprising the slope of interest, namely angle of shearing resistance, ϕ, cohesion, c, tensile strength, unit weight, together with knowledge of the relevant seepage scenarios, strength degradation processes, and seismic events likely to occur.

Results show that that earthquake loading and seepage can substantially decrease slope stability, increase the volume of material sliding away during each landslide event and alter the evolution of the slope over time, while tensile strength exhibits a less strong influence especially as strength degradation progresses. Dimensionless ready-to-use charts are provided for the benefit of practitioners.

Keywords: Limit analysis; Slope stability; Successive failures; Tension crack; Seismic; Slope morphology

1. INTRODUCTION

Morphological evolution of cliffs is a traditional subject in engineering geology and geomorphology (Carson and Kirkby, 1972; Hutchinson, 1969; Hutchinson, 1970; Hutchinson, 2001; Selby, 1982 ; Parsons, 2002). Modelling the progressive retreat of cliffs has recently received considerable attention by the engineering community due to increasing coastal erosive processes caused by climate change and increased environmental awareness at national and European level (Bray and Hooke, 1997). Also the insurance industry needs reliable models for the predictions of the amount of cliff retreat over time for residential buildings located in exposed areas whereas local authorities and decision makers need to know the level of risk faced by residential buildings and public infrastructure (e.g. coastal roads, pedestrian footpaths, car parks, etc.).

Cliff evolution is caused and/or accelerated by several physical agents (Arkin and Michaeli, 1985; de Lange and Moon, 2005; Briaud, 2008 ; Collins and Sitar, 2011). Key drivers of slope instability are seismic action (Chen and Liu, 1990; Ling and Leshchinsky, 1995; Loukidis *et al.*, 2003; Wasowski *et al.*, 2011; Rathje and Antonakos, 2011 ; Yang and Chi, 2014Tsai and Chien, 2016), rainfall and climatic variations (Leroueil, 2001; Frayssines and Hantz, 2006; Take and Bolton, 2011; Conte and Troncone, 2012 ; Springman *et al.*, 2013), weathering (Yokota and Iwamatsu, 2000 ; Hachinohe *et al.*, 2000), crack formation (Baker, 1981 ; Hales and Roering, 2007) and wave action for seacliffs (Benumof *et al.*, 2000 ; de Lange and Moon, 2005). In this paper we shall consider all these actions but the last one. The first stability analyses of slopes were based on limit equilibrium methods (Espinoza et al., 1994). However, they present several shortcomings (Duncan, 1996) one of which is the fact that the solution is neither a lower nor an upper bound of the true collapse load. The use of numerical methods such as finite element method (Zheng *et al.*, 2005; Huang and Jia, 2009 ; Potts and Zdravković, 2001) and material point method (Yerro et al., 2015) to provide approximate solutions to the slope stability problem is also increased in the latest decades. The Discrete Element Method has also been employed (Utili and Crosta, 2011b ; Camones *et al.*, 2013Utili et al., 2015) and also for 3D analyses of the stability of rock slopes (Boon et al., 2014) which has been made possible by computational advances in the DEM contact detection algorithms to deal with polyhedral blocks (Boon *et al.*, 2012 ; Boon *et al.*, 2013) and new algorithms for rock slicing (Boon et al., 2015). However, in cases of rather uniform slopes subject to a few cracks, such a numerical approach is not justified.

This paper aims to develop an analytical solution describing the morphological evolution of natural cliffs with limited tensile strength subject to progressive retreat induced by ground strength degradation, for instance weathering induced, for static and seismic scenarios and various hydrological conditions. The limit analysis upper bound method (Chen, 1975) and the pseudo-static approach (Terzaghi, 1943) are used to derive the analytical law describing the evolution of homogeneous c, ϕ slopes subject to strength degradation, seismic action, formation of tension cracks and various seepage conditions. Perhaps the greatest limitation of the analytical solution presented resides in assuming the slopes homogeneous. In fact natural heterogeneity, layering, different material properties of the layers etc. tend to produce geomorphic features unique to each particular slope that are not captured by the solution here presented. However, the other side of the coin is the fact that such a strong assumption allows for the derivation of a comprehensive analytical solution that can be used to achieve a first rough estimation of the past or future evolution of a slope knowing a limited amount of information and to explore the relative influence on slope evolution of the various physical phenomenon considered in general terms.

The ground strength is here characterised by the Mohr-Coulomb failure criterion therefore only three parameters are needed to describe its properties (unit weight, angle of shearing resistance and cohesion). The limit analysis upper bound method was applied to determine each discrete landslide event occurring over time for successive destabilization and complete removal of the failed mass after each event (Utili and Crosta, 2011a). The material accumulated at the slope toe cannot be taken into account in our model since limit analysis is not able to give any information about the final geometry of the debris accumulated after each landslide. Therefore, it is assumed that the debris accumulating at the slope toe is removed by atmospheric agents and erosion (e.g. fluvial or marine), before a new landslide develops. This condition is known in the literature as a strong erosion condition and is typical of weathering-limited processes (Hutchinson, 1973).

2. LIMIT ANALYSIS MODEL

The failure mechanisms assumed in our analysis are 2D single wedge rigid rotational mechanisms (see Fig. 1). The failing wedge EBCD rotates rigidly around point P_1 with the ground lying on the right of the log-spiral CD and the vertical crack BC remaining at rest. The equation of the log-spiral CD is:

$$r = r_0 \exp[tan\phi(\theta - \theta_0)]$$

with r being the distance of a generic point of the spiral to its centre (see Fig. 1).

For sake of simplicity, the calculations presented below refer to the case of a horizontal slope crest. Since the limit analysis method does not provide information about the final geometry of the debris accumulated after each landslide, it is therefore assumed that the debris is entirely removed by atmospheric agents and/or erosion before the occurrence of the next slope failure.

The following geometrical relationships will be employed in the derivation of the semi-analytical solution:

$$r_{y_1} = r_{x_1} \exp\left[tan\phi\,(y_1 - x_1)\right] \tag{1}$$

$$r_{z_1} = r_{x_1} \exp\left[tan\phi\,(z_1 - x_1)\right] \tag{2}$$

$$r_{y_2} = r_{x_2} \exp\left[tan\phi\,(y_2 - x_2)\right] \tag{3}$$

$$r_{z_2} = r_{x_2} \exp\left[tan\phi\,(z_2 - x_2)\right] \tag{4}$$

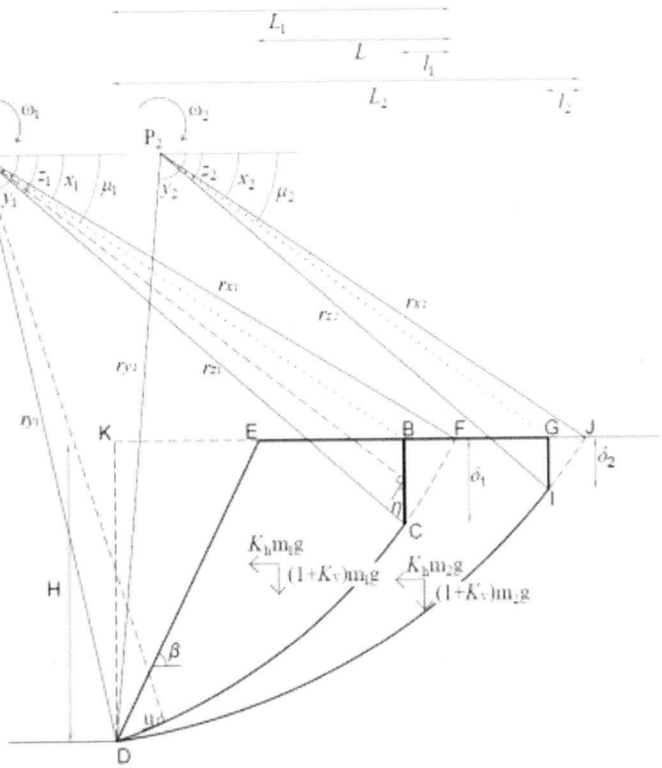

Fig. 1. Second failure mechanism: Region of soil GIDCB by rotates around point P_2. Region of soil EBCD has already slipped away when the first failure occurred.

$$H = r_x\{\exp[tan\phi\,(y_1 - x_1)]\,\sin y_1 - \sin x_1\} \tag{5}$$

$$L = r_{x_1}\left\{\frac{\sin(x_1+\beta)}{\sin\beta} - \exp[tan\phi\,(y_1 - x_1)]\frac{\sin(y_1+\beta)}{\sin\beta}\right\} \tag{6}$$

$$L_1 = r_{x_1}\cos x_1 - \cos y_1 \exp[tan\phi\,(y_1 - x_1)] \tag{7}$$

$$l_1 = r_{x_1}\cos x_1 - \cos z_1 \exp[tan\phi\,(z_1 - x_1)] \tag{8}$$

$$L_2 = r_{x_2} \cos x_2 - \cos y_2 \exp\left[tan\phi\,(y_2 - x_2)\right] \quad (9)$$

$$l_2 = r_{x_2} \cos x_2 - \cos z_2 \exp\left[tan\phi\,(z_2 - x_2)\right] \quad (10)$$

where r_x, r_y and r_z are the radii of the spiral at the angles x, y and z respectively, L, L_1, $l_1 L_2$, l_2 are the horizontal lengths (see Fig. 1) and H is the height of the slope.

2.1. First failure (landslide)

The calculations for the first failure mechanism are here illustrated. The upper bound is derived by imposing energy balance for the failing wedge EBCD:

$$\dot{W}_d + \dot{W}_{cf} = \dot{W}_{ext} \quad (11)$$

where $\dot{W}d$ is the rate of energy dissipated along the log-spiral surface (shear failure), $\dot{W}cf$ is the rate of energy dissipated by crack formation and $\dot{W}ext$ the rate of external work done by the failing wedge.

The calculation of $\dot{W}d$ accounting for the energy dissipated along the log-spiral segment CD is reported in (Utili, 2013). Cracks may develop from the slope face (DE) and/or from the upper part of the slope (EF) (see Fig. 1). Cracks or fissures can be the result of a variety of phenomena, for instance exceedance of the ground tensile strength (Baker, 1981), the occurrence of differential settlements (Vanicek and Vanicek, 2008), desiccation (Konrad and Ayad, 1997; Dyer *et al.*, 2009 ; Péron *et al.*, 2009) and freezing (Hales and Roering, 2007). Cracks may cause a significant decrease in the stability of a slope both in static (Michalowski, 2013; Utili, 2013; Gao *et al.*, 2015 ; Utili, 2015) and seismic conditions (Utili and Abd, 2016).

Here cracks are treated as no-tension non-cohesive perfectly smooth (no friction) interfaces, therefore the angle η between the velocity vector and the crack surface (seeFig. 1) is $0° < \eta < 180°$. Following (Michalowski, 2013), two types of cracks can be considered: cracks existing in the slope before the formation of any failure mechanism, called pre-existing cracks by Michalowski, and cracks forming as part of the failure process due to the exceedance of the ground tensile strength, here called tension cracks, that take place contemporaneously to the formation of localized deformations leading to the failure of the slope. A pre-existing crack may have been formed in the past by tensile stresses that are no longer acting on the slope, for instance due to tectonic movements. Tension cracks instead indicate cracks that are generated by tensile stresses exceeding the current ground tensile strength leading to the formation of a failure mechanism (Terzaghi, 1943 ; Baker, 1981). Although the presence of pre-existing cracks can be easily accounted for in limit analysis (Michalowski, 2013 ; Utili, 2013), here only tension cracks are considered. This is because the evolution of a slope subject to deterioration/weathering is ruled by the deterioration of the ground tensile strength, leading to the onset of tension cracks whereas pre-existing cracks generated in the course of the geological history of the slope formation are likely to affect the first failure mechanism only.

To calculate the energy dissipated by the formation of a crack with limit analysis, Michalowski (2013) has considered limiting the Mohr Coulomb linear envelope by the stress circle of an unconfined uniaxial tensile strength test with the circle being tangent to the linear envelope (see Fig. 2). This failure criterion is realistically non-linear in the tension zone and on the other hand lends itself to limit analysis calculations. The energy expended for the formation of a tension crack, $\dot{W}_{cf}$ turns out to be (Michalowski, 2013):

$$\dot{W}_{cf} = r_{x_1}{}^2 \omega \left(\frac{\sin x_1}{\tan \mu_1} \right)^2 \left(\frac{f c^{M-C}}{2} \int_{\mu_1}^{z_1} \frac{1 - \sin \theta}{\cos^3 \theta} d\theta + \frac{f_t}{1 - \sin \phi} \int_{\mu_1}^{z_1} \frac{\sin \theta - \sin \phi}{\cos^3 \theta} d\theta \right) \quad (12)$$

with μ_1 being the angle made by the segment P_1B with the horizontal (see Fig. 1), fc^{M-C} being the Mohr-Coulomb unconfined compressive strength of the ground and f_t the unconfined tensile strength as measured from laboratory experiments (see Fig. 2). It is convenient to introduce a dimensionless coefficient, t, defined as the ratio of the unconfined tensile strength measured in laboratory experiments, f_t, over the full unconfined tensile strength predicted by the Mohr-Coulomb criterion, f_t^{M-C} (see Fig. 2a):

$$t = \frac{f_t}{f_t^{M-C}} \quad (13)$$

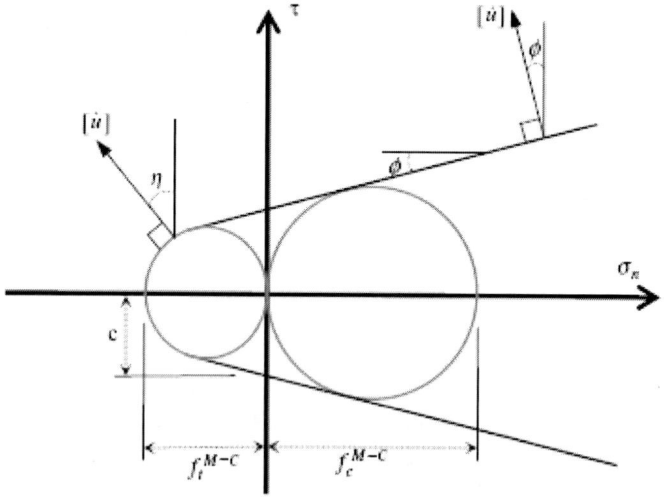

(a)

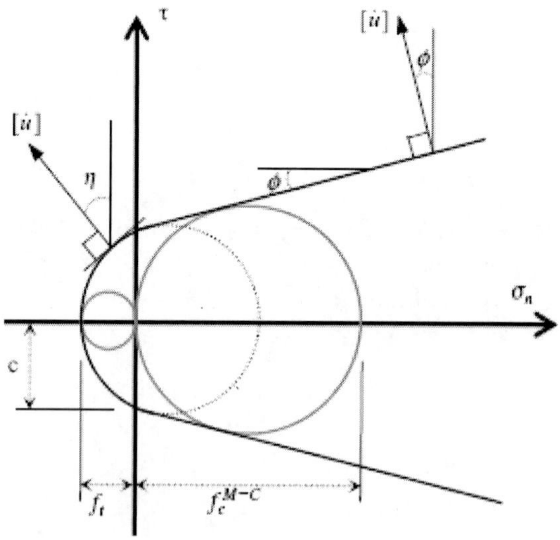

(b)

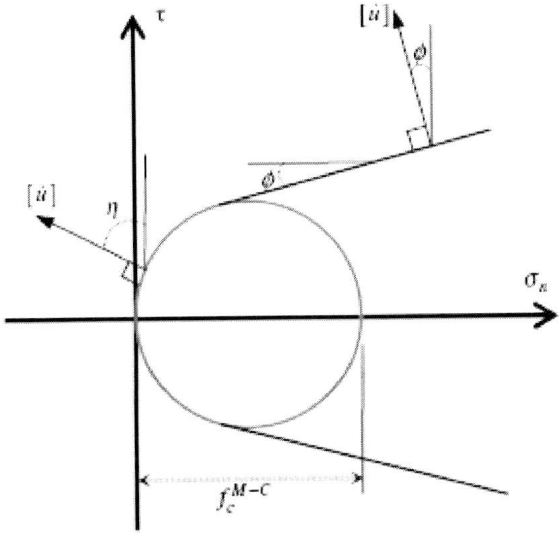

(c)

Fig. 2. Mohr-Coulomb failure criteria illustrating the failure mechanisms associated with (a) Soil of full unconfined tensile strength ($t = 1$). (b) Soil of limited tensile strength ($t = 0.5$). (c) Soil of zero tensile strength ($t = 0$), based on Michalowski (2013).

It is straightforward to observe that $0 < t < 1$. Also both f_c^{M-C} and f_t^{M-C} are uniquely related to c and ϕ:

$$f_c^{M-C} = 2c\left(\frac{\cos\phi}{1-\sin\phi}\right) \tag{14}$$

$$f_t^{M-C} = 2c\left(\frac{\cos\phi}{1+\sin\phi}\right) \tag{15}$$

Three different conditions controlling crack formation are tackled here:

i.tension cut-off ($t = 0$);

ii.limited tensile strength, e.g. $t = 0.2$ and $t = 0.5$;

iii.full tensile strength, i.e. the possibility of tensile failure of the ground is neglected

The rate of external work for the sliding wedge EBCD, $\dot{W}_{ext}$, is made by two contributions:

$$\dot{W}_{ext} = \dot{W}_{\gamma} + \dot{W}_{w} \tag{16}$$

with $\dot{W}_{\gamma}$ representing the external work done by the weight of the wedge and $\dot{W}w$representing the work done by the water pressure.

$\dot{W}_{\gamma}$ is calculated as the work of block EFD minus the work of block BFC. In turn the work of block EFD is calculated by algebraic summation of the work of blocks P_1FD, P_1FE and P_1ED (Chen et al., 1969) and the work of block BFC is calculated by summation of the work of blocks P_1FC, P_1FB and P_1BC (Utili and Nova, 2007; Utili and Crosta, 2011a ; Utili, 2013). Note that here, in addition to the weight force, a horizontal pseudo-static force, $FPS_h = mK_hg = \gamma K_hA$, with g being the gravitational acceleration and m the mass of the wedge, and a vertical one, $FPS_v = mK_vg = \gamma K_vA$, are added to account for seismic action (Chen and Liu, 1990 ; Utili and Abd, 2016).

To calculate $\dot{W}_w$, a uniform value of the dimensionless coefficient ru is assumed over the whole slope with ru defined as (Bishop and Morgenstern, 1960):

$$r_u = \frac{u}{\gamma_{sat} h} \tag{17}$$

with u being the total pore water pressure in the considered point of the failure line, $\gamma_{sat}$the ground bulk unit weight and h the depth of the point considered from the ground surface. The assumption of uniform ru is a strong one (Barnes, 2010), however this assumption is still commonly made in slope stability analysis (e.g. Michalowski, 2013). The work of water along the cracked part BC and the log-spiral part CD (Fig. 3) was calculated as a surface integral over the whole surface (Michalowski, 1995 ; Michalowski, 2013):

$$\dot{W}_w = \dot{W}_{BC} + \dot{W}_{CD} = \int_{0r_{u_1}}^{z_1} u[v]_n dS_{BC} + \int_{z_1}^{y_1} u[v]_n dS_{CD} \tag{18}$$

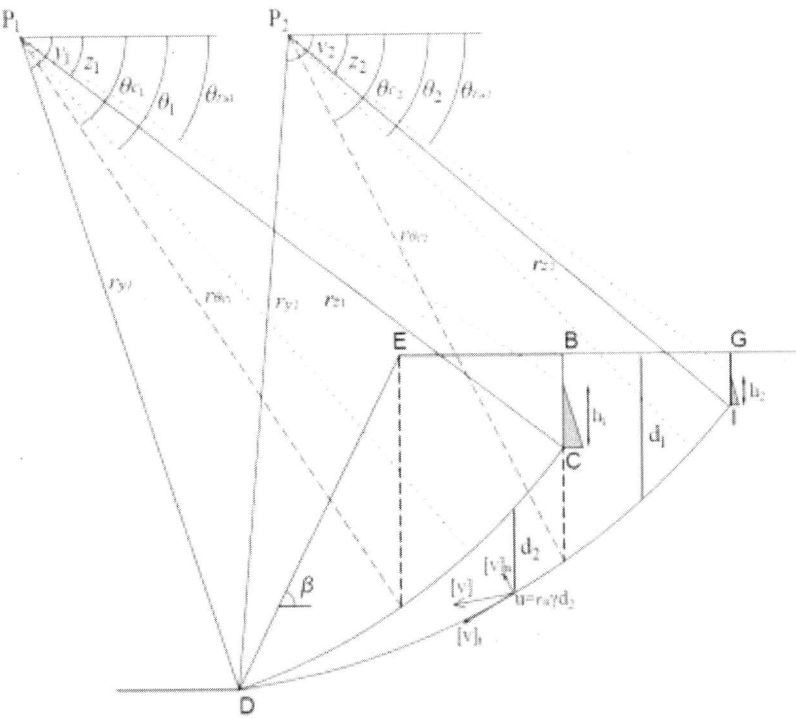

Fig. 3. Presence of pore pressure in the calculation of the most critical failure mechanism.

where $[v]_n$ is the normal component of the boundary velocity (Fig. 3) and

$$\tan\theta r_{u_1} = \frac{r_z \sin z_1 - h_1}{r_z \cos z_1} = \frac{\exp(\tan\phi(z_1 - x_1))\sin z_1 - \frac{r_u \gamma}{\gamma_w}(\exp(\tan\phi(z_1 - x_1))\sin z_1 - \sin x_1)}{\exp(\tan\phi(z_1 - x_1))\cos z_1}.$$

Substituting Eqs. (12), (16) ; (18) into Eq. (11), the final equation to calculate the stability factor, $N = \gamma H/c$, is obtained:

equation(19)

$$\frac{\gamma H}{c} = g(x_1, y_1, z_1, \phi, \beta, K_h, \lambda, r_u) = \frac{f_{d_1} + f_{cf}[\exp(\tan\phi(y_1 - x_1))\sin y_1 - \sin x_1]}{(1+\lambda K_h)(f_{1v} - f_{2v} - f_{3v} - f_{4v} + f_{5v} + f_{6v}) + K_h(f_{1h} - f_{2h} - f_{3h} - f_{4h} + f_{5h} + f_{6h}) + r_u f_w} \quad (19)$$

with $\lambda = K_v/K_h$ (consistently with Fig. 1, the + sign indicates vertical downward acceleration, whereas the – sign indicates vertical upward acceleration). The global minimum of $g(x_1, y_1, z_1, \phi, \beta, K_h, \lambda)$ over the three geometrical variables x_1, y_1, z_1 provides the least (best) upper bound on the stability factor for the case that has been considered. The static case is a particular case obtained setting $K_h = K_v = 0$. The calculations of the expressions for $\dot{W}_{ext}$ for each block are provided in Appendix A.

Unlike the case of intact slopes, failure mechanisms may in principle daylight on the slope face above the slope toe; however in (Utili and Abd, 2016) no potential mechanism passing above the slope toe turned out to be the most critical, therefore no potential failure mechanisms passing above the toe were considered in this analysis.

2.2. Second and successive failures (landslides)

The analytical expressions for the second failure also apply to every successive failure. After the region EBCD (Fig. 1) has slipped away and due to further ground degradation, at some point a second landslide will occur. The double logarithm spiral shaped area GIDCB will rigidly rotate around the center of rotation P_2, as yet undefined, with the material below the logarithmic spiral ID and right of the vertical crack GI remaining at rest. The mechanism is now defined by six variables $x_1, z_1, y_1, x_2, z_2, y_2$, where x_1, y_1, z_1 are the angles defining the first log spiral failure line (i.e., the current slope profile produced by the previous failure), that will be called 'old' landslide and will be denoted by the superscript *o* and x_2, y_2, z_2 are the angles defining the second log spiral failure line (see Fig. 1), that will be called 'new' landslide and will be denoted by the superscript *n*.

Two contributions constitute the rate of the external work for the sliding wedge GIDCB; the work done by the weight of the wedge that slides away $\dot{W}_\gamma$ and the work done by the water pressure $\dot{W}_w$. The weight of the wedge that slides away $\dot{W}_\gamma$ is calculated this time as the work by block KJD minus the work of blocks KBCD and JIG. The work of block KJD is calculated as the summation of three contributions: $\dot{W}_1^n$-$\dot{W}_2^n$-$\dot{W}_3^n$, with $\dot{W}_1^n$, $\dot{W}_2^n$ and $\dot{W}_3^n$ being work rates done by blocks P_2JD, P_2JK and P_2KD respectively. The rate of work of block KBCD is calculated as the summation of six contributions: $\dot{W}_1^o$-$\dot{W}_2^o$-$\dot{W}_3^o$-($\dot{W}_4^o$-$\dot{W}_5^o$-$\dot{W}_6^o$), with $\dot{W}_1^o$ to $\dot{W}_6^o$ being the work rates of regions P_1FD, P_1FK, P_1KD, P_1FC, P_1FB and P_1BC respectively. The work rate of JIG is calculated as $\dot{W}_4^n$-$\dot{W}_5^n$-$\dot{W}_6^n$, with $\dot{W}_4^n$, $\dot{W}_5^n$ and $\dot{W}_6^n$ being the work rates done by P_2JI, P_2JG and P_2GI respectively. Therefore, $\dot{W}_\gamma$ is calculated as:

$$\dot{W}_\gamma = \dot{W}_1^n - \dot{W}_2^n - \dot{W}_3^n - \dot{W}_4^n + \dot{W}_5^n + \dot{W}_6^n - \dot{W}_1^o + \dot{W}_2^o + \dot{W}_3^o + \dot{W}_4^o - \dot{W}_5^o - \dot{W}_6^o \tag{20}$$

with n and o referring to the new and the old landslide respectively.

The calculations of the expressions for $\dot{W}_1^n$, $\dot{W}_2^n$, $\dot{W}_3^n$ etc. and for $\dot{W}_w$ are provided inAppendix A. Substituting Eqs. (12), (18) ; (20) into Eq. (11) and dividing all terms by $\dot{\theta}$ and rx_2, and rearranging, the stability factor, $N_s = \gamma H/c$, is obtained:

$$\frac{\gamma H}{c} = g(x_1, y_1, z_1, x_2, y_2, z_2, \phi, K_h, \lambda, r_u) = \frac{f_{d_2} + f_{cf}[\exp(\tan\phi(y_2 - x_2))\sin y_2 - \sin x_2]}{\begin{bmatrix}(1+\lambda K_h)(f_{1v}{}^n - f_{2v}{}^n - f_{3v}{}^n - f_{1v}{}^o + f_{2v}{}^o + f_{3v}{}^o + f_{4v}{}^o - f_{5v}{}^o - f_{6v}{}^o - f_{4v}{}^n + f_{5v}{}^n + f_{6v}{}^n) + \\ K_h(f_{1h}{}^n - f_{2h}{}^n - f_{3h}{}^n - f_{1h}{}^o + f_{2h}{}^o + f_{3h}{}^o + f_{4h}{}^o - f_{5h}{}^o - f_{6h}{}^o - f_{4h}{}^n + f_{5h}{}^n + f_{6h}{}^n) + r_u f_w\end{bmatrix}} \quad (21)$$

The global minimum of $g\,(x_1, y_1, z_1, x_2, y_2, z_2, \phi, \beta, K_h, \lambda)$ over the three geometrical variables x_2, y_2, z_2, provides the least (best) upper bound on the stability factor for the second and any successive landslide. Note that the second (and every successive) mechanism could pass through any point since the current slope profile is no longer straight, as presented in Fig. 4. Therefore failure mechanisms daylighting at any point of the slope profile left after the first failure has occurred have to be considered. To this end, the slope profile was divided into a discrete number of points (n) and each point has been assumed as the toe of a sub-slope whose height, h_i, is smaller than the overall height H (see Fig. 4). The most critical mechanism among all the possible mechanisms has to be found. The critical cohesion value, c_i, and angles, x_i, y_i and z_i, associated with the critical failure mechanism, were determined for all n sub-slopes of different height, h_i, with the parameter y_i assuming a different value for each sub-slope analysed. The most critical failure mechanism among the n potential mechanisms is the one associated to the highest cohesion value. As long as a sufficiently large value of n is considered, n does not affect the obtained results.

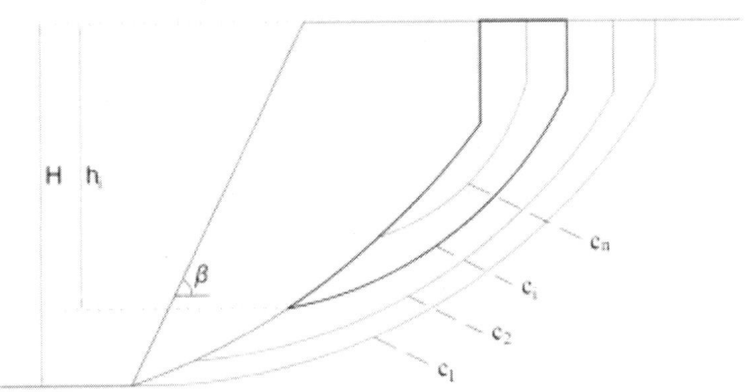

Fig. 4. Potential failure mechanisms for the second failure, relative to different mechanisms considered (for different critical heights h_i).

2.3. Weathering induced cohesion decrease

By applying the procedure described, it is possible to determine as many failure mechanisms as needed to follow the slope evolution until full degradation of the ground strength has taken place. Here, the friction angle is assumed to remain constant in time with only cohesion decreasing over time, since experimental evidence from the weathering of rocks and cemented soils show that weathering causes mainly a decrease of cohesion and to a much lesser extent to the angle of shearing resistance (see Fig. 5). However, extension of the model to account for a decrease of both c and ϕ is straightforward. The case of both c and ϕ decreasing is reported in (Utili and Crosta, 2011a) for geomaterials with full tensile strength.

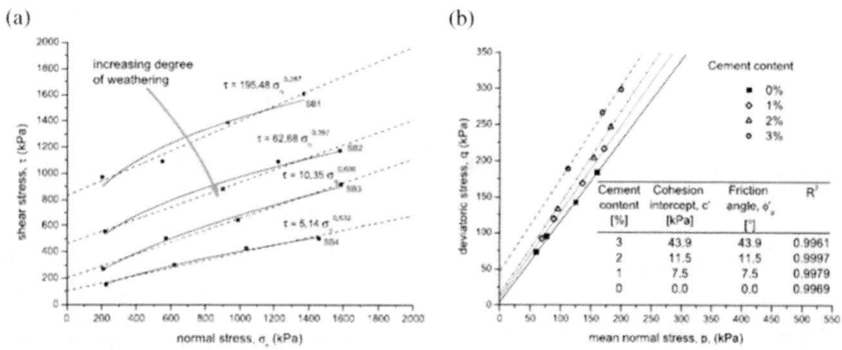

Fig. 5. (a) Weathering of granite (after Kimmance, 1988), (b) failure loci of cemented sands for different cement contents (after Wang and Leung, 2008).

Computations for the evolution of slopes have been carried out using Matlab for a wide range of parameters (ϕ and initial slope inclination β), for intact slopes ($t = 1$), slopes with soil tensile strength limited to $t = 0.2$ and $t = 0.5$ and slopes with no tensile strength, $t = 0$ (tension cut-off) under different seismic and seepage scenarios.

3. MORPHOLOGICAL EVOLUTION OF SLOPES

In Fig. 6 the typical evolution undergone by a slope subject to weathering induced strength degradation is shown. The features of the morphological evolution exhibited by the slope are similar to what shown in (Utili and Crosta, 2011a) apart from the upper part of the failure mechanisms which is always vertical due to the presence of cracks. A number of successive failures of decreasing area and depth of mechanism occur until a deeper mechanism of much larger area takes place (see Fig. 6). To calculate the deeper mechanism, the slope profile, which is composed of several log-spiral pieces, was approximated by a straight line obtained as the linear envelope of the piecewise log-spiral profile (see Fig. 6). This approximation is acceptable since the region delimited by the log spiral and the vertical line is small in comparison with the size of the failure mechanisms.

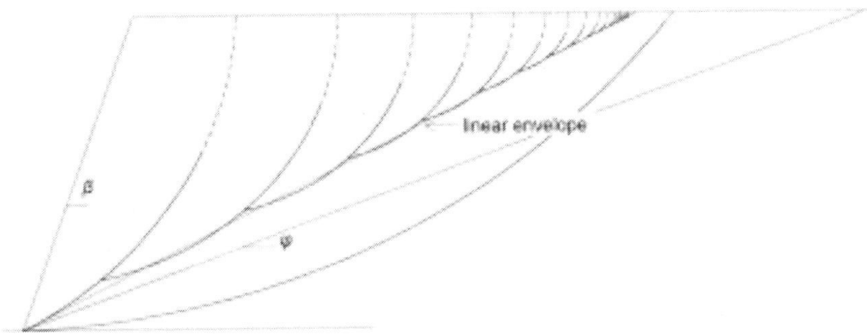

Fig. 6. Deep seated failure mechanism for a slope with $\beta = 70°$ and initial friction angle $\phi = 20°$.

In Fig. 7, the evolution of initially steep slopes, with very limited or completely absent tensile strength, is shown. It emerges that this type of slope exhibits a peculiar type of evolution: several successive small failure mechanisms made of thin slices occur (from vertical line FC to vertical line GI) until a much larger failure mechanism takes place (mechanism JMK). The occurrence of thin slice mechanisms, which in the limit case of an initially vertical slope ($\beta = 90°$) become infinitesimally thin slices, is described and physically explained in (Utili, 2013).

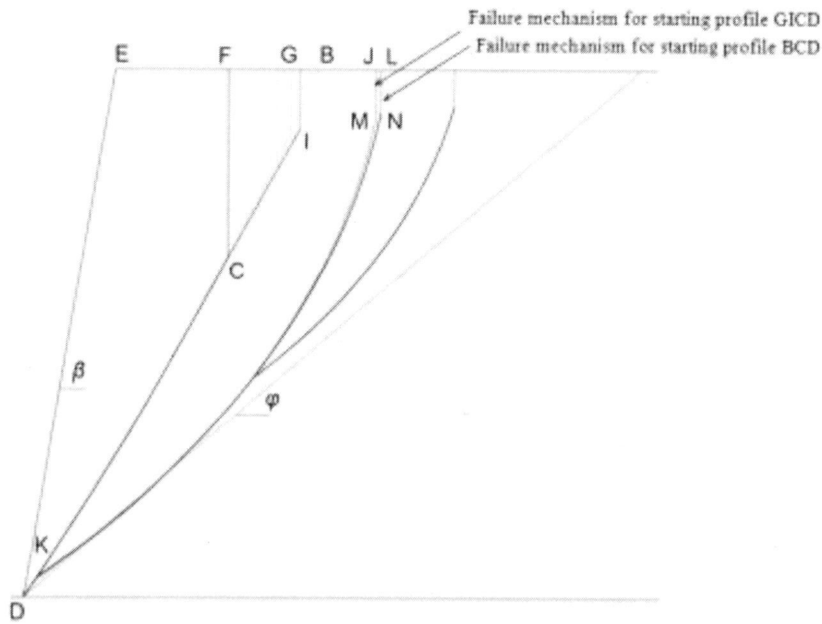

Fig. 7. Most critical failure mechanism for the second 'significant' failure for slope with initial slope inclination $\beta = 80°$ and friction angle $\phi = 40°$ (tension cut off).

4. PARAMETRIC ANALYSIS OF SLOPE EVOLUTION FOR VARIOUS HYDROGEOLOGICAL SCENARIOS

In Fig. 8 the effect of the seismic action is analysed for various values of tensile strength. For each failure, numbered in chronological order, the area of the sliding mass is plotted. It emerges that the influence of the soil tensile strength on the sliding area is small, but seismic action has the effect of amplifying these small differences. To investigate the influence of seepage instead, the critical height for a slope with an initial inclination of $\beta = 60°$ and $\phi = 20°$ is plotted in Fig. 9 for 6 successive failures.

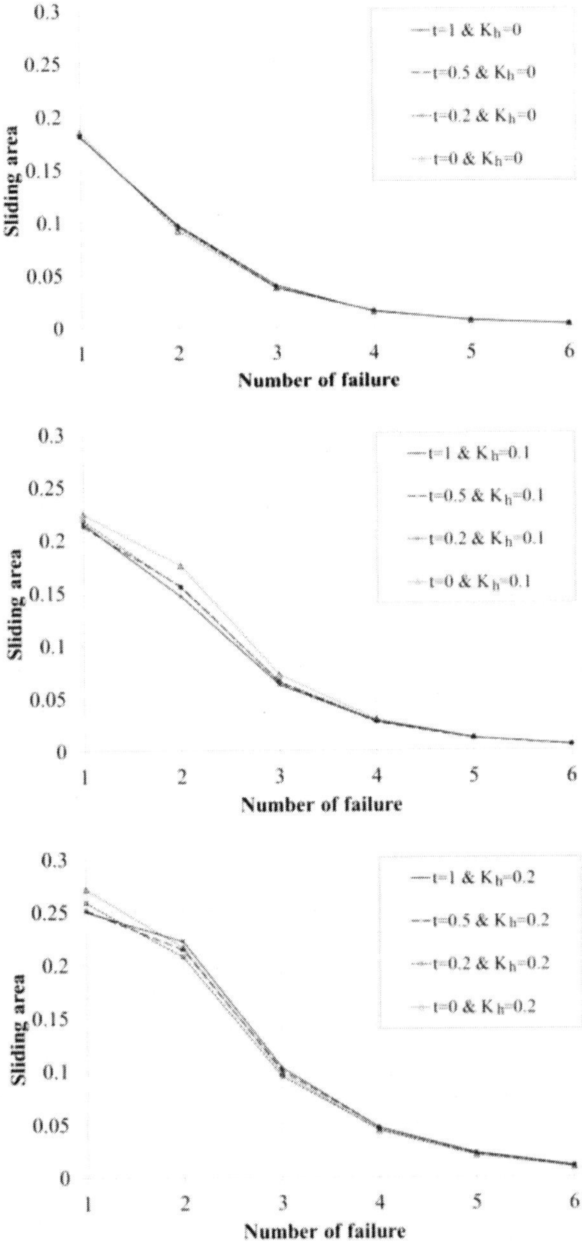

Fig. 8. Dimensionless normalised sliding area for inclination $\beta = 70°$ and for friction angle $\phi = 40°$, showing the influence of the tensile crest on the evolution of slopes for different K_h, for 6 successive failures.

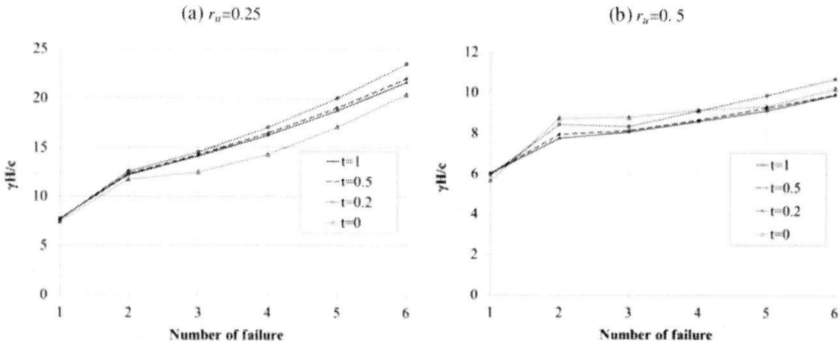

.Fig. 9. Critical height (γH/c) for six successive failures for slope inclination $\beta = 60°$ and $\phi = 20°$ for soil of full unconfined tensile strength ($t = 1$), soil of limited tensile strength ($t = 0.5$ and $t = 0.2$) and soil of zero tensile strength ($t = 0$), (a) for $ru = 0.25$ and (b) for $ru = 0.50$.

In Fig. 10, the step-like relationship between dimensionless normalised cohesion and crest retreat is plotted for 8 successive failures for various seismic scenarios ($K_h = 0$, $K_h = 0.1$ and $K_h = 0.2$). It can be observed that if the first two failures are excluded, the values of critical cohesion and crest retreat lie on straight lines in agreement to what observed in (Utili and Crosta, 2011a) for static dry slopes not subject to crack formation. From the figure it also emerges that the tensile strength does not affect significantly either the cohesion or crest retreat. However this is less the case for slopes subject to substantial seismic action.

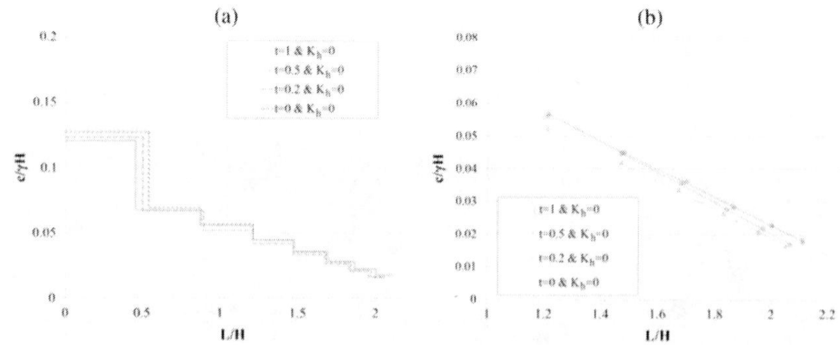

Fig. 10. Dimensionless normalised cohesion versus crest retreat for inclination $\beta = 70°$, (a) step-like relationship between cohesion and crest retreat obtained for $\phi = 20°$, showing the influence of the tensile crest on the evolution of slopes for $K_h = 0$, $K_h = 0.1$ and $K_h = 0.2$, for 8 successive failures (b) If the first two failures are excluded, the values of critical cohesion and crest retreat lie on straight lines only for intact slopes.

The evolution of the stability number (γH/c) and the dimensionless cliff retreat (L/H) as a function of the friction angle represented in the same graph as curves from the analysis of six successive failures, for initial slope inclinations of $\beta = 60°$, $\beta = 70°$ and $\beta = 80°$ are plotted in Fig. 11. Solid black lines are used for the stability number and dashed lines for the dimensionless crest retreat, while the corresponding number of failure is reported on each curve. It is evident that the influence of tensile strength and of the presence of cracks on the stability of the slopes is more significant in steep slopes. When seismic acceleration is taken into consideration the influence on both stability number and cliff retreat is significant (Fig. 12). It can be seen in Fig. 13 and as explained by (Michalowski, 2013), that the adverse influence of pore-water pressure on the stability number increases for larger values of ϕ.

Note that for some cases of slopes with low ϕ, after a number of failures, the model predicts a much larger landslide in terms of area (volume), but not in terms of crest retreat (Fig. 6) so this is more of interest for anthropic constructions at the toe of the slope. If the slope is a cliff by the ocean – an environment where continuous wash away actions take place – it is reasonable to expect the erosion of the failed material before the next mechanism occurs. Instead, in case of less exposed cliffs (e.g. inland) we can expect that this deep mechanism will not take place due to the stabilising action of the weight of the debris and therefore it is reasonable to disregard it.

In Table 1 values of crest retreat and area of the failure mechanism are reported for various initial slope inclinations ($\beta = 60°$, 70° and 80°), angle of shearing resistance ($\phi = 20°$, 30° and 40°) for the three scenarios of crack formation considered in the paper (tension cut-off, limited tensile strength and full tensile strength). The results of the parametric investigation for initially intact slopes subject to seismic action and various seepage seepage scenarios are reported in Table 2.

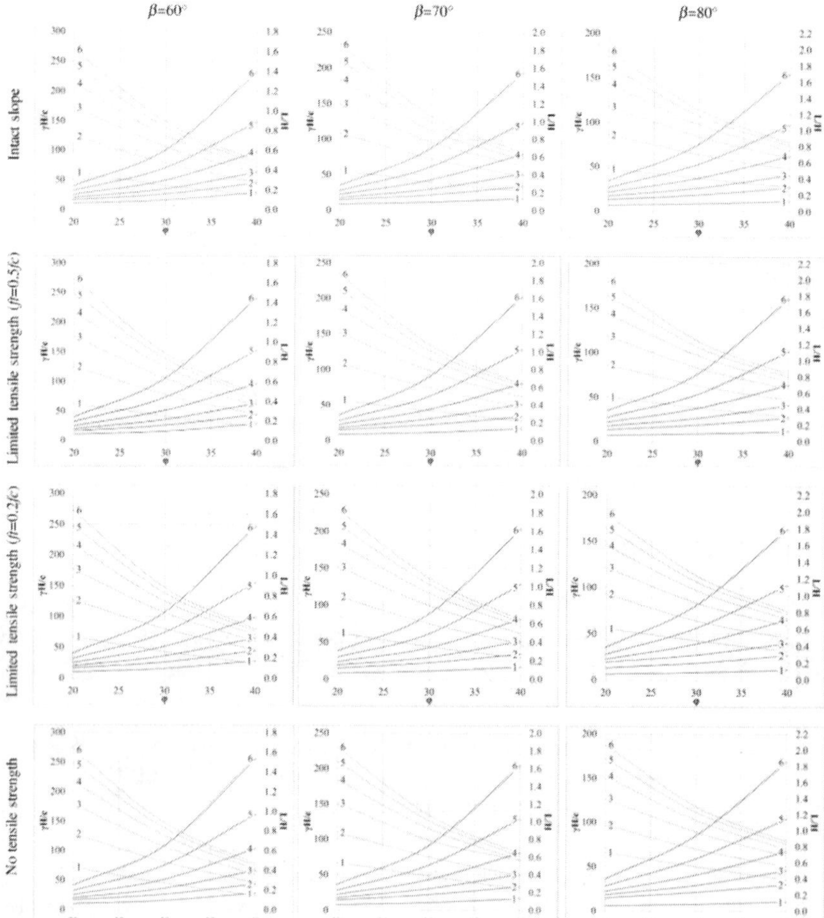

Fig. 11. Critical height (γH/c) and dimensionless cliff retreat (L/H) as a function of the friction angle represented in the same graph as curves for six successive failures for slope inclinations $\beta = 60°$, $\beta = 70°$ and $\beta = 80°$ for the case of intact slope, for soil of full unconfined tensile strength ($t = 1$), soil of limited tensile strength ($t = 0.5$ and $t = 0.2$) and soil of zero tensile strength ($t = 0$). Solid black lines are used for the critical height and dashed lines for the dimensionless crest retreat. The corresponding failure number is marked on the curves.

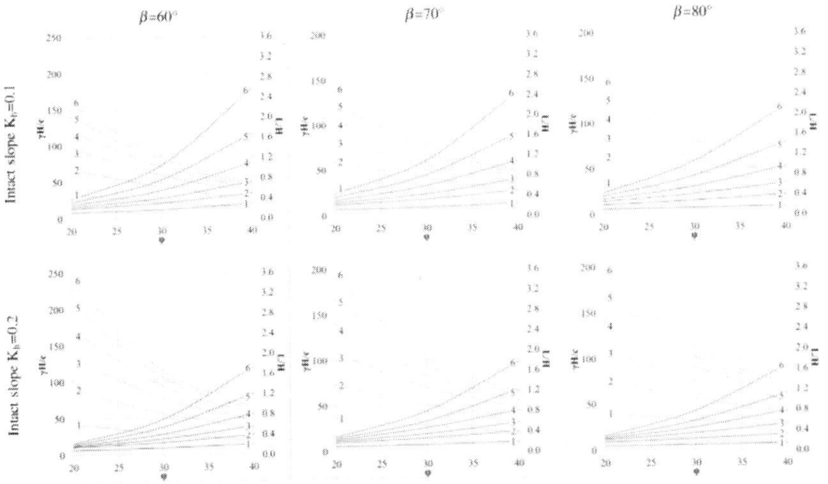

Fig. 12. Critical height (γH/c) and dimensionless cliff retreat (L/H) as a function of the friction angle represented in the same graph as curves for six successive failures for slope inclinations $\beta = 60°$, $\beta = 70°$ and $\beta = 80°$ for intact slopes and for seismic scenarios with coefficient $K_h = 0.1$ and $K_h = 0.2$. Solid black lines are used for the critical height and dashed lines for the dimensionless crest retreat. The corresponding failure number is marked on the curves.

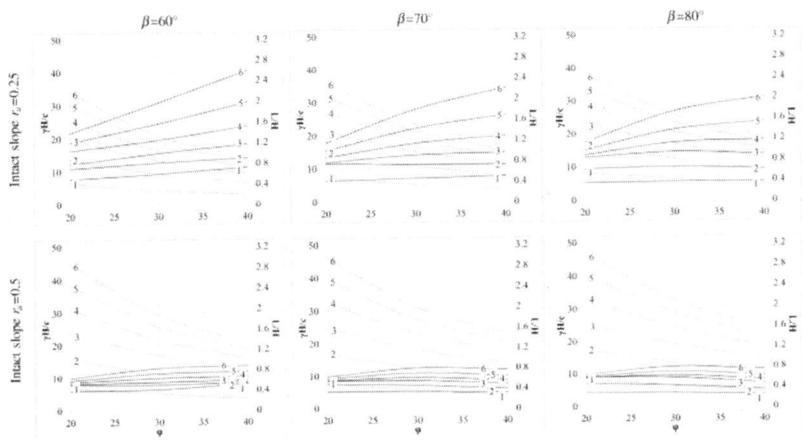

Fig. 13. Critical height (γH/c) and dimensionless cliff retreat (L/H) as a function of the friction angle represented in the same graph as curves for six successive failures for slope inclinations $\beta = 60°$, $\beta = 70°$ and $\beta = 80°$ for the case of intact slope and for pore pressure with coefficient $ru = 0.25$ and $ru = 0.5$. Solid black lines are used for the critical height and dashed lines for the dimensionless crest retreat. The corresponding failure number is marked on the curves.

Table 1. Values for associated crest retreat and sliding area for slopes with initial inclination β and angle of shearing resistance ϕ, for slopes with $t = 1.0$, $t = 0.5$, $t = 0.2$ and $t = 0$.

	β = 60°						β = 70°						β = 80°					
	φ = 20°		φ = 30°		φ = 40°		φ = 20°		φ = 30°		φ = 40°		φ = 20°		φ = 30°		φ = 40°	
	L/H	Area	L/H	Area	L/H	Area	L/H	Area	L/H	Area	L/H	Area	L/H	Area	L/H	Area	L/H	Area
Failure	Intact slope $t = 1.0$																	
1	0,377	0,356	0.244	0.244	0.145	0.162	0,453	0,345	0.317	0,255	0,218	0,181	0,525	0,345	0,407	0,269	0,301	0,207
2	0,742	0,270	0,468	0,141	0,277	0,072	0,867	0,353	0,578	0,178	0,386	0,097	0,947	0,364	0,702	0,239	0,518	0,154
3	1,039	0,171	0,632	0,066	0,369	0,027	1,202	0,216	0,766	0,085	0,497	0,040	1,300	0,241	0,922	0,117	0,659	0,064
4	1,279	0,105	0,748	0,033	0,428	0,011	1,473	0,138	0,898	0,042	0,568	0,016	1,580	0,151	1,076	0,058	0,748	0,025
5	1,466	0,067	0,830	0,016	0,466	0,004	1,689	0,088	0,990	0,021	0,613	0,006	1,806	0,093	1,184	0,028	0,808	0,010
6	1,616	0,041	0,887	0,008	0,490	0,002	1,862	0,056	1,055	0,010	0,642	0,003	1,981	0,059	1,260	0,014	0,844	0,004
Failure	Limited tensile strength $t = 0.5$																	
1	0,378	0,349	0,248	0,246	0,137	0,159	0,463	0,344	0,334	0,257	0,217	0,181	0,552	0,350	0,419	0,269	0,310	0,206
2	0,753	0,287	0,469	0,134	0,269	0,068	0,874	0,362	0,591	0,182	0,385	0,097	0,955	0,359	0,697	0,232	0,521	0,156
3	1,054	0,175	0,633	0,065	0,360	0,027	1,215	0,224	0,782	0,088	0,495	0,038	1,306	0,238	0,914	0,113	0,660	0,060
4	1,297	0,108	0,747	0,031	0,418	0,011	1,484	0,140	0,915	0,042	0,565	0,015	1,583	0,149	1,066	0,055	0,748	0,024
5	1,486	0,069	0,827	0,015	0,456	0,004	1,701	0,086	1,008	0,021	0,609	0,006	1,808	0,092	1,172	0,027	0,804	0,010
6	1,639	0,043	0,882	0,007	0,479	0,002	1,870	0,055	1,072	0,010	0,638	0,002	1,981	0,059	1,246	0,013	0,839	0,004
Failure	Limited tensile strength $t = 0.2$																	
1	0,404	0,354	0,260	0,251	0,148	0,163	0,501	0,354	0,353	0,259	0,233	0,182	0,614	0,364	0,470	0,278	0,346	0,214
2	0,760	0,284	0,481	0,133	0,278	0,066	0,893	0,328	0,599	0,191	0,388	0,095	1,025	0,343	0,702	0,213	0,505	0,136
3	1,060	0,171	0,638	0,062	0,364	0,025	1,214	0,203	0,793	0,091	0,497	0,038	1,337	0,232	0,917	0,110	0,644	0,055
4	1,298	0,107	0,750	0,030	0,421	0,010	1,471	0,127	0,929	0,044	0,567	0,015	1,598	0,138	1,065	0,053	0,730	0,023
5	1,485	0,066	0,828	0,015	0,457	0,004	1,674	0,076	1,023	0,021	0,612	0,006	1,810	0,084	1,168	0,026	0,784	0,009
6	1,633	0,041	0,883	0,007	0,479	0,002	1,831	0,047	1,089	0,010	0,640	0,002	1,976	0,052	1,240	0,012	0,818	0,004
Failure	Tension cut-off $t = 0$																	
1	0,431	0,359	0,268	0,248	0,148	0,163	0,539	0,357	0,386	0,266	0,243	0,185	0,726	0,389	0,552	0,299	0,393	0,218
2	0,770	0,280	0,483	0,133	0,276	0,062	0,876	0,344	0,603	0,177	0,389	0,091	1,068	0,391	0,771	0,237	0,554	0,150
3	1,072	0,171	0,639	0,062	0,359	0,023	1,213	0,220	0,790	0,088	0,499	0,037	1,403	0,239	0,981	0,114	0,686	0,064
4	1,301	0,103	0,751	0,029	0,412	0,009	1,473	0,133	0,923	0,041	0,567	0,015	1,678	0,142	1,129	0,053	0,773	0,025
5	1,485	0,062	0,826	0,014	0,446	0,004	1,682	0,080	1,012	0,020	0,610	0,006	1,896	0,087	1,231	0,025	0,830	0,010
6	1,625	0,039	0,879	0,007	0,468	0,001	1,842	0,050	1,075	0,009	0,637	0,002	2,065	0,053	1,302	0,012	0,865	0,004

Table 2. Values for associated crest retreat and sliding area with initial inclination β and angle of shearing resistance ϕ, for intact slopes under seismic action or with the existence of water ($t = 1$).

	β = 60°						β = 70°						β = 80°					
	φ = 20°		φ = 30°		φ = 40°		φ = 20°		φ = 30°		φ = 40°		φ = 20°		φ = 30°		φ = 40°	
	L/H	Area	L/H	Area	L/H	Area	L/H	Area	L/H	Area	L/H	Area	L/H	Area	L/H	Area	L/H	Area
Failure	$K_h = 0.1$																	
1	0,490	0.415	0.330	0,293	0.199	0.194	0.553	0.393	0.400	0,293	0,286	0,216	0,628	0,390	0,485	0,302	0,380	0,238
2	0,977	0.407	0,626	0,203	0.379	0.101	1.078	0,478	0,744	0,274	0,504	0,147	1,139	0,458	0,841	0,289	0,635	0,194
3	1,391	0,288	0,850	0,106	0,501	0,042	1,531	0,344	1,004	0,143	0,652	0,062	1,595	0,349	1,118	0,164	0,809	0,084
4	1,738	0,196	1,013	0,056	0,582	0,018	1,910	0,234	1,191	0,076	0,751	0,028	1,987	0,244	1,321	0,088	0,922	0,037
5	2,026	0,136	1,132	0,030	0,635	0,008	2,226	0,163	1330	0,041	0,817	0,012	2,311	0,176	1,470	0,046	0,998	0,016
6	2,267	0,095	1,218	0,016	0,669	0,003	2,490	0,114	1,431	0,023	0,860	0,005	2,590	0,123	1,577	0,025	1,048	0,007
Failure	$K_h = 0.2$																	
1	0,618	0,483	0,420	0,337	0,281	0,238	0,667	0,455	0,505	0,340	0,366	0,250	0,731	0,442	0,586	0,349	0,456	0,270
2	1,292	0,649	0,817	0,301	0,516	0,146	1,317	0,616	0,920	0,356	0,652	0,222	1,366	0,570	1,011	0,355	0,750	0,229
3	1,913	0,535	1,121	0,174	0,676	0,068	1,945	0,524	1,262	0,209	0,850	0,102	1,993	0,541	1,365	0,224	0,969	0,115
4	2,469	0,421	1,356	0,099	0,790	0,031	2,503	0,423	1,514	0,124	0,989	0,046	2,550	0,423	1,627	0,133	1,114	0,055
5	2,968	0,339	1,530	0,059	0,865	0,015	3,004	0,342	1,714	0,072	1,085	0,023	3,050	0,340	1,835	0,077	1,215	0,025
6	3,415	0,273	1,668	0,034	0,917	0,007	3,455	0,276	1,863	0,043	1,148	0,010	3,498	0,274	1,989	0,046	1,283	0,012
Failure	$r_u = 0.25$																	
1	0,415	0,402	0,276	0,304	0,188	0,233	0,143	0,236	0,338	0,291	0,233	0,228	0,537	0,364	0,406	0,292	0,286	0,228
2	0,873	0,390	0,593	0,228	0,422	0,147	0,354	0,131	0,691	0,275	0,507	0,195	1,059	0,388	0,796	0,276	0,578	0,197
3	1,280	0,287	0,847	0,139	0,598	0,083	0,514	0,070	0,968	0,169	0,714	0,108	1,472	0,346	1,087	0,201	0,810	0,140
4	1,625	0,211	1,045	0,086	0,732	0,048	0,630	0,038	1,185	0,105	0,872	0,065	1,829	0,241	1,317	0,120	0,984	0,080
5	1,926	0,158	1,203	0,054	0,834	0,027	0,720	0,021	1,360	0,066	0,993	0,038	2,150	0,179	1,502	0,073	1,117	0,046
6	2,187	0,119	1,329	0,034	0,913	0,016	0,788	0,012	1,498	0,041	1,085	0,022	2,427	0,134	1,645	0,045	1,218	0,027
Failure	$r_u = 0.5$																	
1	0,436	0,444	0,313	0,357	0,280	0,406	0,490	0,404	0,355	0,327	0,246	0,268	0,533	0,378	0,403	0,310	0,277	0,250
2	0,997	0,531	0,723	0,343	0,676	0,327	1,083	0,511	0,790	0,347	0,605	0,269	1,119	0,450	0,857	0,323	0,636	0,248
3	1,515	0,445	1,077	0,248	1,014	0,232	1,597	0,468	1,152	0,276	0,891	0,201	1,624	0,474	1,233	0,304	0,937	0,221
4	1,993	0,382	1,382	0,183	1,305	0,166	2,082	0,386	1,471	0,199	1,123	0,131	2,110	0,395	1,560	0,218	1,182	0,146
5	2,443	0,324	1,644	0,136	1,555	0,123	2,526	0,330	1,744	0,147	1,315	0,088	2,557	0,331	1,842	0,156	1,382	0,095
6	2,868	0,292	1,871	0,098	1,770	0,089	2,944	0,280	1,979	0,106	1,475	0,061	2,974	0,280	2,083	0,115	1,548	0,066

5. CONCLUSIONS

Analytical solutions have been presented to investigate the effect of weathering induced strength degradation, seismic action, crack formation and various seepage scenarios on the geomorphological evolution of homogeneous slopes, employing the upper bound limit analysis method. Slopes are characterised by uniform cohesion, angle of shearing resistance, and finite tensile strength. The solutions were obtained considering a succession of discrete failure events (landslides) progressively altering the slope morphology over time and can be used to achieve a first rough estimation of the past or future evolution of a slope knowing a limited amount of information.

A comprehensive parametric analysis has been carried out to investigate the relative influence on slope evolution of the physical phenomena considered, i.e. weathering induced strength degradation, seismic action, crack formation and seepage. It emerges that strength degradation, seismic action and seepage exhibit a stronger influence on the morphologic evolution of slopes than the formation of cracks. High seismic acceleration and/or seepage forces cause larger slope failure mechanisms and therefore larger slope crest retreats. Tension cracks forming due to limited tensile strength influence the first slope failures more than the subsequent ones.

ACKNOWLEDGMENTS

The research carried out in this paper was partly funded by the H2020 Marie Skłodowska-Curie Actions RISE 2014 'Geo-ramp' grant number 645665.

REFERENCES

1. Arkin, Y., Michaeli, L., 1985. Short-and long-term erosional processes affecting the stability of the Mediterranean coastal cliffs of Israel. Eng. Geol. 21, 153–174.
2. Baker, R., 1981. Tensile strength, tension cracks, and stability of slopes. Soils Found. 21, 1–17.
3. Barnes, G.E., 2010. Soil Mechanics: Principles and Practice. Palgrave macmillan. Benumof, B.T., Storlazzi, C.D., Seymour, R.J., Griggs, G.B., 2000. The relationship between incident wave energy and seacliff erosion rates: San Diego County, California. J. Coast. Res. 16, 1162–1178.
4. Bishop, A., Morgenstern, N., 1960. Stability coefficients for earth slopes. Geotechnique 10, 129–153.

5. Boon, C., Houlsby, G., Utili, S., 2012. A new algorithm for contact detection between convex polygonal and polyhedral particles in the discrete element method. Comput. Geotech. 44, 73–82.
6. Boon, C., Houlsby, G., Utili, S., 2013. A new contact detection algorithm for three-dimensional non-spherical particles. Powder Technol. 248, 94–102.
7. Boon, C., Houlsby, G., Utili, S., 2014. New insights into the 1963 Vajont slide using 2D and 3D distinct-element method analyses. Geotechnique 64, 800–816.
8. Boon, C., Houlsby, G., Utili, S., 2015. DEM modelling of a jointed rock beam with emphasis on interface properties. Géotech. Lett. 5, 49–55.
9. Bray, M.J., Hooke, J.M., 1997. Prediction of soft-cliff retreat with accelerating sea-level rise. J. Coast. Res. 13, 453–467.
10. Briaud, J.L., 2008. Case histories in soil and rock erosion: Woodrow Wilson Bridge, Brazos River Meander, Normandy Cliffs, and New Orleans Levees. J. Geotech. Geoenviron. 134, 1425–1447.
11. Camones, L.A.M., Vargas Jr, E.D.A., De Figueiredo, R.P., Velleso, R.Q., 2013. Application of the discrete element method for modeling of rock crack propagation and coalescence in the step-path failure mechanism. Eng. Geol. 153, 80–94.
12. Carson, M.A., Kirkby, M.J., 1972. Hillslope Form and Process. Chen, W.F., 1975. Limit Analysis and Soil Plasticity. Elsevier. Chen, W.F., Liu, X., 1990. Limit Analysis in Soil Mechanics. Elsevier. Chen, W.F., Giger, M., Fang, H.Y., 1969. On the limit analysis of stability of slopes. Soils Found. 9, 23–32.
13. Collins, B.D., Sitar, N., 2011. Stability of steep slopes in cemented sands. J. Geotech. Geoenviron. 137, 43–51.
14. Conte, E., Troncone, A., 2012. A method for the analysis of soil slips triggered by rainfall. Geotechnique 62, 187–192.
15. de Lange, W.P., Moon, V.G., 2005. Estimating long-term cliff recession rates from shore platform widths. Eng. Geol. 80, 292–301.
16. Duncan, J.M., 1996. State of the art: limit equilibrium and finite-element analysis of slopes. J. Geotech. Eng. 122, 577–596.
17. Dyer, M., Utili, S., Zielinski, M., 2009. Field survey of desiccation fissuring of flood embankments. Proceedings of the ICE-water Management. 162, pp. 221–232.
18. Espinoza, R., Bourdeau, P., Muhunthan, B., 1994. Unified formulation for analysis of slopes with general slip surface. J. Geotech. Eng. 120, 1185–1204.
19. Frayssines, M., Hantz, D., 2006. Failure mechanisms and triggering factors in calcareous cliffs of the subalpine ranges (French Alps). Eng. Geol. 86, 256–270.
20. Gao, Y., Song, W., Zhang, F., Qin, H., 2015. Limit analysis of slopes with cracks: comparisons of results. Eng. Geol. 188, 97–100.
21. Hachinohe, S., Hiraki, N., Suzuki, T., 2000. Rates of weathering and temporal changes in strength of bedrock of marine terraces in Boso Peninsula, Japan. Eng. Geol. 55, 29–43.
22. Hales, T., Roering, J.J., 2007. Climatic Controls on Frost Cracking and Implications for the Evolution of Bedrock Landscapes. Earth Surface, Journal of Geophysical Research, p. 112.
23. Huang, M., Jia, C.Q., 2009. Strength reduction FEM in stability analysis of soil slopes subjected to transient unsaturated seepage. Comput. Geotech. 36, 93–101.
24. Hutchinson, J.N., 1969. A reconsideration of the coastal landslides at Folkestone Warren, Kent. Geotechnique 19, 6–38.
25. Hutchinson, J.N., 1970. A coastal mudflow on the London Clay cliffs at Beltinge, North Kent. Geotechnique 20, 412–438.
26. Hutchinson, J.N., 1973. The response of London Clay cliffs to differing rates of toe erosion. Geol. Appl. Idrogeol. 7, 221–239.
27. Hutchinson, J.N., 2001. The fourth Glossop lecture reading the ground: morphology and geology in site appraisal. Q. J. Eng. Geol. Hydrogeol. 34, 7–50.
28. Kimmance, G., 1988. Computer aided risk analysis of open pit mine slopes in kaolin mined deposits. Ph. D thesis. Imperial College London (University of London).

29. Konrad, J.M., Ayad, R., 1997. A idealized framework for the analysis of cohesive soils undergoing desiccation. Can. Geotech. J. 34, 477–488.
30. Leroueil, S., 2001. Natural slopes and cuts: movement and failure mechanisms. Geotechnique 51, 197–243.
31. Ling, H.I., Leshchinsky, D., 1995. Seismic performance of simple slopes. Soils Found. 35, 85–94.
32. Loukidis, D., Bandini, P., Salgado, R., 2003. Stability of seismically loaded slopes using limit analysis. Geotechnique 53, 463–479.
33. Michalowski, R.L., 1995. Slope stability analysis: a kinematical approach. Geotechnique 45, 283–293.
34. Michalowski, R.L., 2013. Stability assessment of slopes with cracks using limit analysis. Can. Geotech. J. 50, 1011–1021.
35. Parsons, A.J., 2002. Hillslope form. Routledge. Péron, H., Hueckel, T., Laloui, L., Hu, L., 2009. Fundamentals of desiccation cracking of finegrained soils: experimental characterisation and mechanisms identification. Can. Geotech. J. 46, 1177–1201.
36. Potts, D.M., Zdravković, L., 2001. Finite Element Analysis in Geotechnical Engineering: Application (Thomas Telford). Rathje, E.M., Antonakos, G., 2011. A unified model for predicting earthquake-induced sliding displacements of rigid and flexible slopes. Eng. Geol. 122, 51–60.
37. Selby, M.J., 1982. Hillslope materials and processes. Hillslope Materials and Processes. Springman, S., Thielen, A., Kienzler, P., Friedel, S., 2013. A long-term field study for the investigation of rainfall-induced landslides. Geotechnique 63, 1177.
38. Take, W., Bolton, M., 2011. Seasonal ratcheting and softening in clay slopes, leading to first-time failure. Geotechnique 61, 757–769.
39. Terzaghi, K., 1943. Theoretical Soil Mechanics. John Wiley & Sons Inc., New York. Tsai, C.C., Chien, Y.C., 2016. A general model for predicting the earthquake-induced displacements of shallow and deep slope failures. Eng. Geol. 206, 50–59.
40. Utili, S., 2013. Investigation by limit analysis on the stability of slopes with cracks. Geotechnique 63, 140–154.
41. Utili, S., 2015. Discussion of "limit analysis of slopes with cracks: comparisons of results". Eng. Geol. 197, 306–307.
42. Utili, S., Abd, A., 2016. On the stability of fissured slopes subject to seismic action. Int. J. Numer. Anal. Methods Geomech. 40, 785–806.
43. Utili, S., Crosta, G., 2011a. Modeling the evolution of natural cliffs subject to weathering: 1. Limit analysis approach. J. Geophys. Res. Earth Surf. 116, F01016.
44. Utili, S., Crosta, G., 2011b. Modeling the evolution of natural cliffs subject to weathering: 2. Discrete element approach. J. Geophys. Res. F: Earth Surf. 116, F01017.
45. Utili, S., Nova, R., 2007. On the optimal profile of a slope. Soils Found. 47, 717–729.
46. Utili, S., Zhao, T., Houlsby, G., 2015. 3D DEM investigation of granular column collapse: evaluation of debris motion and its destructive power. Eng. Geol. 186, 3–16.
47. Vanicek, I., Vanicek, M., 2008. Earth Structures: In Transport, Water and Environmental Engineering (Springer Science & Business Media). Voulgari, C., 2016. Engineering Modelling of the Progressive Retreat of Cliffs (PhD Thesis at University of Warwick (UK)).
48. Wang, Y., Leung, S., 2008. Characterization of cemented sand by experimental and numerical investigations. J. Geotech. Geoenviron. 134, 992–1004.
49. Wasowski, J., Keefer, D.K., Lee, C.-T., 2011. Toward the next generation of research on earthquake-induced landslides: current issues and future challenges. Eng. Geol. 122, 1–8.
50. Yang, X.G., Chi, S.C., 2014. Seismic stability of earth-rock dams using finite element limit analysis. Soil Dyn. Earthq. Eng. 64, 1–10.
51. Yerro, A., Alonso, E., Pinyol, N., 2015. The material point method for unsaturated soils. Geotechnique 65, 201–217.

52. Yokota, S., Iwamatsu, A., 2000. Weathering distribution in a steep slope of soft pyroclastic rocks as an indicator of slope instability. Eng. Geol. 55, 57–68.

53. Zheng, H., Liu, D., Li, C., 2005. Slope stability analysis based on elasto-plastic finite element method. Int. J. Numer. Methods Eng. 64, 1871–1888.

CHAPTER 2

Application of Geotechnical and Geophysical Field Measurements in an Active Alpine Environment

D.R. Lucas, K. Fankhauser, S.M. Springman

ETH Zürich, Institute for Geotechnical Engineering, Wolfgang-Pauli-Str. 15, 8093 Zürich, Switzerland

ABSTRACT

A gravelly scree slope in the Meretschibach catchment, a location in the Swiss Alps in the vicinity of Agarn, canton Valais, has been observed to deform downslope at up to 0.5 m p.a. The potential instabilities at this site include surficial landslides, some of them originally thought to be triggered by an increase in pore water pressure with a subsequent loss of shear strength as a consequence of rainfall infiltration and rockfalls. A programme consisting of monitoring, laboratory testing and investigation was developed, to perform a thorough soil characterisation needed in order to produce a realistic ground model. The long-term geotechnical monitoring included in situ soil temperature, suction as well as volumetric water content measurements using dielectric permittivity and time domain reflectometry (TDR) sensors. This data was complemented by electrical resistivity tomography (ERT) to provide extensive knowledge on the depth to bedrock and to validate the volumetric water contents in specific locations. The datasets are completed by recordings from two nearby weather stations. Seasonal changes of precipitation and temperature were reflected in corresponding trends in all measurements. A comparison of volumetric water content records was obtained using capacitance and time domain reflectometry (TDR) sensors with ERT, yielding reasonable agreement. The resulting ground model, which integrates all currently available parameters, delivers the essential information and boundary conditions for predicting and validating slope instabilities in the future, using numerical and physical modelling.

1. INTRODUCTION

The Meretschibach catchment (Fig. 1), situated in the Swiss Alps, canton Valais, near the village of Agarn (620 m.a.s.l.), has been investigated due to dynamic processes of erosion, deposition and remobilisation of debris, which can evolve into different kinds of mass movement. Debris flows have reached Agarn from an active channel and caused damage to infrastructure in the past. This has been well documented (Rickenmann and Zimmermann, 1993 ; Oggier, 2011). However, little has been published about slope instabilities triggered in gravelly soils in the adjacent scree slope, where surface movements of up to 0.5 m p.a. have been obtained from InSAR measurements (Dr. Hugo Raetzo, 2012, FOEN, pers. Comm.).

Three prevalent types of downslope movement were expected, either based on past research (Springman et al., 2012), or observed during the monitoring programme on the steep scree slope:

- shallow landslides triggered by an increase in pore water pressure that reduces soil suction and shear strength;
- boulders that topple, fragment, bounce and roll downslope from weathered rock walls at the top of the slope;
- ongoing surficial erosion, including destabilisation of the top ground layer through rainfall runoff and flow, movement of dry debris as well as small avalanches in winter.

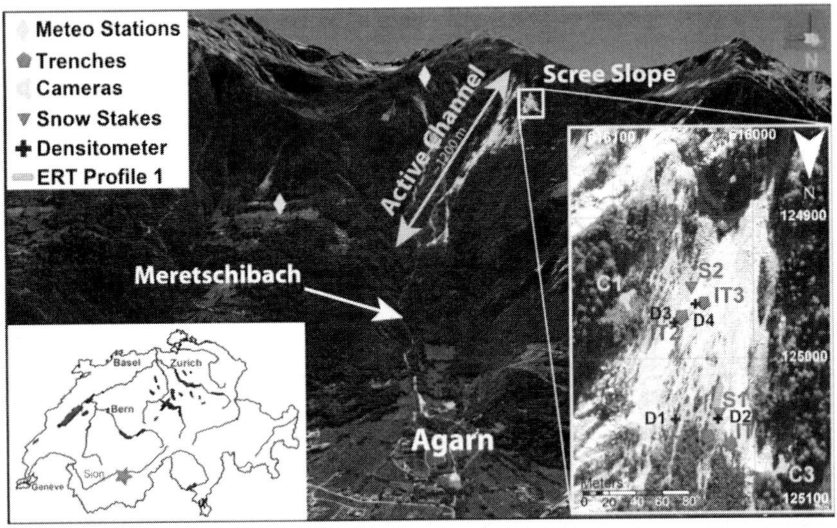

Fig. 1. Overview of the field area, located in canton Valais, Switzerland (star on small map). The image looking south shows a view of Agarn, situated on the Rhone valley floor and the Meretschibach catchment on the mountain slopes behind. The most active area within the catchment can be divided into an active channel and a scree slope. The weather stations (yellow diamonds) are located at 1370 m.a.s.l. (IGT) and 2220 m.a.s.l. (WSL). The right image shows an enlargement of the scree slope (yellow rectangle) with the locations of the instrumented trenches (IT1-IT4), the ERT profile on which the monitoring was performed, snow stakes (S1–S2), densitometer measurements (D1–D4) and cameras (C1–C4).

Characterisation and long-term monitoring of this slope are necessary in order to gain more understanding of these processes, although this is complicated by slope inclinations of 33–43°, heterogeneous gravelly soil, patches of native vegetation, seasonal limitations and difficult accessibility due to the alpine location at 1840–2000 m.a.s.l. A realistic ground model can be established, defining extent of soil layers and depth to bedrock, as well determining relevant soil properties at different depths.

A long-term monitoring campaign, integrating geotechnical and geophysical methods together with laboratory analyses, allows a progressively more detailed calibration and validation of the ground model and relevant parameters, while observing and measuring the response of the scree slope to seasonal processes. Instrumentation was installed in the scree slope at shallow depths (< 1 m) within four trenches (IT1-IT4), to perform real time monitoring of the volumetric water content (VWC), temperature and suction (Lucas *et al.*, 2015 ; Springman *et al.*, 2015). The VWC and the suction measurements provide valuable information to assess changes in soil shear resistance, particularly for soil in a partially saturated state and to enable a comparison between the VWC acquired by two different types of moisture sensors. Additionally, the precipitation in the area was measured with two weather stations located within a distance of 2 km, at elevations of 2220 and of 1370 m.a.s.l., respectively.

Geophysical measurements complemented the long term soil monitoring, providing a convenient way to extend the knowledge of the subsurface to larger areas in the scree slope (Bogoslovsky and Ogilvy, 1977; Hack, 2000; Bichler *et al.*, 2004; Otto and Sass, 2006; Sass, 2007 ; Sass *et al.*, 2008). Electrical resistivity tomography (ERT) measurements were conducted repeatedly throughout the spring and summer months of 2014, on a profile located close to IT1 (Fankhauser, 2014), allowing the in situ VWC recordings to be validated in specific locations using Archie's law (Archie, 1942 ; Friedel *et al.*, 2006). Effective validation can demonstrate complementarity of geotechnical and geophysical methods and provide a more widespread interpretation of the volumetric water content (VWC), leading to a more detailed characterisation of the scree slope.

The data presented here was acquired over a period of two years, and is analysed by seasons. Trends in summer and winter, and thawing/freezing processes in spring and autumn, are discussed further.

2. SOIL GROUND CHARACTERISTICS

2.1. Soil classification

Fig. 1 shows the location of the research site and, specifically, the instrumented trench sites IT1–IT4. Samples were taken from test pits and trenches during dry summer weather conditions. The following characteristics were noted: the ground surface was considered to be 'depth zero' for each test pit; at least 4 kg of soil were extracted for each disturbed sample. Cobbles larger than 70 mm, and any boulders, were left at the site. Roots were found in almost all of the samples.

Grain size analyses were performed later in the laboratory. The soil is mostly gravel, with diverse percentages of sand and fines, and it is classified (Fig. 2 ; Fig. 3 and Table 1) according to Swiss standard classification (SN 670004-2NA), as poorly-graded gravel with silt and sand (GP-GM), silty gravel (GM), well-graded gravel with silt and sand (GW-GM), and well-graded gravel (GW). Fig. 2 shows the soil profiles in the instrumented trenches.

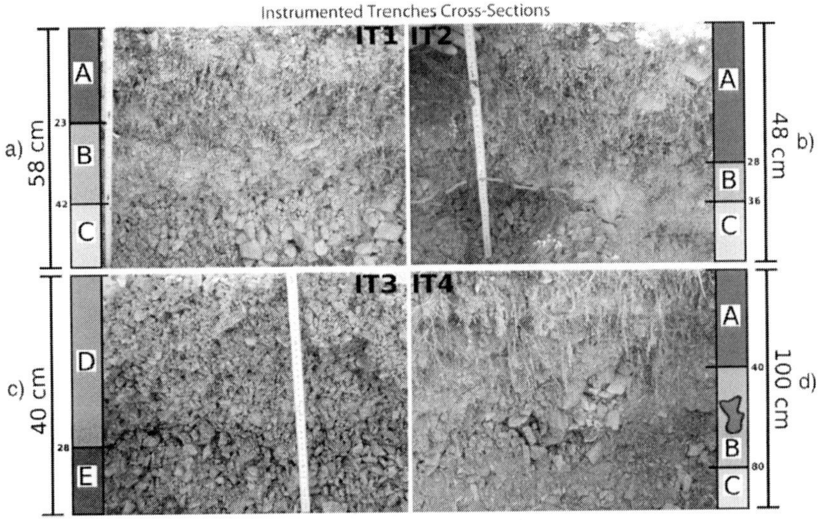

Fig. 2. Soil profiles at instrumented trenches. a) IT1, b) IT2, c) IT3 and d) IT4. For IT1, IT2 and IT4: A (depth ~ 0–40 cm), gravel with silt, sand and some roots; B (depth ~ 25–40 cm), gravel with silt and sand; C (depth below ~ 40 cm), coarser gravel with less fines; IT3: D (depth ~ 0–25 cm), gravelly soil with little sand and silt content; E (depth below ~ 25 cm) change to a coarser gravel.

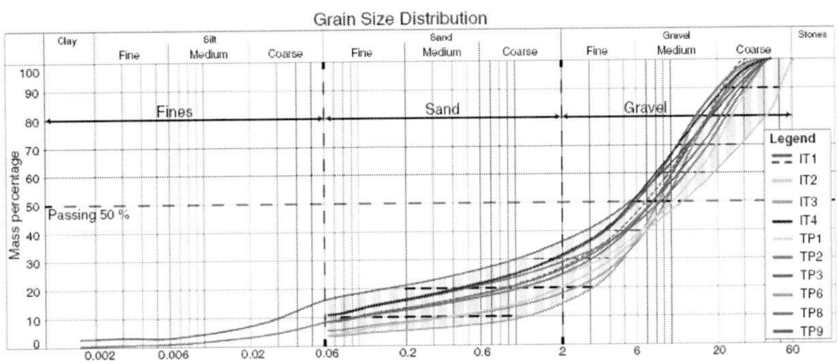

Fig. 3. Grain size distribution: a selection from Meretschibach scree slope test pits TP 1–9 are plotted, together with IT1–IT4. The range is shaded in grey to illustrate the characteristics and content of fines for the entire set of samples.

Table 1.Soil classification from test pits and instrumented trenches.

Test pit	Swiss standard classification (SN 670004-2b NA)	Moisture content [%]	Percent of fines [%]	C_u [−]	C_c [−]	G_S [−]
TP1	GP-GM	2.9–3.7	5.7–8.2	73.6–139.3	8.3–14.4	
TP2	GP-GM	3.0–5.4	5.5–10.8	33.3–105.1	7.9–11.2	
TP3	GM	3.7–4.3	6.7–16.4	72.8	9.1	
TP4/IT1	GP-GM	–	7.7–9,2	98.2–115.3	6.2–8.2	2.68
TP5	GP-GM	–	5.7	37.6	5.0	
TP6	GP-GM	–	5.4	28.7	5.1	
TP7	GP-GM	–	8.4	106.1	7.7	
TP8	GP-GM	–	8.5	110.2	10.1	
TP9		–	11.0	–	–	
TP10	GW-GM		11.7	94.9	2.2	
IT2	GW	2.2	4.0	46.9	2.32	
IT3	GW		3.4	8.4	1.9	
IT4	GP-GM		10.9	47.5	2.32	

C_u: Coefficient of uniformity, C_c: Coefficient of curvature; G_S: Specific gravity.

Three common soil units are found in IT1, IT2 and IT4, but with different thicknesses: (A) gravel with fines and high root content; (B) gravel with fines, and less or no roots, (C) coarser gravel. An orange dashed line separates the gravel from a soil layer with greater fines content, considered here as more stable and suitable for instrumentation placement, from either coarser gravel layers or isolated lenses of coarser gravel that were more unstable for excavation (IT4). In contrast, IT3 consists of two different units described as finer gravel (D) and a transition to slightly coarser gravel (E).

2.2. In situ unit weight

The in situ unit weight was determined with a soil densitometer (Magdeburger Prüfgeräte GmbH (HMP)), which uses the balloon method (DIN 18125-2:2011-03) from which void ratio and porosity can be derived. It could be transported to alpine sites to carry out measurements in gravelly soil.

The unit weights calculated from the tests are presented in Table 2 and the test locations are shown in Fig. 1. These values were in agreement with the void ratio range of 0.34–0.69, which was determined in the laboratory from test samples.

Table 2. Dry unit weight obtained with the soil densitometer method.

Measurement	Dry unit weight [kN/m^3]	Void ratio [—]	Relative density [%]	Porosity [—]
D1	17.63	0.52	48.57	0.34
D2	16.98	0.58	31.43	0.37
D3	17.69	0.51	51.43	0.34
D4	16.09	0.66	8.57	0.40

2.3. Slope inclination

The slope inclination in the scree slope (Fig. 4) was obtained using the geographical information system (GIS). The most predominant angle of inclination in the scree slope is between 33 and 43°, although some steeper areas (mostly upslope) can be identified > 43°.

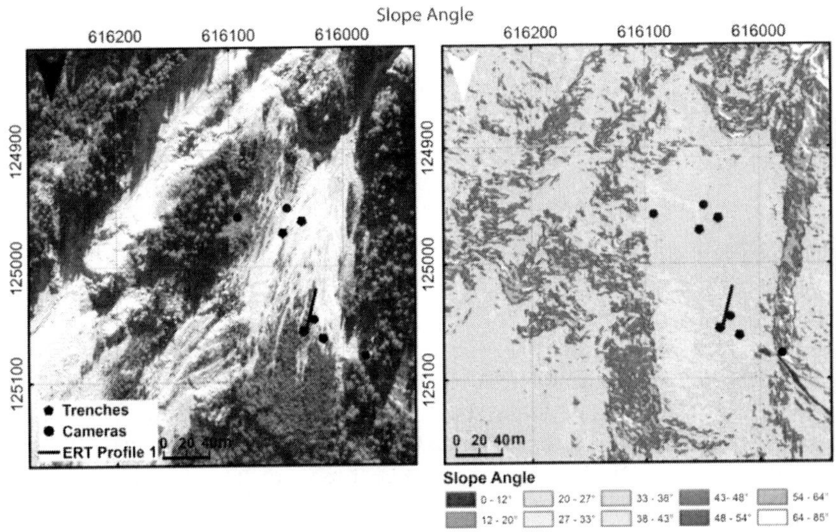

Fig. 4. Left: Aerial photograph (SWISSIMAGE) of the scree slope and top part of the active channel; right: slope angle in °: of the same area. Black symbols show the location of instrumentation (see Fig. 1 for comparison).

2.4. Instrumented trenches

Selected spots on the scree slope (i.e. landslide source areas) are monitored in situ to measure water infiltration-suction characteristics in the ground, to provide time-series data on infiltration quantities and rates, as well as the influence on the capillary suction, which directly affects the effective stress. Four monitoring profiles were installed (IT1–IT4, Fig. 1 ; Fig. 2), consisting of sets of dielectric permittivity sensors (EC-5 and 10HS), time domain reflectometry (TDR) soil moisture and temperature sensors (Table 3). Two trenches, IT1 and IT4, are supplemented by co-located tensiometers. All of the sensors were installed in trenches up to 1 m depth and of minimum 1 m width, to minimise interference between sensors at similar depths.

Table 3. Technical information about the sensors installed.

Sensor	Name	Brand	Length	Measuring range	Volume of influence cm^3	Accuracy	Temperature range [°C]
Capacitance	EC-5	Decagon Devices*	0.089 m	0–100% VWC	240	± 1–2% VWC with soil specific calibration	−40 to +50
	10 HS (only IT2)		0.145 m	0–57% VWC with polynomial	1320	using soil specific calibration, ±0.02 m^3/m^3 (±2% VWC) in any soil	Survival temperature: −40–50 Operating temperature: 0–50
Time domain reflectometry	TDR100	Campbell Scientific**	0.15 m	0–100% VWC	188.5	± 1.5% VWC	−40 to +55
Tensiometers	Jetfill 2725 ARL	Keller AG Company***	0.15, 0.30, 0.45 m	0–85 kPa	Direct contact	± 1.5%	Avoid freezing conditions
	Remote 2100F	Keller PR23-S****	Installed at depths from 0.2 to 1 m.	0–85 kPa			
Temperature	RTD PT100	Heraeus	Installed at depths from 0.2 to 1 m.–	−50 to 100 °C	Direct contact	±0.05 °C	−50 to 100 °C

* http://manuals.decagon.com/Manuals/13876_EC-5_Web.pdf.
** https://s.campbellsci.com/documents/us/manuals/tdr100.pdf.
*** http://www.soilmoisture.com/RESOURCE_INSTRUCTIONS-all_products/Resource_Instructions_0898-2725_2725%20Jet-Fill%20Tensiometers.pdf.
**** http://www.soilmoisture.com/RESOURCE_INSTRUCTIONS-all_products/Resource_Instructions_0898-2100_2100F%20Soilmoisture%20Probe.pdf.

Two weather stations were installed and are monitored within the study area, a SR50A Campbell weather station (precipitation, ambient air temperature, humidity) controlled by the Swiss Federal Institute for Forests, Snow and Landscape research (WSL), and an OTT Pluvio² device (precipitation, ambient air temperature, humidity and wind direction) installed by ETH (IGT). While the Campbell station is used to measure precipitation in the form of rain when the ambient air temperature is higher than zero (no heater), the OTT Pluvio² is able to measure precipitation in the form of rain, snow, or a mixture of both during all seasons.

Environmental changes, including freezing/thawing conditions, particularly during the crucial autumn/spring snow-melt period, can be identified through ambient and soil temperature measurements. The TDRs and dielectric permittivity sensors (EC-5, 10HS) are calibrated to measure the volumetric water content (VWC).

2.4.1. Sensors and calibration

2.4.1.1. Sensors

Two types of sensor were installed to measure VWC (Table 3): time domain reflectometry (TDR100; Campbell Scientific) and capacitance sensors (EC-5 and 10HS; Decagon Devices), via the apparent dielectric constant/permittivity of the soil surrounding the probes.

The volume of sensitivity of the sensors (soil volume influenced during VWC measurement) becomes relevant while interpreting data in terms of first response to a saturation event. A representative VWC of gravel in the scree slope should reflect the range of particle sizes and the measurement volumes, with 181 (Onset, n.d.) and 1160 cm^3 (ICT international, 2008) respectively for EC-5s and 10HSs (method after Sakaki et al., 2008), and 1.4 times the diameter of the rod spacing (2.1 cm) for TDRs (O'Connor and Dowding, 1999), leading to a volume of 101 cm^3 (Table 3).

The suction was recorded with tensiometers (Jetfill 2725 in IT1 and Remote 2100F in IT4; Keller AG), paused in October before winter (to avoid freezing conditions) and restarted again in May. While the Jetfill were installed a few metres upslope of the trench, to a maximum depth of 0.45 m, the Remote were installed in the trench with the other instruments at different depths (up to 1 m).

Further information on all sensors can be found in supplementary material A.

2.4.1.2. Calibration

The capacitance and TDR sensors were calibrated in the laboratory using poorly-graded gravel with silt and sand (GP-GM) from the field, over the range of VWCs estimated for the unsaturated conditions expected in situ. The specimens were prepared by moist tamping, with void ratios in the range of $e = 0.53 - 0.69$, simulating suitable installation conditions with both medium and loose densities.

A linear calibration was fitted to the measured values in most cases, in which the coefficient of determination was $r^2 = 0.56–0.98$. In some cases, this could be improved by fitting a polynomial equation, and this was adopted, for instance, for IT2 capacitance sensor 10HS at 13 cm depth.

Although Topp's equation (Topp et al., 1980) allows prediction of VWC for grain sizes up to 2 mm, the VWC obtained were between factors of 2–3 smaller than for a site-specific linear calibration at 20 °C using the soil from the field with a $d_{max} \leq 31.5$ mm (albeit with a 8% content of fines). This significant difference could be due to effect of the texture, structure, and density on the electrical response in soils, as mentioned by Topp et al. (1980). These results confirm that a site specific calibration function is necessary when using poorly-graded gravel with silt and sand. While Topp's approach could not be completely applicable to the grain size distribution of the scree slope, the adjustment of the permittivity due to changes in temperature derived by Bogena et al. (2007), using a liquid with known dielectric properties (2-isopropoxyethanol (i-C3E1) water mixtures) was tested against Bogena's equation for loamy silt. This has a significant effect on the calculated VWC, which in many cases maps closely to the values of the site-specific calibration results. As the use of a site-specific calibration function has been recommended by other authors (e.g. Take et al., 2007 and Mittelbach et al., 2011), this has been adopted. A site-specific equation for the correction of permittivity due to temperature (applicable to capacitance sensors installed in scree soil) is advised for future studies.

Ultimately, the performance of a sensor type is directly related to the soil conditions in the field. Large stones and boulders, sometimes comparable to the size of the capacitance sensor prongs, complicate effective insertion of the probes into the soil. Heterogeneity contributes natural variability in density, void ratio and grain sizes that have impacted on the process of site-specific calibration and functionality of TDRs and capacitance sensors during monitoring.

2.5. Electrical resistivity tomography (ERT)

A 2D electrical resistivity survey was carried out on a fixed profile line next to instrumented trench IT1 (Fig. 1) to complement the point measurements from the TDRs and dielectric permittivity sensors, and achieving a more widespread characterisation of the subsurface. The basic principle of geoelectrical resistivity measurements, as well as the equipment and configurations used in this project, are briefly summarised in Supplementary material A. Further details on acquisition, processing etc. can be found in Fankhauser (2014), Günther and Rücker (2013), Günther et al. (2006), Günther (2011), Rücker et al. (2006) and Lowrie (2007). A total of six ERT measurements (three of which are presented here) were poorly-graded gravel with silt and sand (GP-GM) conducted in a monitoring phase from May to July 2014.

2.6. Archie's law

ERT measurements are highly susceptible to changes in the subsurface water content/saturation as the current injected during a geoelectrical survey flows mostly through the pore water. An empirical relationship between the resistivity of a porous medium to the amount of pores, their connectivity and saturation, as well as the resistivity of the pore filling fluids, was found by Archie (1942):

$$\rho = a\rho_w \Phi^{-m} S^{-n} \qquad \text{with} \quad S = \frac{\theta}{\Phi} \qquad (1)$$

where ρ is the resistivity of the porous soil, ρw is the resistivity of the pore filling fluid, Φ is the soil porosity (typically n in a geotechnical context) and S the saturation, which can also be expressed as the ratio between the volumetric water content θ and the porosity. Furthermore, a is the tortuosity factor set to 1 for granular soils, m is the cementation factor and n is the coefficient of saturation. The latter two factors are assumed to be constant and, if taken from literature (e.g. Archie, 1942; Lowrie, 2007 ; Schön, 1983), are usually given as $m \cong 1.3$ and $n \cong 2$ for granular soil. Archie's law is considered to be valid in this case (without the need for any correction factors), as the soil consists mainly of gravel and sand and contains almost no fines, especially no clay (Figs. 2 and 3, Table 1). The relationship can be used for effective comparisons between the geotechnical soil moisture measurements and ERT acquisitions (e.g.Brunet *et al.*, 2010; Lehmann *et al.*, 2013 ; Springman *et al.*, 2013). For this purpose, the key parameters, Φ and ρw, were estimated from soil and water samples (small creeks and surface run-off near to the scree slope) extracted from the field, and analysed through laboratory and field experiments as $\Phi = 0.33 \mp 0.08$ and $\rho w = 155 \mp 28\ \Omega\text{m}$ (at 25 °C; for temperature corrections, the equation and the geophysical temperature compensation factor stated in Hayashi (2004) were used).

A convenient way to use the temporally varying soil resistivities (repeatedly measured ERT profile) is to calculate the saturation relative to a so-called baseline model (chosen here as the acquisition on 15th May 2014) and so avoid having to estimate many (uncertain) parameters in Archie's law. The resulting trend (Fig. 16) can be compared relative to the readings obtained from TDR and dielectric permittivity sensors. This process is further described in Supplementary material A.

3. RESULTS AND DISCUSSION

3.1. VWC and temperature monitoring in instrumented trenches IT1–IT4

An overview of the data time series recorded since October 2013 is presented in Fig. 5, where a set of two plots display the volumetric water content (VWC) and temperature versus time (in months at the bottom and the top, separated by dashed lines), in chronological order of installation for each of the instrumented trenches (from top to bottom: IT1–IT4).

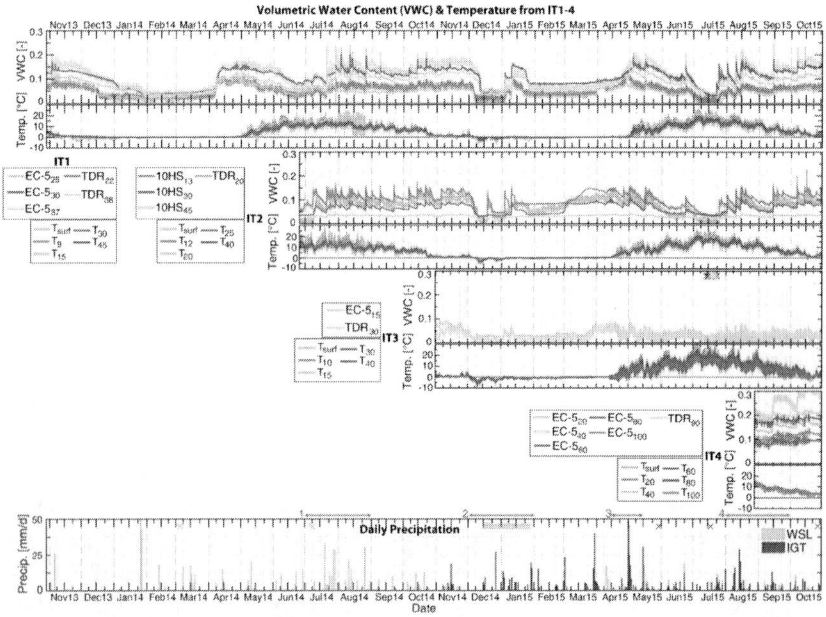

Fig. 5. Overview of measurements at instrumented trenches. From top to bottom for each trench IT1–IT4: VWC, temperature; precipitation from WSL and IGT stations on the bottom.

The VWCs displayed in all of the plots range from 0 to 0.3 and the temperatures from – 10 to 30 °C. The precipitation from the two weather stations (WSL and IGT) is shown at the bottom, over a range of 0 to 50 mm per day. Each coloured line represents a sensor installed at a given depth, as shown in each legend with the corresponding sensor name (Dec = dielectric permittivity sensor; T = temperature), as well as the depth (in centimetres below the surface) as a sub-index. The VWC from the three types of moisture sensors (dielectric permittivity sensors and TDRs) can be compared when they are at similar depths. *Tsurf* is the temperature measured in the data logger at the surface, usually a few metres away from the trench. The background shading represents a definition of summer (yellow) and winter (blue), as the season during which fieldwork can be carried out and a time during which access to the site is not possible, respectively. Missing data due to loss of signals or eliminated data from a specific device are marked at the top of the plot by a cross of the same colour.

Clear and expected seasonal differences are observed between summer, winter and spring/autumn in terms of temperature and volumetric water content (VWC).

3.2. Seasonal response of the scree slope

3.2.1. Summer

Fig. 6 ; Fig. 7 represent the summers of 2014 and 2015, respectively at different locations of the scree slope (Fig. 1, IT1–IT3). High temperatures in summer, combined with precipitation in the form of rainfall, which infiltrates into the ground, produces almost immediate increases in VWC near the surface. The variation in magnitude of VWC between trenches IT1–3 (Fig. 7) can be attributed to the different grain size distributions (GW: IT2–3; GP-GM: IT1,4). Two types of infiltration behaviour can be identified in trenches IT1-IT2, which can be seen when comparing the VWC in July 2014. The increase of VWC can either be gradual with a convex drop-off as the ground drains, or reach immediate peaks, followed by linear to concave decreases. The former were associated with periods of high suctions (especially in deeper tensiometers, and during periods of little to no rainfall), when the unsaturated hydraulic conductivity decreased, while immediate peaks occurs at higher degree of saturation. Furthermore, suctions reduced either in the same timeframe as the rain infiltration or marginally slower than the VWCs had increased. This is due to the radius of influence of the sensors, EC-5 has a greater radius of 20 mm and the tensiometers around 1 mm, therefore capacitance sensors react faster during infiltration (Beck, 2010). Sun exposure is higher in IT3, with temperature ranging from 0 to 27 °C, and daily peaks close to ambient air temperature.

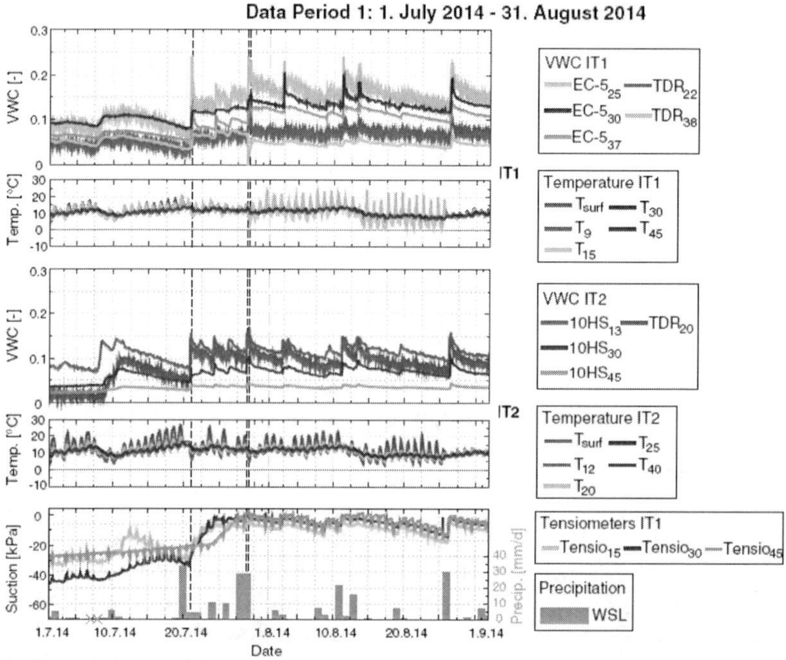

Fig. 6. VWC, temperature and precipitation at IT1–IT2 and suction data at IT1 are displayed from 1st July 2014 to 31st August 2014. The debris flow events in the active channel (Fig. 1) are marked with a dashed line each. The sub-index in the legend indicates the sensor depth in cm. Missing precipitation data are marked with a cross.

3.2.2. Winter

Temperatures are near or below 0 °C (Fig. 8), so that precipitation occurs in the form of snowfall or mixed with rainfall. The pore water fluid is expected to freeze when temperatures are below 0 °C, decreasing the permeability of the unsaturated soil even further. Snow can accumulate on, and insulate, the slope, reducing the rate and volume of water infiltration (Fig. 8, camera 2, 28.12.2014, 10:00). These two processes explain the degree of saturation dropping to a minimum and the lack of variation in VWC with ongoing winter precipitation (Fig. 8, December, beginning January), while infiltrating precipitation and snow-melt processes cause rising VWC when there is an increase from negative to positive temperatures (Fig. 8, end December, beginning January). Episodes of freezing and snow melting are observed later in some trenches, because the ground temperature is dependent on sun exposure, which differs depending on the location on the scree slope (up- or downslope).

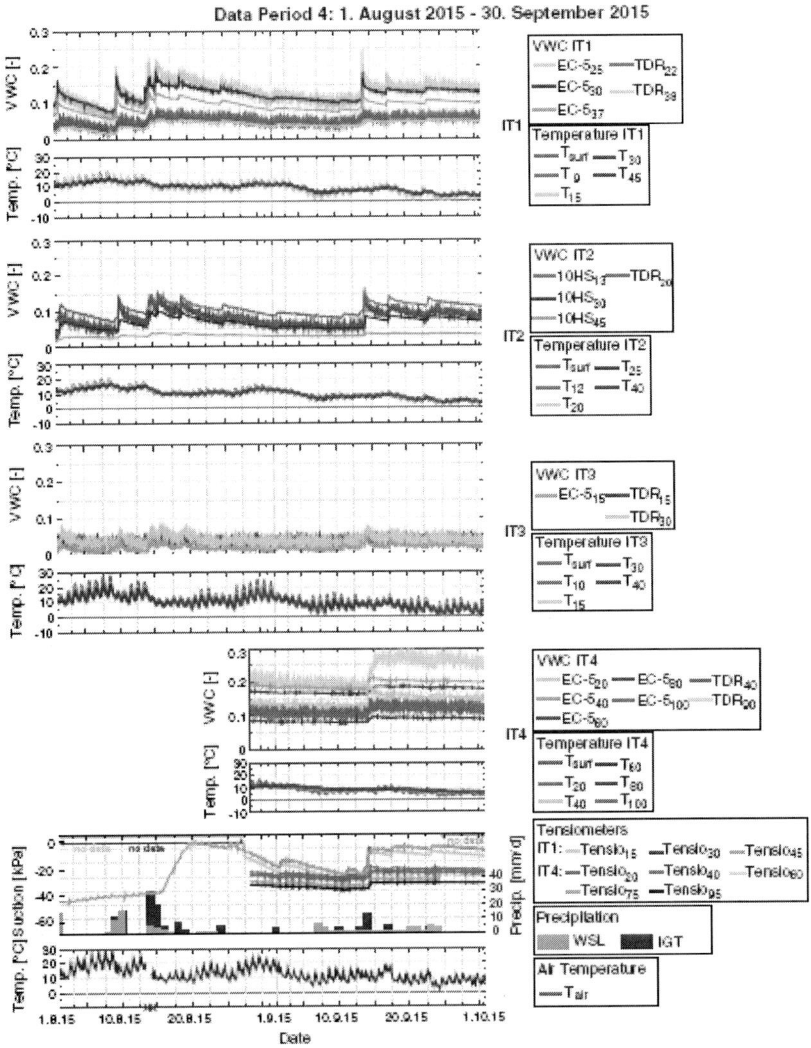

Fig. 7. VWC, temperature and precipitation at IT1–IT4 and suction data from IT1 and IT4 are displayed from 1st August 2015 to 30th September 2015. In addition, the ambient air temperature recorded at the IGT weather station is shown on the bottom. The sub-index in the legend indicates the sensor depth in cm. Missing suction data are marked with an arrow in corresponding colour and missing air temperatures with crosses.

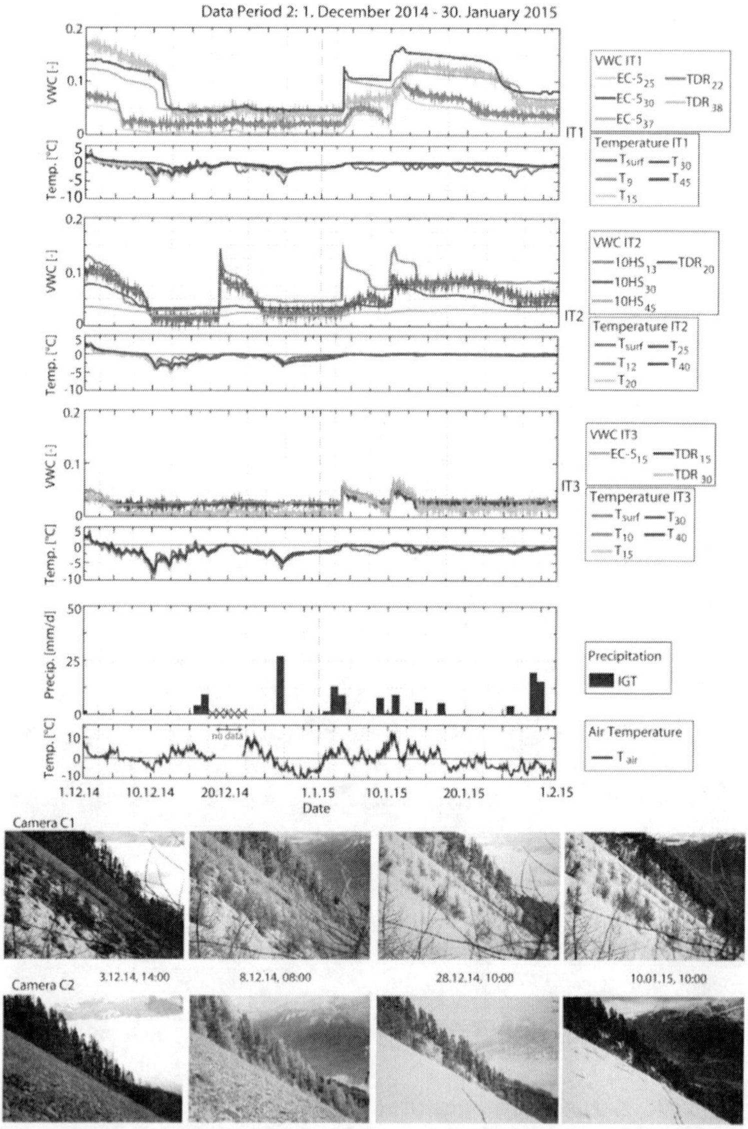

Fig. 8. VWC, temperature and precipitation at IT1–IT3 from 1st December 2014 to 31st January 2015. The ambient air temperature recorded at the IGT weather station is shown below the other graphs. The sub-index in the legend indicates the sensor depth in cm. Missing data are marked with crosses. Photos from selected dates (cameras C1 and C2; Fig. 1) are shown on the bottom.

Patterson and Smith (1981) proposed that K_a is not highly sensitive to ice content in partially saturated coarse sand samples, since the slight increment measured is a product of replacing air ($K_a = 1$) by ice ($K_a = 3–4$). Moreover, they measured K_a at negative temperatures in fine-grained soils (clay loam, silt loam), showing that the VWC for ice-water mixtures was within (± 2.5%) of values calculated by Topp et al.'s (1980) equation, suggesting that it could be applicable, at least, to unfrozen and frozen fine-grained soils.

Considering Patterson's approach, with Topp's values of VWC compared to soil-specific calibration, and VWCs measured at sub-zero temperatures by capacitance and TDR sensors at different depths (Fig. 8):

- IT1 TDR_{22} and $EC\text{-}5_{25}$ show similar ranges of VWC, with EC-5 slightly lower,
- TDR_{38} and $EC\text{-}5_{37}$ match well,
- as do IT2 TDR_{20}, $10HS_{13}$ and IT3 TDR_{15} and $EC\text{-}5_{15}$.

This issue has been further studied by Spaans and Baker (1995), who proposed improving VWC estimation by calibrating TDR sensors in frozen soil, dependent on VWC and temperature.

Frost heave causes change in structure and expansion of saturated fine-grained soils through ice lens formation in the frozen zone under sub-zero temperatures, accompanied by water migration from the unfrozen zone to the ice lenses (Miller, 1973; Taber, 1929 ; Taber, 1930). As thermodynamic equilibrium is required at the interface of the ice lens and the water, suction builds up, changing the VWC and therefore the hydraulic conductivity of the soil. Furthermore, frost heave in a steep slope can contribute to surficial movement, with surface debris rolling downslope (Matsuoka, 1998). This process is unlikely to have happened at the study site, even when soil reached sub-zero temperatures (Fig. 5 ; Fig. 8), although there would have been more likelihood to occur, very locally in silty soil around roots of stunted vegetation in IT1 and IT4, due to the higher content (around 10%) of fines (Fig. 3, Table 1).

3.2.3. Spring/autumn

The spring snow-melt (Bayard et al., 2004) happens over conditions of sun exposure and higher temperatures during the day, producing a significant steady supply of water with more gradual infiltration. In combination with precipitation in the form of rainfall, water infiltrates through the partial insulation (e.g., remaining snow patches) into the ground, raising the VWC

(Fig. 9 ; Fig. 10). The decrease of temperatures in autumn set the conditions for snowfall and potential soil freezing, which results in reduced VWCs.

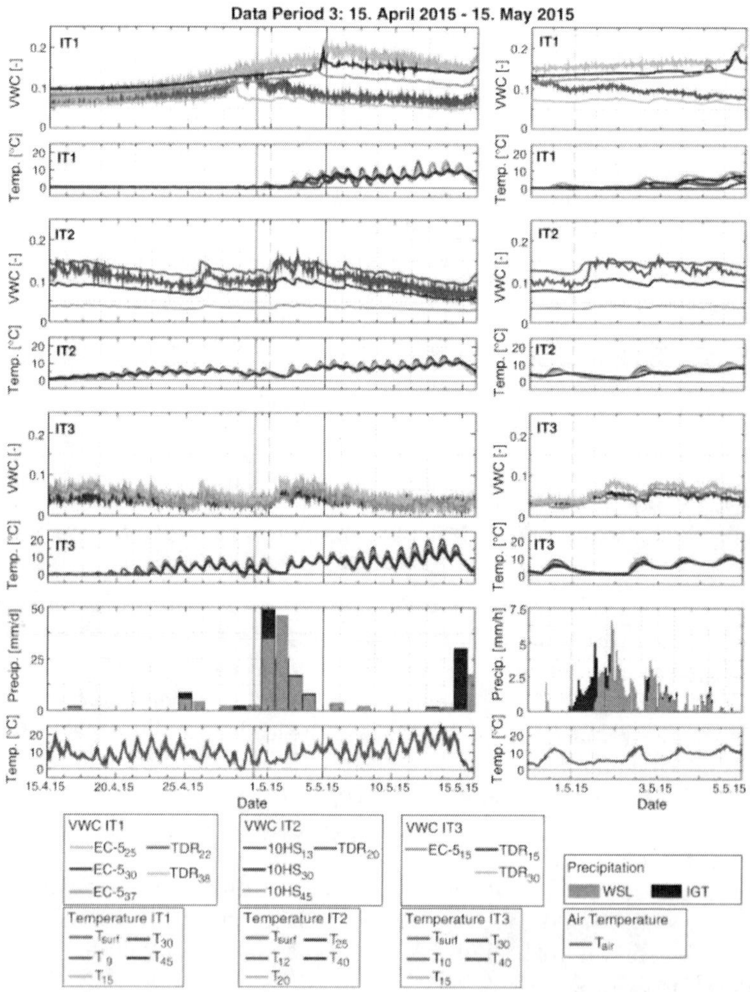

Fig. 9. VWC, temperature and precipitation at IT1–IT3 from 15th April 2015 to 15th May 2015 on the left hand side and from 30th April 2015 to 5th May 2015 on the right hand side. The ambient air temperature recorded at the IGT weather station is shown on the bottom. The sub-index in the legend indicates the sensor depth in cm. Corresponding photographs is shown in Fig. 10.

Fig. 10. Photographs from cameras C1 and C2, illustrating the time series in Fig. 9.

Fig. 9 shows data from mid-April to mid-May 2015, including a larger scale plot of the most extreme precipitation event (2nd May 2015). The spring season shows the transition in soil temperature rising from minimum values, while the air temperature is already positive, with a daily variation of ~10 °C. The temperatures rise earlier in IT2 and IT3 (upslope) respectively, whereas IT1 (downslope) follows days later, confirming the effect of sun exposure. The daily temperature variation in IT1 through IT3 is greater for shallower depths (15–20 cm; yellow), compared with deeper sensors (40–45 cm; red). However, the magnitude of this variation in IT3 is much greater at depth than in any other trench. (For interpretation of the references to colour in this figure legend, the reader is referred to the web version of this article.)

3.3. Tensiometers

The suction versus time was monitored at depth ranges equivalent to VWC measurements for IT1 and IT4 (Fig. 11 ; Fig. 12). The data are plotted to a monthly scale, separated by dashed lines, corresponding to the different periods. Missing or excluded data were marked by a continuous solid arrow at the top of each graph, using the colour trace for the malfunctioning sensor.

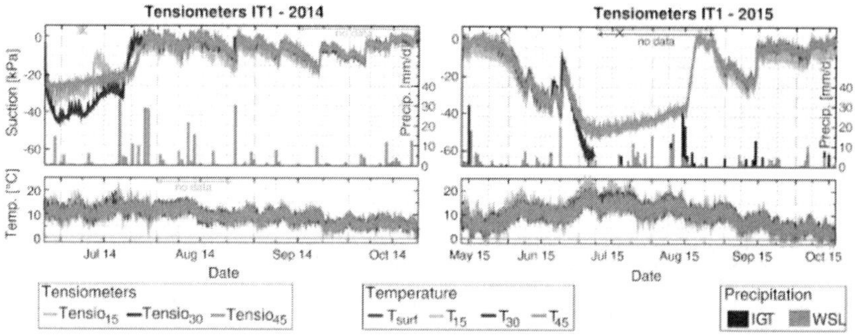

Fig. 11. Tensiometer measurements at IT1, temperature and precipitation data for two summer seasons (25th June 2014 to 23rd October 2014 and 13th May 2015 to 18th October 2015).

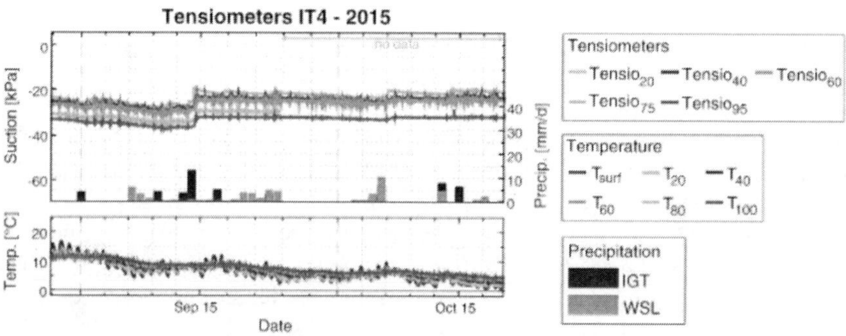

Fig. 12. Tensiometer measurements at IT4: suction, precipitation and temperature data (28th August 2015 to 20th October 2015).

Fig. 11 shows the suction measured in IT1 at 15, 30 and 45 cm depth, complemented by daily total precipitation in mm/day, plotted on a parallel y-axis and temperatures at different depths in a separate plot below. Periods run from the end of June to October 2014 and May to October 2015.

The three tensiometers at IT1 were installed in soil units, A, B, and C, (roots with gravel and fines, gravel with sand and fines, and coarser gravel respectively, GP-GM, Section 2.1). Correlation between the loss in suction after rainfall events can clearly be observed, after all tensiometers show a significant increase in pore water pressure during periods of high precipitation in July 2014. In addition, shallower tensiometers react faster after a rainfall event than the deeper tensiometers, due to top-down infiltration.

The unsaturated hydraulic conductivity of the soil will decrease during July 2014 and 2015 (maximum annual temperatures) due to the higher suction level, resulting in a small variation of the suction, regardless of the rain events recorded. The soil only approaches saturation after a major rainfall event, leading to higher permeability and a more dynamic response to rain infiltration.

Fig. 12 shows the suction measured in IT4 (GP-GM) at five depths (20–95 cm), complemented by daily total precipitation and temperatures. The period extends from the end of August to October 2015. Suctions increase with depth, indicating top-down infiltration through macro-voids after rain (September 15th), with gradual recovery of suction, which is surprising given the lenses of coarser gravel exposed in IT4 (Fig. 2).

3.4. Temperature and VWC with depth

Fig. 13 shows the temperature distribution with depth from 5th to 25th September 2015 for IT1–IT4, from top to bottom, respectively. The temperatures are indicated by colours (blue for colder, red for warmer), displayed in a colour bar on the right hand side. Dashed and solid (black and green) contour lines represent the boundaries at 10 °C, 5 °C and 2 °C respectively. Each depth label corresponds to a depth where sensors are located. The maximum plot depth represents the deepest sensor in each trench, ranging from 40 (IT2–IT3) to 100 cm (IT4). This 20-day autumn period (Fig. 5) is characterised by a decreasing trend in temperatures, with fluctuations between 1 and 13 °C, while the volumetric water content responds dynamically to rain precipitation. Periods of warmer, colder and transition temperatures can be identified in the soil in IT1 (Fig. 13), showing a significant response of the soil to variations in ambient air temperature, even though the general trend is decreasing.

Reddish or blueish vertical 'strips' alternate daily. The response is more apparent at shallower depths (< 20–25 cm) corresponding to GP-GM soils (Table 1), with nearby e = 0.52 (D1, Table 2, Fig. 1), the influence is less and temperatures lag behind in deeper, generally coarser, soil. The temperatures during warm and cold periods are about 7–13 °C and 1–5 °C, respectively with transition zones of 3–7 °C. The temperatures plotted as ground level (0 cm depth) correspond to a sensor installed in the data logger, situated on the surface a few metres from the trench and can be warmer than the temperatures measured in the soil particularly for the cold and transition times. The lowest daily temperatures over all trenches were recorded at 12 cm depth at IT1.

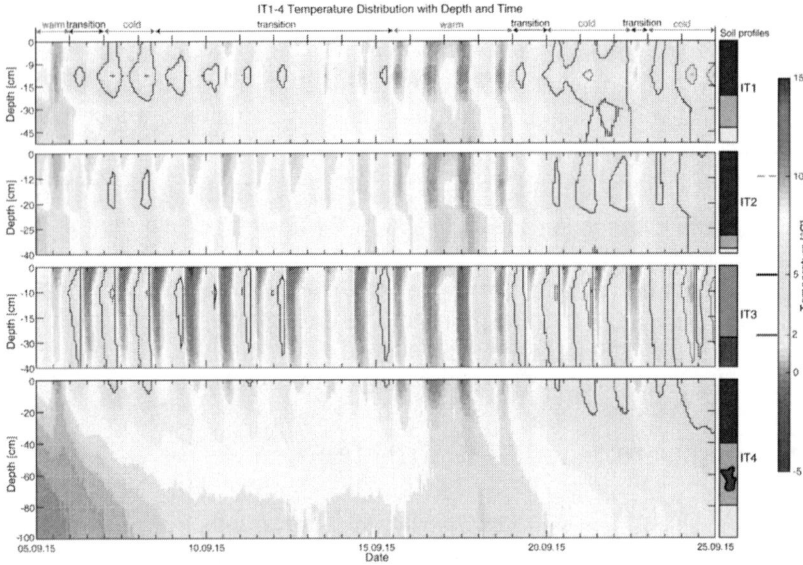

Fig. 13. Temperature distribution with depth in September 2015. IT1–IT4 are shown from top to bottom with sensor depths labelled for each trench. Contour lines are drawn at 10 °C, 5 °C and 2 °C.

The soil temperatures are generally warmer in IT2 than IT1, because the soil is coarser (GW; Table 1), albeit with a similar e = 0.51 (D3, Table 2, Fig. 1). As in IT1, the influence of ambient air temperatures extends to the trench bottom, with greater variation in the top 25 cm. Periods of transition and cold are less evident due to greater thermal conductivity; this can be related to less vegetation and greater sun exposure than for IT1 (Fig. 1), warmer periods are similar in both trenches. Ambient temperature variations in IT3 affect all depths at the same time. Heat is transferred into the soil more immediately and uniformly than for other trenches, due to coarser (GW, Table 1) more permeable soil (Fig. 2c), and higher e = 0.66 (D4, Table 2). Greater sun exposure (Fig. 1) with generally lower values of VWC (Fig. 5), contribute to these greater daily variations in soil temperature, with temperatures of 12 °C and significant drops to 2 °C over 24 h (Fig. 13, 7–9th July).

Poorly graded gravel (GP-GM) in IT4, with a similar content of fines to IT1 (Table 1), less vegetation and e = 0.58 measured nearby (D2, Table 2), could explain the higher temperatures during cold and transition periods than for IT1 at similar depths. Soil temperatures exhibit daily variations until depths of 40 cm, whereas sensors below this depth are less influenced; they exhibit a gradual transition to a colder season instead, with particularly cold trends noticeable between 5–13th and 18–23rd, starting at depths of 10 cm to the trench bottom (Fig. 13).

Fig. 14 shows the VWC distribution with depth for the same time period as for Fig. 13, for IT1–IT4 from top to bottom, respectively. Colours ranging from yellow to blue (colour bar on the right hand side) represent increasing VWC. Each depth label corresponds to the depth where sensors are located. Maximum depths vary from 30 (IT3) to 100 cm (IT4) and the plots are scaled accordingly.

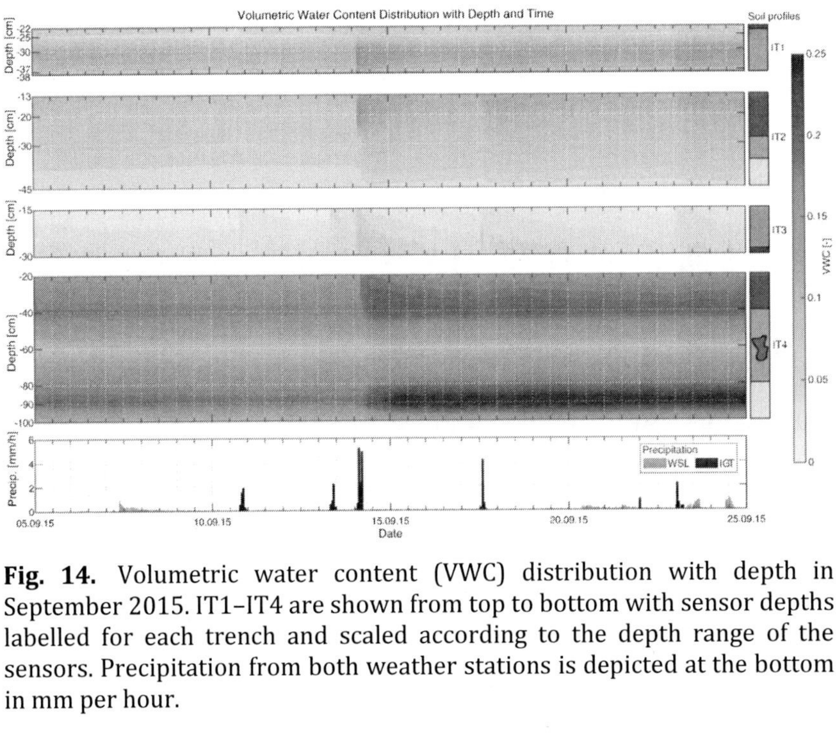

Fig. 14. Volumetric water content (VWC) distribution with depth in September 2015. IT1–IT4 are shown from top to bottom with sensor depths labelled for each trench and scaled according to the depth range of the sensors. Precipitation from both weather stations is depicted at the bottom in mm per hour.

This VWC representation enhances the understanding of water infiltration into the soil. Precipitation events are clearly observable as 'wet fronts' in each of the trenches (e.g. on 14th September 2015). IT1 shows lower VWCs (0.05) over the top 28 cm and higher VWCs (0.1–0.2) below, during all periods of observation, with an increase during the rain event (14th September) to 0.015, when 4 h of rain intensity (1–5 mm/h) occurred. Since there is no visible sign of water infiltration and pore water pressure dissipation into the shallow depths, it is possible that lateral flow occurred in the scree slope through the coarser soil layer.

In all other trenches, water infiltrates during significant rainfall events, increasing the VWC to different degrees and depths. Drainage occurred from the top 30 cm in IT2 into the coarser underlying gravel. The narrowest range and lowest VWCs are exhibited in IT3, due to fast infiltration and dissipation of pore water pressure from the much coarser material, with exponentially decaying trends after each rain event (14th, 17th and 23rd September).

Higher VWCs and drainage at 60 cm depth into a coarser gravel layer typify data from IT4, located nearby IT1 (Fig. 1). While a significant increase in saturation was observed in a 10 cm thick layer (depths of 0.9–1 m) after 15th September, and maintained for at least 10 days until the end of the period (25th September), the tensiometers above and below this layer (Tensio75, Tensio95; Fig. 12) maintained suctions of between 30 and 40 kPa. This can be interpreted as pore water pressure build-up in a potential failure surface.

3.5. Sensor comparison: TDR vs. capacitance sensors

Fig. 15 shows the comparison between TDRs and capacitance sensors (EC-5, 10HS) at similar depths, which is complemented by measurements of temperature, suction, precipitation, soil and ambient air temperature. All VWC sensors consistently measure within a range of 0 to 0.3, which is expected, considering the given range of porosity (2.2 ; 2.6). Resulting time series from capacitance sensors and TDRs generally exhibit a similar dynamic response in terms of VWC to rain infiltration.

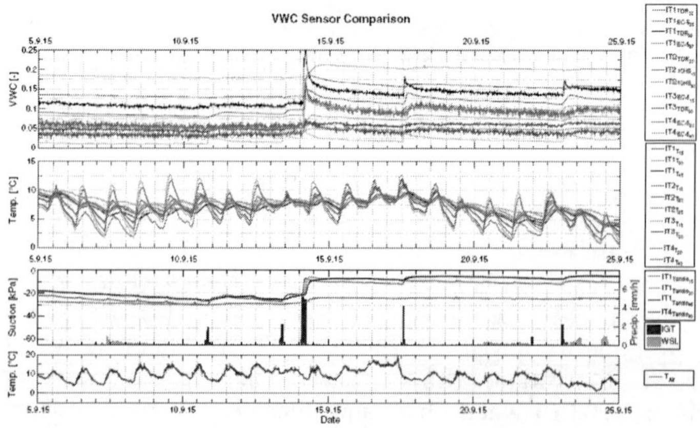

Fig. 15. Comparison between TDRs and dielectric permittivity sensors for IT1–IT4 at similar depths. From top to bottom: VWC, temperatures, suction and precipitation, and ambient air temperature in a period from 5th to 25th September 2015.

Sensors at similar depths in the same trench, or similar depths in different locations, agree in terms of peak occurrence, but not in magnitude of variation, nor in increase and decay patterns.

Comparing sensors TDR_{38} and EC-5_{37} in IT1, the response on 14th September is immediate in both cases with a significant, sharp peak, and drainage in the coarser gravel represented by concave decay pattern due to high hydraulic conductivity. However, the magnitude of variation in VWC for the TDR is greater than for the EC-5. This can be attributed mostly to the heterogeneity of the soil, the volume of sensitivity, because the EC-5 averages the VWC over a larger volume than the TDR and possibly to the dependency of the EC-5 sensors on temperature (Section 2.4.1.2). Even though, the daily temperature variation recorded at IT1 in T_{30} and T_{45} on 14th is actually smaller than in IT3 and other sensors in IT1 at shallower depths, the trend in temperatures varies between 4 and 8 °C during these two weeks. When comparing the rain events on 17th and 23rd for the same sensors, the difference in variation of magnitude of VWC is less, but the patterns of decay in VWC change after rainfall. Sharp peaks develop with a concave decay profile (IT1, TDR_{38}) in contrast to softer shape peaks and convex decays (IT1, EC-5_{37}), that can be explained by the soil heterogeneity or a larger volume of sensitivity for the capacitance sensors. Two other sensors TDR_{22} and EC-5_{25} in IT1 at similar depth, show similar values of VWC and variations in VWC after a rain event, with slightly higher VWC recorded at TDR than EC-5 sensors, for the reasons explained before.

Capacitance sensors $10HS_{13}$ and $10HS_{30}$ and TDR_{20}, located at different depths in IT2, reaction times were compared as a reference. The sensors have similar reaction, with a slight lag from $10HS_{30}$ on 14th September (also observed for deeper sensors in other trenches), due to the initial smaller rain intensity (mm/h) (Fig. 15) that probably infiltrated only to shallower depths. $10HS_{30}$ and TDR_{20} registered similar VWCs before the rain, with greater magnitude response for TDR_{20}, which remains at a higher VWC after that. Sensors installed nearby at the same reference depth (Fig. 1) $10HS_{30}$ (IT2) and TDR_{30}(IT3) were also compared. Both sensors indicate similarly low VWCs (GW, Table 1), and VWC in TDR_{30} (IT3, $e = 0.66$) is lower than $10HS_{30}$ (IT2, $e = 0.51$), as would be expected. Mittelbach et al. (2011) also compared TDR and 10HS in clayey loam, and have preferred 10HS in medium to low VWC using site-specific calibration, because of the greater sensitivity and accuracy in this specific range.

Finally, corrections for temperature effects could be applied to account for some of the differences, whereas the remaining discrepancies can be due to laboratory calibration, imperfect installation, varying volumes of influence as well as the heterogeneity of the gravelly soil.

3.6. Electrical resistivity tomography (ERT)

The resulting geoelectrical tomograms in Fig. 16a–c show the subsurface resistivities obtained from inversion of the ERT data acquired on 15th May, 16th June and 25th July 2014. All models depict a two-layered subsurface: an upper, relatively low resistive layer, which can be attributed to the unsaturated gravelly soil and a highly resistive layer underneath, which is assumed to represent the quartzite bedrock.

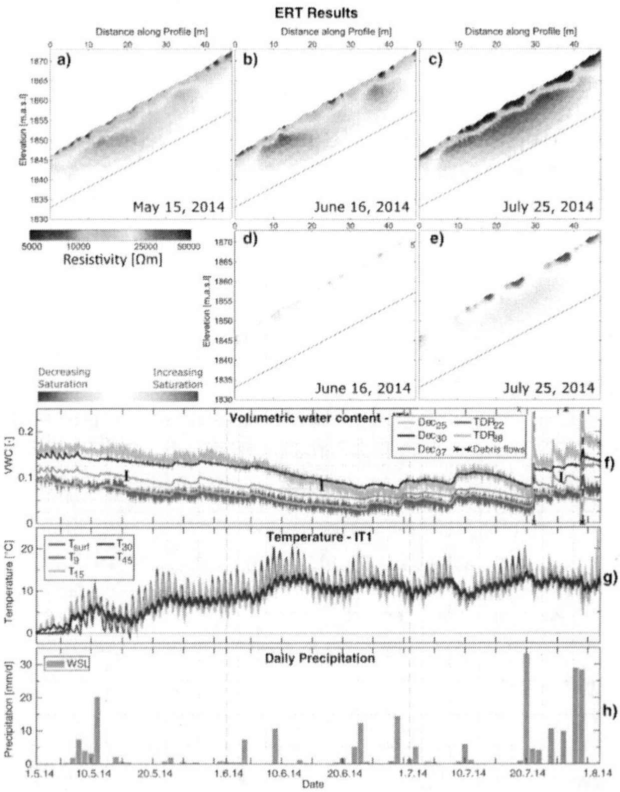

Fig. 16. a–c: Geoelectrical tomograms from 15th May, 16th June and 25th July 2014. The resistivity scale beneath the tomograms is the same for every acquisition. d–e: increasing and decreasing saturation, relative to the subsurface model obtained on 15th May (see Eq. A1 in Supplementary material A for further information). Finally, the three acquisition dates are indicated as vertical error bars within the volumetric water content measurements from TDR and capacitance sensor instrumentation in IT1 (f), which is shown from 1st May to 1st August 2014. The temperature (g) and the precipitation (h; in mm per day) are shown in the lower two time series plots.

The precipitation data before each of the acquisitions show that the conditions were relatively dry and undergoing a drainage phase for the first two campaigns, whereas the data acquired in July was recorded within a period of heavy rainfalls.

These seasonal variations are mostly reflected in resistivity changes within the soil layer, in particular as a significant reduction of resistivities in the July model. The saturation relative to the subsurface model obtained in May (Fig. 16d–e) clearly reflects these conditions; there is a slight net near-surface decrease in saturation in June, whereas an increase in saturation is observed within the soil layer in July. This trend in near-surface soil saturation corresponds very well to the volumetric water content measurements (which are directly proportional to saturation; (Eq. (1)) from TDR and dielectric permittivity sensors installed at shallow depths within IT1.

Numeric VWC values calculated from resistivity measurements using Archie's law (Eq. (1)) are presented in Table 4. It should be noted, that values for VWC_{ERT} were calculated using different cementation factors *m* and coefficient of saturation *n*. For all calculations, the resulting VWC_{ERT} agree very well with the saturation trend and the magnitude measured by either type of moisture sensor: higher VWC values in May, lower values (relative decrease in saturation of the gravelly soil) during the dry period in June to higher values again during the period of heavy rainfalls (relative increase in the saturation of the soil) in July 2014. However, the VWC_{ERT} were systematically overestimated when using literature values for *m* and *n*. Determining *m* and *n* by matching the VWCs from the first acquisition (15th May 2014) and using them in subsequent calculations, a very good fit of the VWC_{ERT} to the VWC_{TDR} and the VWC_{Dec} was found for all of the acquisition dates.

Table 4. Comparison of VWCs from ERT to VWCs from TDRs and capacitance sensors (Dec) in IT1. "No." refers to the number of readings taken by the moisture sensors during the recording of the ERT, as VWC_{TDR} and VWC_{Dec} are averaged over all measurements. ρ is the electrical resistivity measured by ERT.

Date	No.	Temperature [°C]	VWC_{TDR} [−]	VWC_{Dec} [−]	ρ_{ERT} [Ωm]	VWC_{ERT} [−][a], m = 1.3; n = 2	VWC_{ERT} [−][b], m = 1.05; n = 1.75	VWC_{ERT} [−][b], m = 1.5; n = 1.3
15.5.2014	3	4.8	0.11	0.11	7217.9	0.14 ∓ 0.01	0.11 ∓ 0.01	0.11 ∓ 0.01
16.6.2014	4	11.7	0.06	0.07	9003.4	0.11 ∓ 0.01	0.08 ∓ 0.01	0.07 ∓ 0.01
25.7.2014	4	12.1	0.1	0.1	5363.7	0.14 ∓ 0.02	0.10∓ 0.01	0.10∓ 0.01

[a] *m* and *n* taken from literature (e.g. Archie, 1942; Lowrie, 2007; Schön, 1983).

[b] *m* and *n* calibrated using the first measurement (15th May 2014), then used in subsequent calculations of VWC_{ERT}.

Two pairs of m and n achieved this fit very well but even though estimation without soil specific laboratory measurements is difficult, $m = 1.5$ and $n = 1.3$ was discarded, because an increase in the cementation factor (from the literature value of 1.3 estimated for loose materials) would mean an increase in cementation, which is certainly not the case in this gravelly scree. Hence, the VWC_{ERT} are indicated as vertical error bars in Fig. 16f, using the VWC_{ERT} calculated with $m = 1.05$ and $n = 1.75$. The resulting VWC_{ERT} fit best with the VWCs from the dielectric permittivity sensor at 37 cm and the TDR at 38 cm depth. The remaining differences (to other sensors) were most likely due to the following factors:

- both methodologies of 'measuring' the VWC are indirect;
- tomograms consist of model cells larger than the cylinders of influence of individual moisture sensors in IT1 and hence an average (spatial) resistivity of surrounding cells was calculated for the comparison. In addition, the VWCs from TDR and dielectric permittivity sensors (Table 4) are also (temporal) averages, calculated over the duration of the ERT acquisition;
- parameters used in Archie's law, such as the resistivity of the pore water and the porosity, were only inferred from preliminary measurements.
- the ERT profile 1 is representative of the specific area of measurement in IT1, but the heterogeneity of the soil has to be taken into account, when comparing recorded VWCs, because of the slight difference in location (Fig. 1).

3.7. Slope characteristics

3.7.1. Sections of the slope

The variation of VWC by type of moisture sensor versus time and depth, within all of the trenches IT1–IT4, is shown in Fig. 5, Fig. 14 ; Fig. 15. The range of VWC recorded at IT2 and IT3, compared to IT1 and IT4, is remarkably narrow for the dielectric permittivity sensors as well as the TDR measurements. These differences in the range of VWC, in addition to the VWC and temperature trends observed, support splitting the slope into up- and down-slope sections, with the following characteristics:

3.7.1.1. Upper scree slope (upslope)

Volumetric water content:

- low range of VWC (e.g. IT2/IT3), Fig. 5;

- Faster infiltration and drainage of water due to predominance of coarser soil with less fines. (e.g. IT3), Section 3.4, Fig. 14;

Temperature:

- temperatures can rise earlier (by up to half a month) than in downslope locations, due to greater sun exposure, Fig. 5;
- daily variations are higher and more pronounced for all sensor depths (e.g. IT3), Section 3.4, Fig. 13;
- extreme maximum values during winter and summer respectively (IT3) due to a coarser soil with higher permeability, Fig. 5;
- earlier snow-melt caused by greater sun exposure (IT2/IT3).

3.7.1.2. Lower scree slope (downslope)

Volumetric water content:

- higher range of VWC (e.g. IT1/IT4), Fig. 5 ; Fig. 14;
- VWC can rise due to infiltration and possible lateral flow of water downslope (IT1/IT4) Section 3.4, Fig. 14;

Temperature:

- more insulation and less sun exposure due to patches of vegetation, Fig. 1 ; Fig. 5;
- daily variations influenced by ambient air temperature are greater than upslope in the first 20–30 cm and less pronounced for deeper sensors (e.g. IT1/IT4), Section 3.4, Fig. 13.

3.7.2. Ground model

The ground model in Fig. 17 (with parameters in Table 5) was derived from the ERT tomograms (hence showing only the lower part of the scree slope). The boundary between the upper soil layer and the highly resistive bedrock underneath can be seen on each of the recorded tomograms (Fig. 16), however, it is clearest in the results obtained from the recordings on 25th July 2014. This measurement took place at the end of a rainy period and therefore current injected into the ground could penetrate more effectively and deeper into the saturated (more conductive) ground. Hence, the solid black line could be drawn in the ground model to divide the two layers.

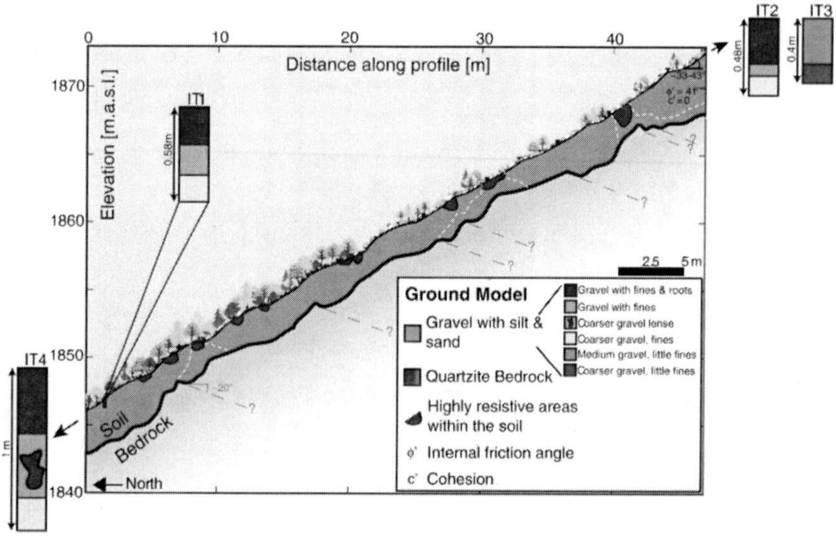

Fig. 17. Ground model derived from ERT measurements on profile 1. A visual description of the soil observed in IT1–IT4 is included, as well as some of the most important parameters, such as the friction angle.

Table 5. Geotechnical soil parameters used in the ground model.

Section	Classification (SN 670004-2b NA)	Percentage of fines [%]	Friction angle, critical state ϕ' [°]	Cohesion [kN/m^2]	G_s [—]
Downslope	GP-GM	5–10	41	0	2.68
Upslope	GW-GM	0–5	41	0	2.68

All ERT surveys, but in particular those that were carried out during dry periods (e.g. 16th June 2014), reveal (local) areas of high resistivity near the surface. Most of these were not consistently observed at the same spots along the profile, were limited in size and could be explained by the difficult ground coupling of an electrode in a dry and highly resistive soil in the respective area. A few of these "patches" were more extensive than others, and could be interpreted to extend down to the bedrock (marked with dashed, light grey lines in the ground model). These were particularly visible on the tomogram from 25th July. However, field observations, as well as GPR measurements on the lower part of the slope, indicate that it is rather unlikely that the bedrock rises to the surface. Furthermore, geological observations favour a series of "steps" at the soil-bedrock interface below ground, as nearby bedrock outcrops above ground exhibit rock layers inclined southward (into the slope) with 10–30° (see also Gabus *et al.*, 2008a ; Gabus *et al.*, 2008b). Assumed bedrock layers are therefore indicated in the ground model with dashed grey lines.

The highly resistive patches on the upper part of the profile, where there is less vegetation (mostly small trees, bushes and shrubs), can be explained as large boulders (estimated up to 0.5 m in diameter), which have been deposited on the surface and which also made electrode insertion difficult. Such boulders are not connected to the bedrock, but might produce a more extensive area of high resistivity. In addition, lenses consisting of larger cobbles, and even boulders, have been discovered within finer gravelly soil, e.g. during the excavation of IT4 (Fig. 2). Such lenses could also be attributed to highly resistive areas.

Due to the shallow depth of the bedrock (1–3 m), it is very difficult to determine resistivity changes within the soil layer. Subtle differences in grain size distributions were seen from trench and soil excavations (Figs. 2 and 3). Although these volumes are too small to be detected by ERT measurements (with an electrode spacing of 1 m), the soil is therefore depicted in the ground model as a uniform layer of gravel with sand and silt. However, a schematic description of the soil profiles in the trenches has been integrated as well, even though only IT1 is located along the ERT profile used for the ground model.

A friction angle of 41° was obtained using a constant shear drained triaxial stress path tests (CSD). A series of CSD triaxial tests were carried out in a medium (150 mm diameter) and large (250 mm diameter) scale apparatus using reconstituted specimens (for further information, see Grob, 2015).

4. CONCLUSIONS

Combining geotechnical and geophysical techniques has led to greater insight about the characteristics of a steep alpine scree slope. A ground model that estimates soil thickness and depth to bedrock, strength parameters, and soil classification, can be used in a posterior analysis of slope response to environmental perturbations through physical and/or numerical modelling. Furthermore, natural hazards can be assessed more effectively to provide input to needs for early warning systems and any mitigation measures.

Long-term monitoring and extensive characterisation using instrumented trenches, geophysical surveys, meteorological data and photographic observation of the slope surface yielded complementary and therefore well integrable results and gave useful indications of processes that could cause three prevalent types of downslope movement and changed perceptions of the most likely events.

Initially, it had been hypothesised that the most likely failure mechanism would be through shallow landslides, triggered by rainfall infiltration and lateral flow in a saturating soil that is heterogeneous in terms of porosity and grading. Given that the slope angle was between 33 and 43° and the friction angle is 41°, there remains some potential for small slips to occur if pore water pressures become positive over a significant layer that runs more or less parallel to the surface (as observed in 1 m depth in one of the trenches). However, the greatest number of downslope movements arise either from boulders falling and toppling downslope, being deposited temporarily in metastable positions, and remobilised either by repeated cycles of freezing and thawing, and snow-melting processes in winter or eroded by rainfall and runoff during the rest of the year.

Seasonal, daily and event driven patterns in changes of VWC, temperature and suction could be identified during the two years of monitoring and there is an agreement in trends of the VWC between dielectric permittivity sensors EC-5, 10HS and TDR in their response to rain infiltration, freezing and thawing processes. Correction factors could be applied in future for capacitance sensors (EC-5, 10HS) to allow for phase change between water and ice. Remaining differences in VWC from neighbouring devices, or similar depths, can be attributed to different installation locations and soil heterogeneity.

VWCs obtained from instrumentation were compared to values independently calculated from ERT measurements, conducted on different dates from May–July 2014. It was found that even though seasonal saturation trends agreed, VWC values were systematically overestimated by ERT, when using literature values for two parameters in Archie's law. A better fit was obtained, when these parameters were calibrated by matching one of the ERT acquisitions to the local trench measurements recorded at the same time. Hence, the agreement of VWCs from both types of measurements led to the conclusion that ERT indeed provides a convenient and fast way to infer VWC values over a larger area (spatial resolution), but has to be carefully calibrated using local VWC measurements, which in turn have the advantageous ability to more frequently record data (temporal resolution). "Small scale" trench instrumentation and "larger scale" ERT surveys complement each other very well and therefore, when combined and calibrated properly, can substantially enhance the characterization and monitoring of (steep) alpine scree slopes.

ACKNOWLEDGEMENTS

The authors are most grateful for funding from the SNF Project No. 200021_144326/1 and Canton Valais, together with the Councils of Agarn and Leuk. The project also contributes to the TRAMM2 programme of the Competence Centre of Environmental Sustainability. Furthermore, we wish to acknowledge, in particular, the support from Pascal Stoebener (WSL), Brian McArdell (WSL), Prof. Dr. Hansruedi Maurer (Exploration and Environmental Geophysics, EGG), Dr. Andrew Kos, Dipl. Ing. Ralf Herzog, Ernst Bleiker, Thomas Buchli, Reto Grob and Nicole Oggier. We would also like to thank Peter Molnar for providing us with insights regarding the recorded data series of rainfall. For the help with sensor-related problems, we would like to thank many employees of Decagon Devices.

REFERENCES

1. Archie, G.E., 1942. The electrical resistivity log as an aid in determining some reservoir characteristics. Trans. AIME 146 (01), 54–62.
2. Bayard, D., Stähli, M., Parraiaux, A., Flühler, H., 2004. The influence of seasonally frozen soil on the snowmelt runoff at two alpine sites in southern Switzerland. J. Hydrol. 309 (1), 66–84.
3. Beck, A., 2010. Measurement of unsarurated hydraulic conductivity with the instantaneous profile method. Semester Project ETH (Zurich, Switzerland).
4. Bichler, A., Bobrowsky, P., Best, M., Douma, M., Hunter, J., Calvert, T., Burns, R., 2004. Three-dimensional mapping of a landslide using a multi-geophysical approach: the Quesnel Forks landslide. Landslides 1 (1), 29–40.
5. Bogena, H.R., Huisman, J.A., Oberdörster, C., Vereecken, H., 2007. Evaluation of a low-cost soil water content sensor for wireless network application. J. Hydrol. 344 (1), 32–42.
6. Bogoslovsky, V.A., Ogilvy, A.A., 1977. Geophysical methods for the investigation of landslides. Geophysics 42 (3), 562–571.
7. Brunet, P., Clément, R., Bouvier, C., 2010. Monitoring soil water content and deficit using electrical resistivity tomography (ERT) — a case study in the Cevennes area France. J. Hydrol. 380 (1), 146–153.
8. Fankhauser, K., 2014. Geophysical Slope Characterization Using GPR and ERT in an Active Debris Flow Catchment. (Master Thesis). EEG, ETH Zurich.
9. Friedel, S., Thielen, A., Springman, S.M., 2006. Investigation of a slope endangered by rainfall-induced landslides using 3D resistivity tomography and geotechnical testing. J. Appl. Geophys. 60 (2), 100–114.
10. Gabus, J. H., Weidmann, M., Burri, M., & Sartori M. (2008a). Feuille 1287 Sierre, Atlas Géologique de la Suisse 1:25'000, Carte 111, swisstopo, Wabern.
11. Gabus, J. H., Weidmann, M., Burri, M., & Sartori M. (2008b). Feuille 1287 Sierre, Atlas Géologique de la Suisse 1:25'000, Note expicative, swisstopo, Wabern.

12. Grob, R., 2015. Measurement of Soil Properties for Gravels in a Scree Slope Using Large Scale Triaxial Testing. (Master Thesis). IGT, ETH Zurich.
13. Günther, T., 2011. Timelapse ERT inversion approaches and their applications. Geoelectric Monitoring, pp. 91–97.
14. Günther, T., Rücker, C., 2013. Boundless Electrical Resistivity Tomography — The User Tutorial (Available at: www.resistivity.net).
15. Günther, T., Rücker, C., Spitzer, K., 2006. Three-dimensional modelling and inversion of DC resistivity data incorporating topography — II. Inversion. Geophys. J. Int. 166 (2), 506–517.
16. Hack, R., 2000. Geophysics for slope instability. Surv. Geophys. 21 (4), 423–448.
17. Hayashi, M., 2004. Temperature-electrical conductivtiy relation of water for environmental monitoring and geophysical data inversion. Environ. Monit. Assess. 96 (1–3), 119–128.
18. ICT International, 2008. 10HS volumen of sensitivity. http://ictinternational .com/content/ uploads/2014/03/10HS-Volume-of-Sensitivity.pdf (Accessed 23.09.16).
19. Lehmann, P., Gambazzi, F., Suski, B., Baron, L., Askarinejad, A., Springman, S.M., Or, D., 2013. Evolution of soil wetting patterns preceding a hydrologically induced landslide inferred from electrical resistivity survey and point measurements of volumetric water content and pore water pressure. Water Resour. Res. 49 (12), 7992–8004.
20. Lowrie, W., 2007. Fundamentals of Geophysics. Cambridge University Press.
21. Lucas, D.R., Askarinejad, A., Herzog, R., Bleiker, E., Springman, S.M., 2015. Volumetric water content determination by TDR sensors and decagons in gravelly soils. XVI European Conference on Soil Mechanics and Geotechnical Engineering (XVI ECSMGE 2015), Edinburgh, UK, 13.-17.9.2015:pp. 3565–3570.
22. Matsuoka, N., 1998. The relationship between frost heave and downslope soil movement: field measurements in the Japanese Alps. Permafr. Periglac. Process. 9 (2), 121–133.
23. Miller, R.D., 1973. Soil freezing in relation to pore water pressure and temperature. 1. International Conference on Permafrost, 2d, Yakutsk, Siberia, 1973. Papers. Mittelbach, H., Casini, F., Lehner, I., Teuling, A.J., Seneviratne, S.I., 2011. Soil moisture monitoring for climate research: evaluation of a low-cost sensor in the framework of the Swiss soil moisture experiment (SwissSMEX) campaign. J. Geophys. Res. Atmos. 116 (D5).
24. O'Connor, K.M., Dowding, C.H., 1999. Geomeasurements by Pulsing TDR Cables and Probes. CRC Press.
25. Oggier, N.C., 2011. Simulierung von Murgängen mit RAMMS am Beispiel des Meretschibachs. (Master Thesis). ETH Zürich und Forschungsanstalt für Wald Schnee und Landschaft (WSL).
26. Onset Computer corporation. EC-5 volume of sensitivity application note. (n.d.). http:// www.onsetcomp.com/files/15926-B%20EC-5%20Volume%20of%20 Sensitivity.pdf (Accessed 23.09.16).
27. Otto, J.C., Sass, O., 2006. Comparing geophysical methods for talus slope investigations in the Turtmann valley (Swiss Alps). Geomorphology 76 (3), 257–272.
28. Patterson, D.E., Smith, M.W., 1981. The measurement of unfrozen water content by time domain reflectometry: results from laboratory tests. Can. Geotech. J. 18 (1), 131–144.
29. Rickenmann, D., Zimmermann, M., 1993. The 1987 debris flows in Switzerland documentation and analysis. Geomorphology 8 (2), 175–189.
30. Rücker, C., Günther, T., Spitzer, K., 2006. Three-dimensional modelling of DC resistivity data incorporating topography — I. Modelling. Geophys. J. Int. 166 (2), 495–505.

31. Sakaki, T., Limsuwat, A., Smits, K.M., Illangasekare, T.H., 2008. Empirical two-point α- mixing model for calibrating the ECH2O EC-5 soil moisture sensor in sands. Water Resour. Res. 44 (4).
32. Sass, O., 2007. Bedrock detection and talus thickness assessment in the European Alps using geophysical methods. J. Appl. Geophys. 62, 254–269.
33. Sass, O., Bell, R., Glade, T., 2008. Comparison of GPR, 2D-resistivity and traditional techniques for the subsurface exploration of the Aeschingen landslide, Swabian Alb (Germany). Geomorphology 93 (1–2), 89–103.
34. Schön, J., 1983. Petrophysik Physikalische Eigenschaften von Gesteinen und Mineralen. Akademie-Verlag, Berlin. Spaans, E.J., Baker, J.M., 1995. Examining the use of time domain reflectometry for measuring liquid water content in frozen soil. Water Resour. Res. 31 (12), 2917–2925.
35. Springman, S.M., Askarinejad, A., Casini, F., Friedel, S., Kienzler, P., Teysseire, P., Thielen, A., 2012. Lessons learnt from field tests in some potentially unstable slopes in Switzerland. Acta Geogr. Slov. 1, 5–29.
36. Springman, S.M., Thielen, A., Kienzler, P., Friedel, S., 2013. A long-term field study for the investigation of rainfall-induced landslides. Géotechnique 63 (14), 1177–1193.
37. Springman, S.M., Lucas, D.R., Oggier, N.C., Kos, A., Fankhauser, K., McArdell, B., 2015. Study of the seasonal response of a scree slope and debris flow catchment in the Swiss alps. XVI European Conference on Soil Mechanics and Geotechnical Engineering (XVI ECSMGE 2015), Edinburgh, UK, 13.-17.9.2015:pp. 1897–1902.
38. Taber, S., 1929. Frost heaving. J. Geol. 428–461.
39. Taber, S., 1930. The mechanics of frost heaving. J. Geol. 303–317.
40. Take, W.A., Arnepalli, D.N., Brachman, R.W.I., Rowe, R.K., 2007. Laboratory and Field Calibration of TDR Probes for Water Content Measurement. Queen's University, GeoEngineering Centre, Kingston, Ontario, Canada.
41. Topp, G.C., Davis, J.L., Annan, A.P., 1980. Electromagnetic determination of soil water content: measurements in coaxial transmission lines. Water Resour. Res. 16 (3), 574–582.

CHAPTER 3

3D DEM Investigation of Granular Column Collapse: Evaluation of Debris Motion and its Destructive Power

S. Utili [a] , T. Zhao [b], G.T. Houlsby [b]

[a] *School of Engineering, University of Warwick, Coventry CV4 7AL, UK*

[b] *Department of Engineering Science, University of Oxford, Oxford OX1 3PJ, UK*

ABSTRACT

This paper presents a numerical investigation of the behaviour of dry granular flows generated by the collapse of prismatic columns via 3D Distinct Element Method (DEM) simulations in plane strain conditions. Firstly, by means of dimensional analysis, the governing parameters of the problem are identified, and variables are clustered into dimensionless independent and dependent groups.

Secondly, the results of the DEM simulations are illustrated. Different regimes of granular motion were observed depending on the initial column aspect ratio. The profiles observed at different times for columns of various aspect ratios show to be in good agreement with available experimental results.

Thirdly, a detailed analysis of the way energy is dissipated by the granular flows was performed. It emerges that most of the energy of the columns is dissipated by inter-particle friction, with frictional dissipation increasing with the column aspect ratio. Also, the translational and rotational components of the kinetic energy of the flows, associated to particle rotational and translational motions respectively, were monitored during the run-out process. It is found that the rotational component is negligible in comparison with the translational one; hence in order to calculate the destructive power of a granular flow slide, only the translational contribution of the kinetic energy is relevant.

Finally, a methodology is presented to calculate the flux of kinetic energy over time carried by the granular flow through any vertical section of interest. This can be related to the energy released by landslide induced granular flows impacting against engineering structures under the simplifying assumption of neglecting all structure-flow interactions. This represents the first step towards achieving a computational tool quantitatively predicting the destructive power of a given flow at any location of interest along its path. This can be useful for the design of engineering works for natural hazard mitigation. To this end, also the distribution of the linear momentum of the flow over depth was calculated. It emerges that the distribution is initially bilinear, due to the presence of an uppermost layer of particles in an agitated loose state, but after some time becomes linear.

This type of analysis showcases the potential of the Distinct Element Method to investigate the phenomenology of dry granular flows and to gather unique information currently unachievable by experimentation.

Keywords: Granular flow; DEM; Landslides; Runout distance; Destructive power; Hazard mitigation

1. INTRODUCTION

Long run-out granular flows (e.g. rock and debris avalanches) can travel distances several times larger than the initial size of their source topography, sweeping away populated areas located far away from the landslide source (Crosta *et al.*, 2005 ; Carrara *et al.*, 2008). Many theories and assumptions have been proposed in the attempt to explain the apparent high mobility of granular material, including the air cushion trapped at the base of moving mass (Shreve, 1968), basal rock melting (Goren and Aharonov, 2007 ; De Blasio and Elverhøi, 2008), sand fluidization (Hungr and Evans, 2004), destabilization of loose granular material at the failure plane (Iverson et al., 2011), acoustic fluidization (Melosh, 1979 ; Collins and Melosh, 2003) or grain segregation-induced friction decrease (Phillips *et al.*, 2006 ; Linares-Guerrero *et al.*, 2007). However, no experimental evidence has been found to validate these theories and assumptions so far.

It has been recognised that debris flows and dry granular flows may behave similarly: for instance, they can sustain shear stresses with very slow deformation due to lasting, frictional grain contacts, and they can flow rapidly, sustaining inelastic grain collisions (Iverson, 1997). Thus, research has mainly focused on small scale laboratory experiments and numerical simulations of dry granular materials (Kerswell, 2005; Saucedo *et al.*, 2005; Mangeney *et al.*, 2007; Mangeney *et al.*, 2010 ; Roche *et al.*, 2011). Although these studies make important simplifications of the problem, they are useful in elucidating the mechanical behaviour of granular flows under simple, well controlled conditions (Crosta et al., 2009).

With regard to the numerical simulations, the Distinct Element Method (DEM) (Cundall and Strack, 1979) has been widely used to simulate dry granular avalanches (Cleary and Campbell, 1993; Staron and Hinch, 2005; Lacaze *et al.*, 2008; Utili and Nova, 2008; Utili and Crosta, 2011a ; Utili and Crosta, 2011b). The application of DEM to the simulation of granular flows, as first proposed by Cleary and Campbell (1993) in 2D, proved to be useful to enhance our understanding of the behaviour of dry granular flows. In the last two decades, it has become increasingly popular to shed light on fundamental mechanical characteristics of landslides (Staron and Hinch, 2007 ; Tang *et al.*, 2009). As stated by Zenit (2005), the use of DEM to simulate landslides is very powerful, since all the numerical data are accessible at any stage of the simulation, including quantities which are difficult, or impossible, to be obtained directly from laboratory experiments, such as the distribution of energy and momentum inside the granular flow and their variation over time and space.

In this paper, we illustrate the results of 3D DEM analyses in plane strain conditions of granular flows originated by the quick release of prismatic granular steps. The paper is organised as follows: in Section 2 the DEM contact model is illustrated; in Section 3 a dimensional analysis of the problem is carried out; then in Section 4 the numerical results obtained are presented. The evolution of the granular column profile over time for various column aspect ratios and inter-granular friction is presented together with the illustration of the flux of the kinetic energy and of the linear momentum of the granular system through vertical planes located at various distances from the initial position of the column and the distribution of linear momentum within the granular flow. Finally, in Section 5, conclusions on the capability of the DEM to model granular flows are illustrated.

2. DEM MODEL AND INPUT PARAMETERS

The DEM open source code ESyS-Particle (Weatherley et al., 2011) was employed to run the simulations here presented. Preliminary runs of granular column collapse tests were run employing non-linear elasticity according to the Mindlin no-slip solution (Johnson, 1985) and linear-elastic contacts. No significant difference was found between these analyses. Hence, since linear elastic contacts require less computational time, linear elasticity was adopted. Accordingly, the normal contact force at the contact plane is linearly proportional to the relative normal displacement between two particles (see Fig. 1(a)), expressed as:

$$F_n = K_n \cdot U_n \tag{1}$$

in which K_n is the normal contact stiffness and U_n is the normal relative displacement between two spheres in contact. The response along the tangential direction (seeFig. 1(b)) is calculated incrementally, as:

$$F_t^n = F_t^{n-1} + K_s \cdot dU_s \tag{2}$$

in which Ftn and Ftn^{-1} are the tangential forces calculated at the current and previous simulation time steps; K_s is the shear stiffness, and dU_s is the incremental tangential displacement. The maximum tangential force is limited by the Mohr–Coulomb criterion (see Figure 1(b)).

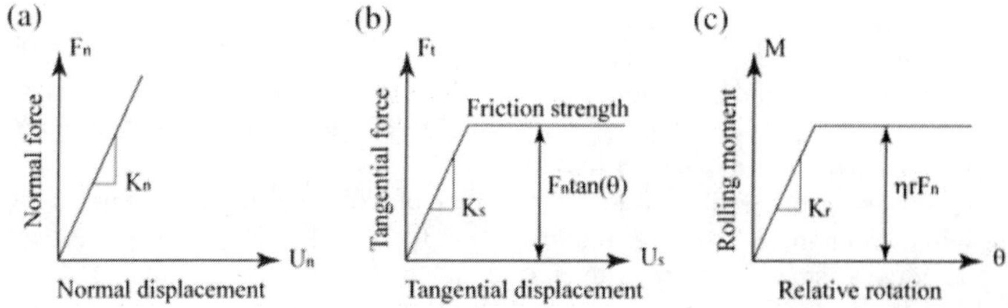

Fig. 1. Particle contact model in DEM (after Belheine et al. (2009)).

A moment–relative rotation law is also present. Here, the rolling law was employed with the only aim of accounting for the shape effect of non-spherical particles. The law accounts for moments arising from the fact that the line of action of the normal contact force in the case of non-spherical particles no longer passes through the centre of mass of the particles and hence generates rotational moments (Belheine et al. (2009)). Without this law, unrealistically low angles of repose would be obtained for the granular assembly. In the model adopted, the magnitude of the elastic rolling moment (M_e) is proportional to the relative rotational angle (see Figure 1(c)), and is calculated incrementally as:

$$M_e^n = M_e^{n-1} + k_r \cdot \Delta\theta_r \tag{3}$$

where *Men* and *Men* $^{-1}$ are rolling moment values calculated at the current and previous simulation time steps; $kr = \beta Ksr^2$ is the rolling stiffness, with β being the coefficient of rolling stiffness, r being the average particle radius; $\Delta\theta r$ is the relative rotational angle between two particles in one iteration step. The maximum rolling moment that can be exchanged is defined as:

$$M plastic = \eta \cdot r \cdot |F_n| \tag{4}$$

in which η is the so-called coefficient of plastic moment.

The particle size distribution (PSD) is one of the most important factors controlling landslide initiation and soil permeability. The PSD of granular flows varies hugely at different locations (see for instance Casagli et al. (2003)). In addition, the grain size distribution may vary significantly within the same landslide mass at different depths (Crosta et al., 2007). Fig. 2 shows examples of particle size distributions from 7 cases of landslides in the Northern Apennines (Casagli et al., 2003) and 6 cases of rock avalanches in Val Pola in the Alps (Crosta et al., 2007). It can be observed that the grain size ranges from 0.001 mm to 1000 mm, with a large percentage of fine and medium sized grains and a small amount of coarse grains. Large discrepancies can be observed between the various site investigations.

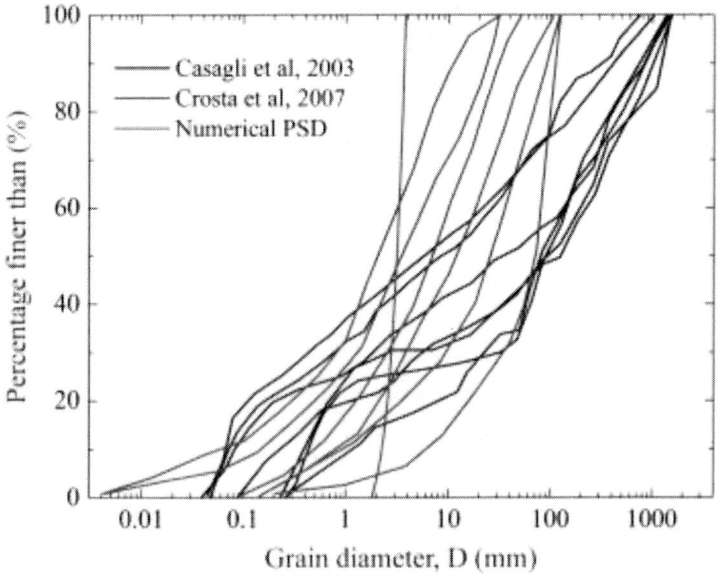

Fig. 2. Particle size distribution of landslides occurred in Northern Apennines (Italy) (after Casagli et al. (2003)) and rock avalanches in the Alps (Italy) (after Crosta et al. (2007)). The particle size distribution adopted in the numerical simulations is plotted as the red curve.

According to Fig. 2, grains with diameters ranging from 0.1 to 10 mm were widely observed in different locations. However, in DEM simulations, due to computational limitations, we used a much narrower particle size distribution with the ratio of maximum to minimum particle sizes equal to 2, as shown in the red curve in Fig. 2. The input parameters are listed in Table 1.

Table 1.Parameters of the granular flow simulations.

DEM parameters	Value	DEM parameters	Value
Particle diameter, D (mm)	See Figure 2	Damping coefficient, α	0.0/0.3
Initial porosity, n	0.43	Coefficient of rolling stiffness, β	1.0
Density, ρ (kg/m^3)	2650	Coefficient of plastic moment, η	0.1
Normal stiffness, K_n (N/m)	3×10^7	**Simulation parameters**	**Value**
Shear stiffness, K_s (N/m)	2.7×10^7	Gravitational acceleration, g (m/s^2)	−9.81/−981
Particle friction angle, ϕ_μ (°)	30	DEM time step size, Δt (s)	10^{-7}

In this study, we carried out numerical experiments of granular columns analogous to the columns of the experimental investigation of Lube et al. (2005). Initially particles were generated within a prismatic space bounded by two rigid and frictionless walls (A and B in Figure 3(a)), a rough base (D in Figure 3(a)), and a periodic boundary in the out-of-plane direction (C in Figure 3(a)) to impose plane strain conditions. The horizontal base (D) of the domain (see Figure 3(a)), is made of particles of the same PSD as the granular column that were kept fixed at all times to simulate a non-erodible base of the same roughness as the flowing material. To generate the granular column, first particles were randomly created in space (with the grain position allocated according to standard algorithms of random number generation), then they were left to settle under gravity until a dense column was obtained. In all the simulations, the granular flow was initiated by the instantaneous removal of wall B.

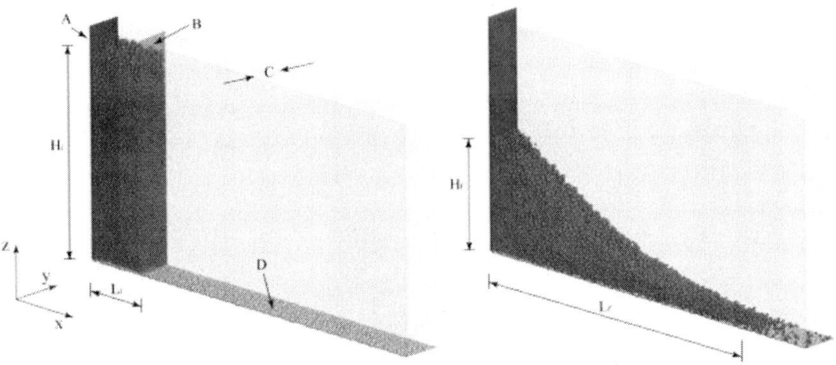

Fig. 3. Model configuration: (a) initial sample; (b) final deposit (A: fixed smooth back wall; B: movable front gate; C: periodic boundary; D: fixed coarse floor; L_i: initial column length; H_i: initial column height; L_f: runout length; H_f: deposit height).

Once the flow has stopped, the final run-out length (L_f) and deposit height (H_f) can be measured (see Figure 3(b)). The presence of single particles detached from the front of the granular mass sometimes makes the calculation of L_f a non-straightforward exercise. Zenit (2005) calculates L_f considering only particles remaining in contact with each other, disregarding individual loose particles detached from the deposit mass centre. In this study, in order to track the position of the front in a way consistent amongst the various simulations (the number of loose single particles moving ahead of the front depends on the aspect ratio), we implemented an algorithm which identifies the front as the boundary between 99% and 1% of the flow mass, i.e. 1% of the mass is travelling ahead of the boundary. This guarantees that the front is not confused with the position of loose single particles jumping ahead the flow.

3. DIMENSIONAL ANALYSIS

The physical parameters ruling the motion and the depositional morphology of the granular column include (Zhao et al., 2012): grain properties, geometrical properties of the domain, flow velocity and duration time. In Table 2, all the parameters are listed together with their units expressed in terms of fundamental dimensions, i.e. mass (M), length (L) and time (T). Depending on their role in the simulation, parameters can be categorised as either independent or dependent. We considered the grain properties and the domain geometry as independent input parameters and final runout distance (L_f), deposit height (H_f), flow velocity (v_f) and flow duration time (t) as the dependent ones.

Table 2. Parameters of the granular flow problem.

	Parameter	Symbol	Unit of measure
Independent parameters	Initial column length	L_i	[L]
	Initial column height	H_i	[L]
	Particle diameter	D	[L]
	Particle density	ρ_s	[ML^{-3}]
	Gravitational acceleration	g	[LT^{-2}]
	porosity	n	[–]
	Normal stiffness	K_n	[MT^{-2}]
	Shear stiffness	K_s	[MT^{-2}]
	Particle friction angle	ϕ_μ	[–]
	Coefficient of rolling stiffness	β	[–]
	Coefficient of plastic moment	η	[–]
Dependent parameters	Final deposit length	L_f	[L]
	Final deposit height	H_f	[L]
	Flow velocity	V_f	[LT^{-1}]
	Granular sliding time	t	[T]

The governing parameters of the mechanical laws ruling particle interactions are the normal and shear particle contact stiffness (K_n, K_s), particle friction (ϕ_μ) and the coefficients of rolling resistance (β, η). Also porosity has a significant influence on the mechanical behaviour of granular flows (Craig, 1997). However, since the primary purpose of this study is to explore the capability of DEM in modelling granular flows, the influence of porosity has not been considered here. The relationship between the independent and dependent quantities can be expressed by a general functional relationship of the form:

$$\left(L_f,\ H_f,\ v_f,\ t\right) = f\left(L_i,\ H_i,\ D,\ \rho_S,\ g,\ K_n,\ K_S,\ n,\ \phi_\mu\right) \qquad (5)$$

Performing dimensional analysis (Palmer, 2008), and taking out porosity which is the same in all the tests carried out (all columns were generated with the same initial porosity), Eq. (5) can be rewritten as:

$$([L],\ [H],\ [V],\ [T]) = f\left(a,\ \varepsilon,\ [S],\ \phi_\mu\right) \qquad (6)$$

where $[L] = (Lf - Li)/Li$ is the normalised run-out distance; $[H] = Hf/Li$ is the normalised maximum final deposit height; $[V] = v_f/\sqrt{gH_i}$ is the normalised flow velocity (v_f can be chosen in a variety of ways, one of which is the velocity of propagation of the flow front, *i.e.* $v_f = dL/dt$); $[T] = t/\sqrt{H_i/g}$ is the normalised flow duration time; $a = Hi/Li$ is the initial column aspect ratio; $\varepsilon = \rho sgHi/(Kn/D)$ is a characteristic compressive strain of the granular column and $[S] = Hi/D$ is the model-to-particle size ratio. Note that [–] indicates a dimensionless variable.

4. NUMERICAL RESULTS

4.1. Calibration of the angle of repose

In studies on non-cohesive granular flows, the material internal friction angle (ϕ) is often approximated by the angle of repose. In simple terms, it can be said that a slope with an angle over the horizontal larger than the angle of repose is unstable, whereas it is stable if the opposite holds true. Typical values for the angle of repose range from 25° for flows made of smooth spherical particles to 40° for rough angular particles (Carrigy, 1970 ; Pohlman *et al.*, 2006). Here, we assumed as baseline reference value, an angle of repose of 31° which is representative of coarse quartz sand (Lube et al. (2005)).

With regard to the calibration of the DEM parameters, concerning the rolling stiffness, β, in Modenese (2013) it is shown that its influence on the angle of repose is negligible, so it was not investigated here. Then, two contact parameters remain to be determined: the intergranular friction angle, ϕ_μ, and the coefficient of rolling resistance, η. In Zhao (2014), it is shown that some arbitrariness exists since the same angle of repose can be obtained from several combinations of them. Here, we opted to choose an intergranular friction angle of $\phi_\mu = 30°$, which is a typical value for quartz sand grains, and to calibrate η against the angle of repose observed when the flow comes to a stop. In Fig. 4, the angles of repose obtained for various values of η are shown. For $\eta > 0$, a linear relationship between η and the angle of repose exists. For $\eta = 0$, an unrealistically low angle of repose is obtained. This is a well-known result reported in Calvetti and Nova (2004), Pöschel and Buchholtz (1993) and Rothenburg and Bathurst (1992)). In all the subsequent analyses, a value of $\eta = 0.1$, corresponding to an angle of repose for the flow of 31°, was adopted.

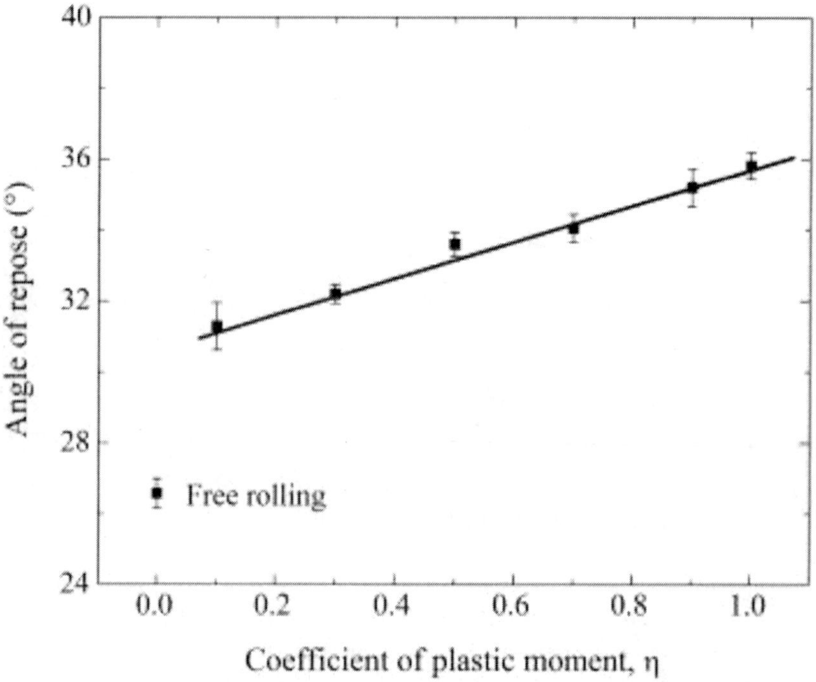

Fig. 4. The angle of repose of a granular deposit for different rolling resistances.

4.2. Kinematics of motion

Depending on the initial aspect ratio of the granular column, different depositional morphologies are obtained (see Figure 5). As it can be expected, the upper surface of the granular flow gradually lowers until the angle of repose is reached. Our results confirm that the flow profiles at various times and the final deposit depend strongly on the initial aspect ratio (Lube et al., 2005). This highlights the need for presenting profiles in dimensionless form. To this end, the final profiles of the granular assembly, normalised by the final runout distance and the deposit height, were plotted in Fig. 6 together with the experimental profiles obtained by Lube et al. (2005). The profiles from our 3D DEM analyses exhibit a good agreement with the experimental results for all the aspect ratios analysed.

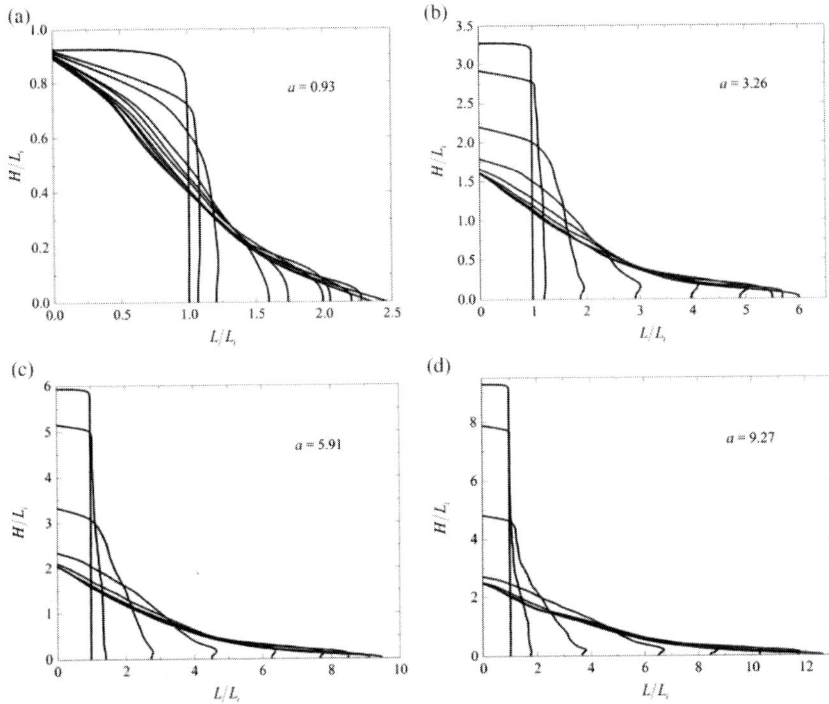

Fig. 5. The evolution of sample profiles for different granular columns (profiles are traced at successive time steps ($\Delta t/\sqrt{H_i/g} = 0.5$)).

A key feature of the DEM is the possibility of obtaining data on the motion of single particles (e.g. trajectories, velocities, etc.) inside the granular flow, allowing performing a detailed analysis of the potential heterogeneities in the flow. For instance, from our analyses, it emerges that in granular columns with small aspect ratios (e.g. $a = 0.93$), only a small portion of material is involved in the motion. To compare the kinematic fields obtained from columns of different aspect ratios, it is convenient to plot the measured velocities in dimensionless form. To this end, we define $\sqrt{gH_i}$ as the characteristic velocity of the system. This represents the maximum

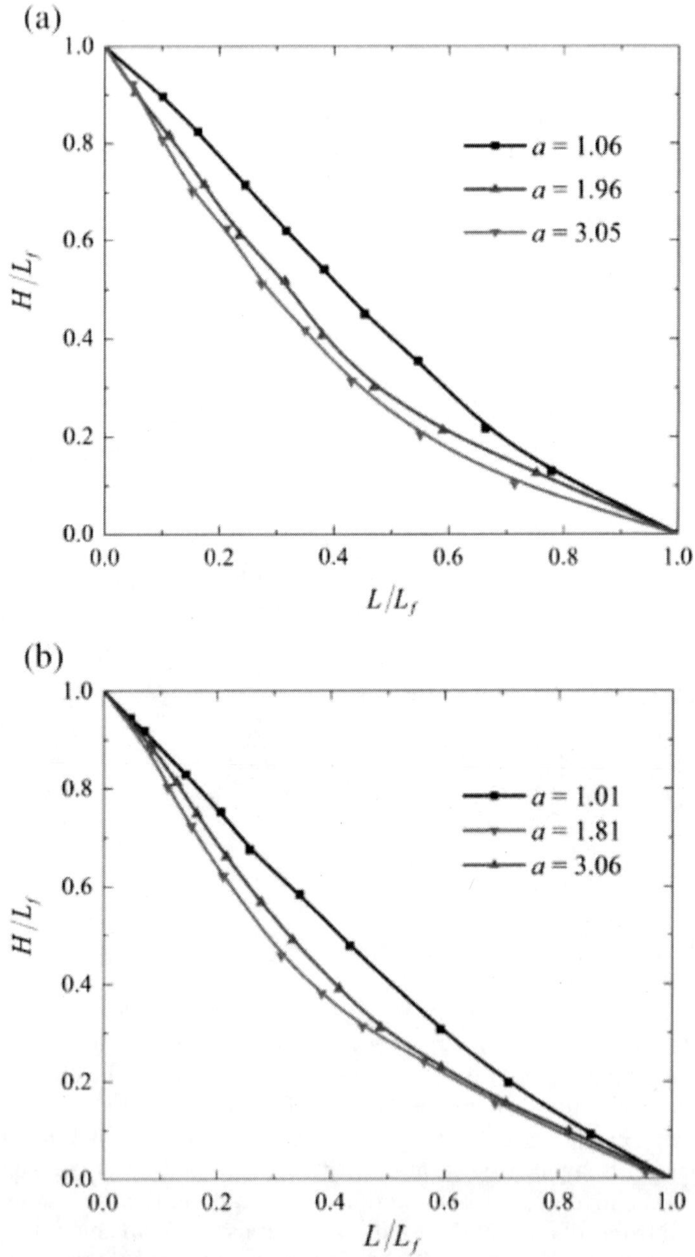

Fig. 6. Final normalised profiles of the granular column deposit: (a) 3D DEM results; (b) experimental results after Lube et al. (2005).

velocity that a single particle in free-fall from a height corresponding to the centre of the column ($Hi/2$) reaches at the end of the fall. In Fig. 7(a), particles with velocities smaller than 1% of the characteristic velocity are plotted in grey, whilst particles moving at higher velocities are plotted in red. From the figure, it can be noted that failure of the granular assembly occurs approximately along a plane, identified by a dashed line in the figure. According to the Rankine's theory of earth pressure (Rankine (1857), assuming that the failure of the granular column occurs when conditions of active thrust are in place, the inclination of the failure plane to the horizontal (θ_f) can be estimated as:

$$\theta_f = 45° + \phi/2 \tag{7}$$

where ϕ is the internal friction angle of the granular material. From Fig. 7(b), it emerges that the inclination angle of the active failure plane is approximately 61° for all the aspect ratios tested. This value is very close to the theoretical value predicted by Eq. (7), which is 60.7°.

4.3. Influence of the column aspect ratio

For rock avalanches and granular flows, the assessment of the final run-out distance is of primary importance, as it determines the extent of the regions affected by the avalanche or landslide. In Fig. 8(a), our simulations have been compared with previous numerical simulations (Staron and Hinch, 2005; Zenit, 2005; Crosta *et al.*, 2009) and experimental observations (Balmforth and Kerswell, 2005; Lajeunesse *et al.*, 2005; Lube *et al.*, 2005) from the literature. We performed three sets of simulations: T1 and T2 refer to simulations run for the same values of ε (ranging from 6.5×10^{-7} to 6.5×10^{-6}), but with $\eta = 0.1$ and $\eta = 0$ respectively, whilst T3 refers to simulations run for ε ranging from 6.5×10^{-5} to 6.5×10^{-4} and $\eta = 0.1$. Some of the T1 and T2 simulations were run with gravitational acceleration scaled up 100 times and $K_n = 3 \times 10^7$ N/m, whereas some others with unscaled gravity but particles 100 times softer, i.e. $K_n = 3 \times 10^5$ N/m. In fact, in light of dimensional analysis, scaling can be introduced by scaling either the value of gravitational acceleration or particle stiffness. The fact that the obtained results are aligned in consistent trends in Fig. 8 can be seen as a verification of the correctness of the performed dimensional analysis. Comparison between T1 and T2 simulations allows examining the influence of particle shape on run-out, whereas comparison between T1 and T3 allows examining the influence of ε. The values of the independent variables

(a)

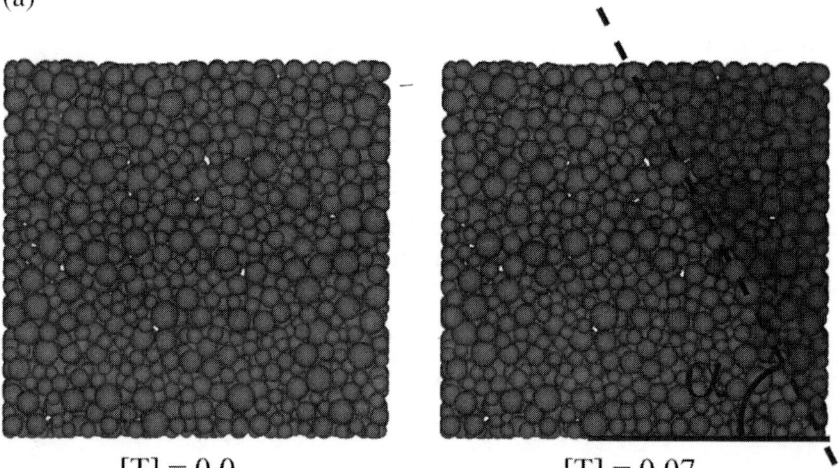

(b)

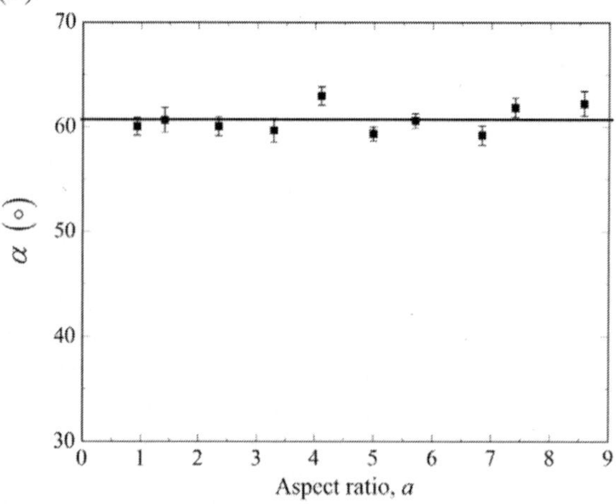

Fig. 7. Active failure state of the granular sample. (a) Motion of solid grains. (b) Inclination angle of the slope failure plane.

employed in our simulations are reported in Table 3 in terms of the dimensionless groups identified in Eq. (6). In the table, the values of the parameters employed in other tests reported from the literature in Fig. 8 ; Fig. 9 for comparison purposes are listed as well. In Fig. 8(a), it can be observed that the final normalised run-out distance obtained from our simulations matches well from a qualitative viewpoint both the experimental observations of Lube et al. (2005) from tests run in plane strain conditions, and the 2D FEM numerical analyses of Crosta et al. (2009). Also it emerges that if rolling resistance is not employed (simulations T2), unrealistically large run-out distances are obtained since particle angularity tends to reduce run-out. In other words, a flow of spherical particles is more prone to sliding than a flow of particles of any non-spherical shape. Equally, if 2D DEM simulations are employed (Staron and Hinch, 2005 ; Zenit, 2005), unrealistically long run outs are obtained. Presumably, this is due to the fact that the 2D kinematics of particle interaction is too different from the real 3D kinematics. Comparison between the simulation series T1 and T3 shows that the characteristic strain of a granular column, ε, has no influence on the observed flow behaviour, at least for the range of values here employed. In conclusion, simulations T1 and T3 show that if particle shape is accounted for, albeit by means of a very crude approximation (i.e. employing a moment–relative rotation law with spherical particles which avoids simulating the real non-spherical geometry of the particles), the obtained run-out and final heights of the simulated flows are in good agreement with the available experimental data.

Table 3. Properties for experimental and numerical tests.

–	Test type	a	ε	[S]	ϕ (°)
Lube et al. (2005)	Experiments in plane strain	[0.5, 20]	$[4.7, 140] \times 10^{-9}$	[9, 1000]	[29.5, 32]
Lajeunesse et al. (2005)	Experiments in plane strain	[0.2, 12]	$[1.2, 8.7] \times 10^{-8}$	[100, 300]	[21.5, 22.5]
Balmforth and Kerswell (2005)	Experiments in plane strain	[0.5, 40]	$[5.2, 420] \times 10^{-9}$	[12, 1000]	[22.5, 26.5]
Crosta et al. (2009)	2D FEM simulations	[0.6, 20]	–	–	[20, 40]
Zenit (2005)	2D DEM simulations	[0.1, 10]	$[2.1, 13.1] \times 10^{-5}$	[28, 173]	30
Staron (2005)	2D DEM simulations	[0.2, 17]	–	[14, 369]	20
This study (T1)[a]	3D DEM simulations in plane strain	[1, 10]	$[6.5, 65] \times 10^{-7}$	[25, 250]	31.7
This study (T2)[b]	3D DEM simulations in plane strain	[1, 10]	$[6.5, 65] \times 10^{-7}$	[25, 250]	26.1
This study (T3)[a]	3D DEM simulations in plane strain	[1, 10]	$[6.5, 65] \times 10^{-5}$	[25, 250]	31.7

[a] Numerical simulation using rolling resistance model ($\beta = 1.0, \eta = 0.1$).
[b] Numerical simulation without using rolling resistance model ($\beta = 0.0, \eta = 0.0$).

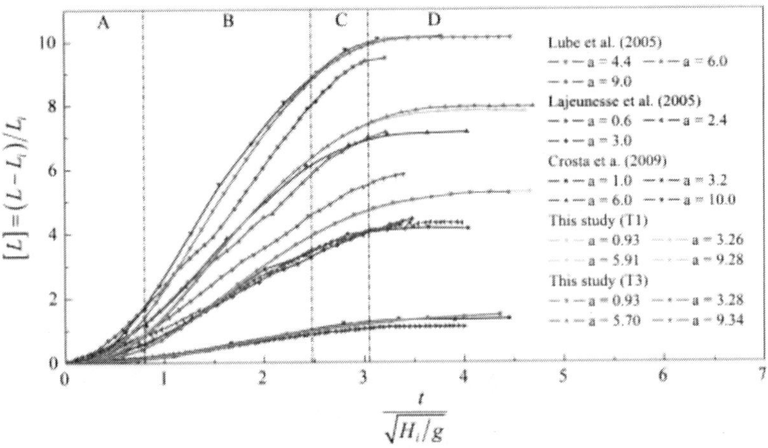

Fig. 9. Normalized granular spreading length versus normalized time from experimental and numerical tests.

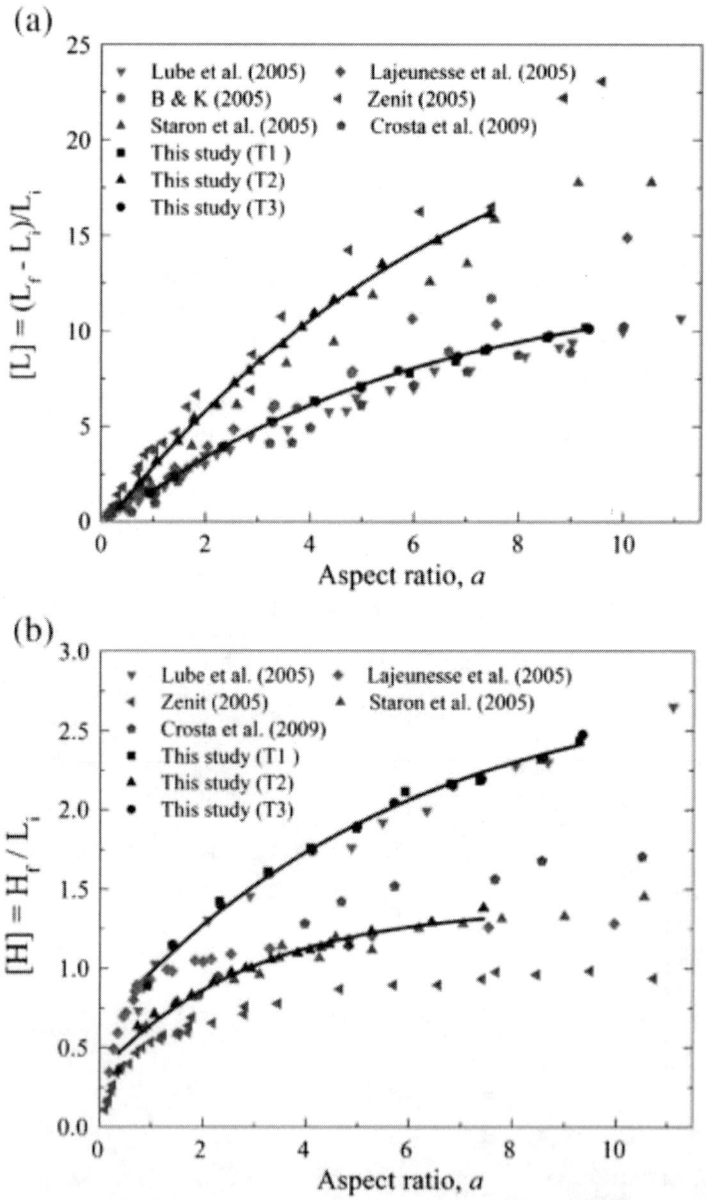

Fig. 8. Relationship between aspect ratio and (a) normalised run-out distance and (b) final deposit height. The red symbols are experimental results; blue symbols are numerical results, while the numerical results of this study are coloured black.

Now let us look at how the front of the flow propagates over time. In Fig. 9, the position of the front is plotted against time for various column aspect ratios. According to previous literature (Lube *et al.*, 2005 ; Crosta *et al.*, 2009), time has been normalised by $\sqrt{H_i/g}$, which can be thought of as the time taken by a single particle in free fall to travel from the centre of the column to the base. Looking at the figure, four typical distinct regimes can be identified: an initial transient acceleration (A), a constant velocity flow (B), a gradual deceleration (C) and a final static deposition (D). It emerges that in terms of DEM simulations, only 3D analyses accounting for the effect of particle shape provide results that are in agreement with experiments.

4.4. Analysis of the energy contributions in the flow

The potential energy of the column at any time is:

$$E_p = \sum_{i=1}^{N} m_i g h_i \tag{8}$$

where m_i and h_i are the mass and height of a single particle *i*, respectively, and *N* is the total number of particles of the column. The kinetic energy of the system at any time is calculated as:

$$E_k = \frac{1}{2}\sum_{i=1}^{N}\left(m_i v_i^2 + I_i \omega_i^2\right) \tag{9}$$

where v_i and ω_i are the translational and angular velocities respectively of a generic particle *i*, and *I* is its moment of inertia (for a spherical particle $I = 2mR^2/5$).

A part of potential energy gets dissipated rather than being transformed into kinetic energy, due to unelastic particle collisions (e.g. unelastic particle rebounds and frictional sliding). In light of the principle of energy conservation, the energy dissipated in the flow at any given time can be calculated as:

$$E_{diss} = E_0 - E_p - E_k \tag{10}$$

where E_0 is the total energy of the system, which can be calculated from the initial potential energy of the column before particles start to move, as:

$$E_0 = MgH_i/2 \tag{11}$$

where $M = \sum_{i=1}^{N} m_i$.

At the particle level, energy is mainly dissipated via frictional sliding at particle contacts and viscous damping along both the normal and tangential directions of contacts when damping is present, but also by the relative rotation between particles once the plastic limit of the rolling moment is reached (see Figure 1(c)). In Fig. 10, the entire temporal evolution of the energy components of the flows, from start of column collapse until end of motion, is plotted. After the instantaneous removal of the confining gate at $T = 0$, particles start to fall downwards, with potential energy being progressively transformed into kinetic energy. The kinetic energy exhibits a peak at around $T = 1.0$ (see the dashed line in the figure), when both the rate of potential energy loss and the rate of cumulative energy dissipation (see the slopes of the curves in the figure) reach their maximum values. After this time, both potential and kinetic energy decrease. The flow comes to a stop at about $T = 4.0$. Considering now columns of different aspect ratios, the dissipated energy in terms of percentage of the initial total energy of the columns has been plotted against their aspect ratio in Fig. 11. From our simulations, it emerges that the higher the aspect ratio, the larger the proportion of energy dissipated during the flow.

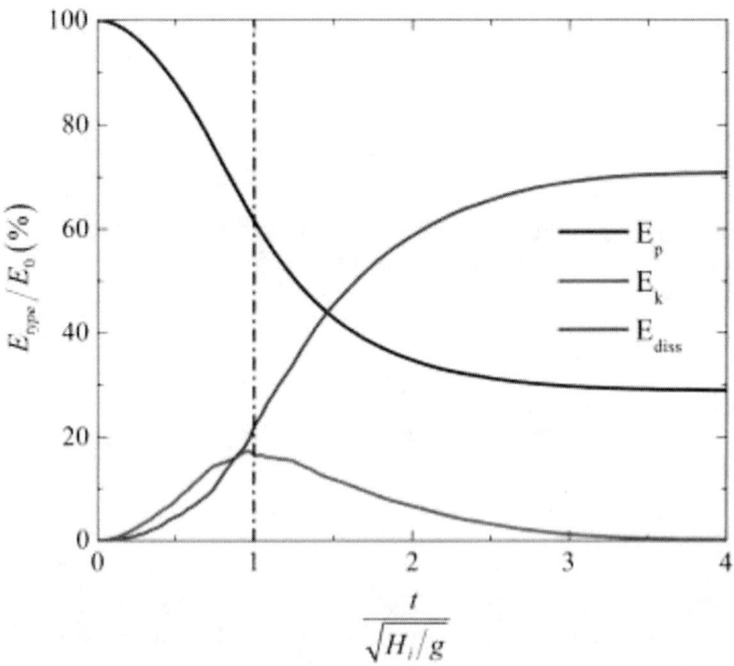

Fig. 10. Variation of energy during granular flow (a = 3.26) (E_0: initial total energy; E_p: normalized potential energy; E_k: normalized kinetic energy; E_{diss}: normalized cumulative energy dissipation).

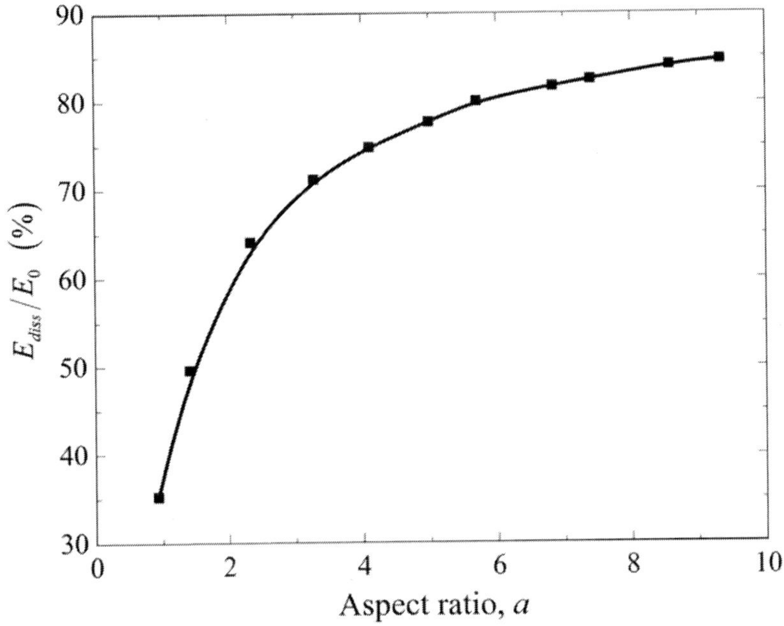

Fig. 11. Total energy dissipated during the flow normalized by the initial potential energy versus column aspect ratio.

Then, we investigated how much of the kinetic energy of the flow is due to the translation of particles and how much is due to their rotations. In Fig. 12, the two sources of kinetic energy are plotted separately against time. Unfortunately, viscous damping coefficients strongly depend on the type of material making the granular flow. Furthermore, a reliable experimental determination of damping coefficients for granular flows is currently not yet available. Therefore, we decided to run two cases only of simulations: one without viscous damping and the other one with a viscous damping coefficient of 0.3, in order to obtain a rough indication of the potential influence of viscous damping on the variables of interest but without trying to model any specific natural granular flow. In the simulations with viscous damping, the damping coefficient of 0.3 was employed in both the normal and tangential directions of particle contacts. This value corresponds to a coefficient of restitution of around 0.5 (Tsuji et al., 1992). In Fig. 12, it can be noted that the kinetic energy stemming from particle rotations remains negligible at all times (i.e. less than 0.5% of the total kinetic energy) in both cases, with and without the presence of viscous damping. In light of this finding, in the next section, where we investigate the spatial distribution of the momentum inside the flow, we have concentrated our attention on the linear momentum since we know the angular momentum to be negligible. Also, from Fig. 12, it emerges that the curves obtained for flows with and without damping, are similar. This implies that the presence of damping decreases the magnitude of the kinetic energy of the system, of the same proportion, throughout the duration of the flow.

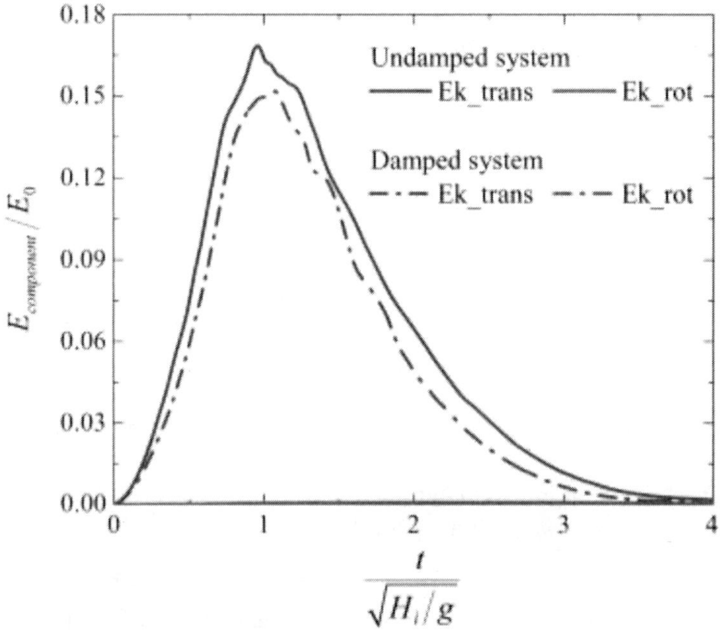

Fig. 12. Evolution of kinetic energy over time (a = 3.26) (E_{k_trans} and E_{k_rot} are the translational and rotational sources respectively of the kinetic energy of the system).

4.5. Linear momentum

The evolution over time of the linear momentum of the flow, $\vec{p} = \sum_{i=1}^{N} m_i \vec{v}_i$, is plotted in Fig. 13, in the case of the presence of viscous dissipation and in its absence. Given the imposition of periodic boundary conditions in the y direction, we expect p_y to be negligible at all times, as it is shown in Fig. 13(a). This confirms the effective presence of plane strain conditions in the x–z plane for the simulated flows. Analogously to the energy analysis of the flow, it is convenient to normalise $\vec{P}$, for the sake of generality in presenting our results. To this end, we introduce a scalar quantity, p_0, with:

$$p_0 \equiv M\sqrt{gH_i} \tag{12}$$

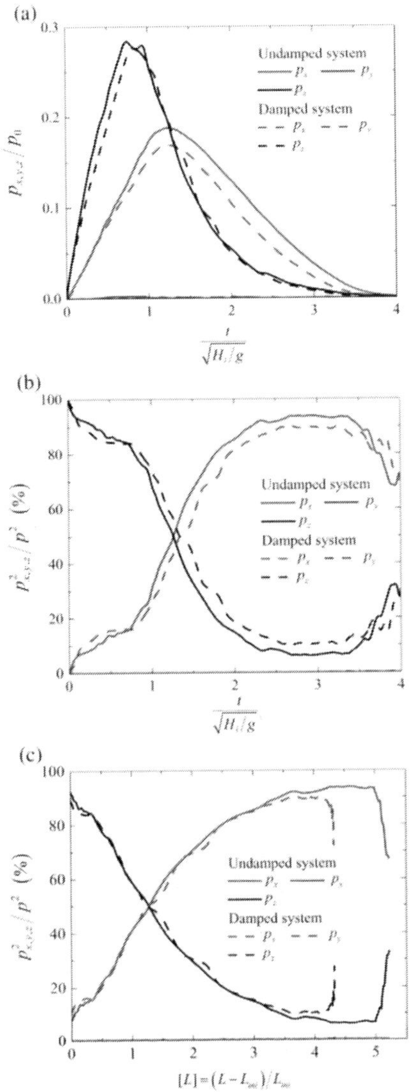

Fig. 13. Evolution of linear momentum over normalised time for flow with and without damping: a) components of momentum normalised by p_0; b) components of momentum as a percentage of the total momentum against normalised time; c) components of momentum as a percentage of the total momentum against normalised run-out; *x* is the direction of flow propagation, *y* is the out of plane direction, and *z* is the vertical direction.

This quantity can be thought of as an approximate average of the momentum of the flow: $p_0 \equiv M\sqrt{gH_i} = M \cdot H_i/\sqrt{H_i/g}$ with $H_i/\sqrt{H_i/g}$ being the characteristic velocity of the flow. In Fig. 13(a), the components of the linear momentum along the *x*, *y* and *z* axes, normalised by p_0, are plotted against dimensionless time. A small difference between the curves for the case with and without damping is noted with damping having the effect of reducing the amount of momentum as it is expected. From the figure, it emerges that the momentum in the vertical direction, p_z, exhibits a higher peak than the momentum in the direction of flow propagation, p_x . The two peaks occur at different times. To better investigate which one is dominant and when, it is convenient to make a relative comparison between the two components. To this end, the square of each component over the square of the magnitude of the vector ($\vec{p}$) is plotted in Fig. 13(b) and (c). The use of the squares allows for plotting the components as percent, since $\frac{p_x^2}{p^2} + \frac{p_y^2}{p^2} + \frac{p_z^2}{p^2} = 100\%$. From the figure, it can be noted that at the beginning of the flow, the vertical component, p_z, dominates due to the fact that the motion of the particles is mainly gravity driven free fall. Then, during the propagation phase, the horizontal component, p_x, becomes dominant, stabilising itself at around 90%. Finally, when the flow is coming to rest, a surge of vertical component appears. This is due to the presence of decelerating particles exhibiting bouncing in the vertical direction, especially near the flow forefront. In comparison with Fig. 13(b), the start point of the chart in Fig. 13(c) looks shifted ahead in time, since it takes some time for the flow front to start moving ahead after gate removal.

Let us now examine the influence of the column aspect ratio on linear momentum. This is an important aspect in order to ascertain how general the observed trends on the linear momentum of the flow are. In Fig. 14, the evolution of the vertical and horizontal components of the linear momentum is plotted (as percentage over the magnitude of the vector $\vec{p}$) against dimensionless time for columns of various aspect ratios. Similar trends amongst the curves can be noted. However, in the initial phase of the flow (for $T < 1.1$), the vertical component of momentum increases with the aspect ratio and obviously the opposite is true for the horizontal one. Instead, after $T = 1.1$, the vertical component of momentum decreases with the aspect ratio. A possible explanation for the observed aspect ratio dependent trend could be that the gravity driven free fall of particles, which gives rise to the particle vertical motion, increases with the height of the column whereas the friction between particles, which opposes the horizontal motion of the particles, is independent of the column aspect ratio. In Fig. 14(b) the components of the momentum are plotted against run-out distance.

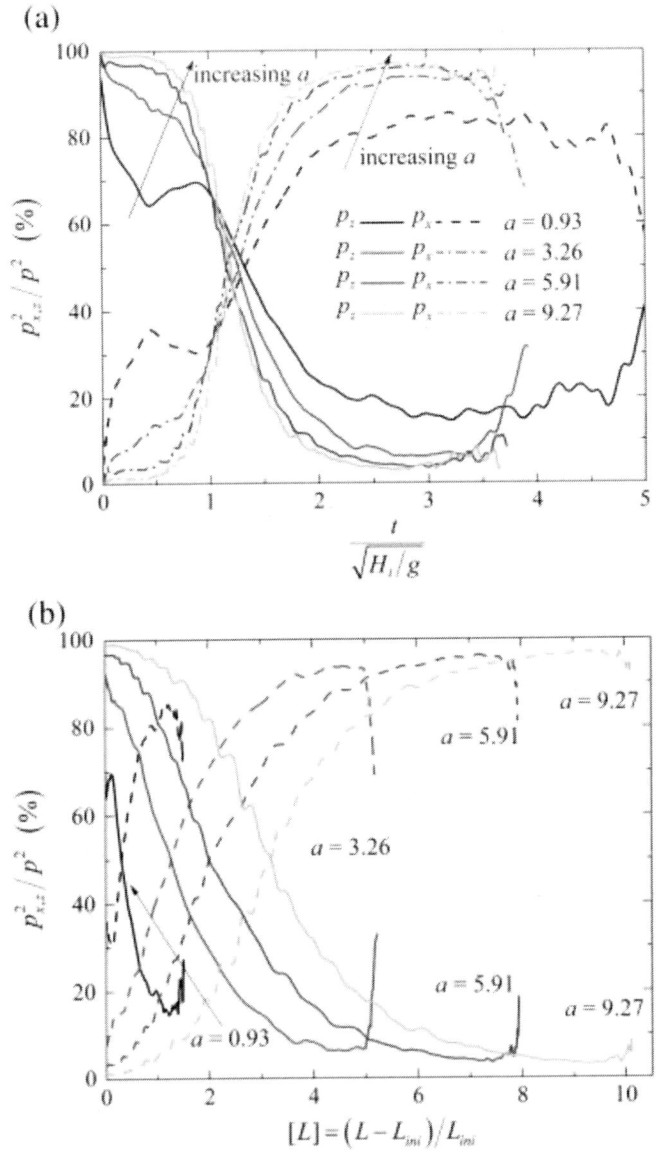

Fig. 14. Evolution of linear momentum over time for various aspect ratios (with viscous damping): a) components of momentum as a percentage of the total momentum against normalised time; b) components of momentum as a percentage of the total momentum against normalised run-out. *x* is the direction of flow propagation, *y* is the out of plane direction, and *z* is the vertical direction.

4.6. Flux of kinetic energy

To assess the vulnerability of existing structures hit by a granular flow/avalanche and to design engineering works for the protection of existing structures, two quantities are of interest: energy and momentum. The kinetic energy of the particle flow can be seen as an upper bound on the destructive energy that could be unleashed on the structure impacted by the flow. The amount of kinetic energy transferred from the flow to the structure depends on how flow and structure interact during the time the structure is impacted by the flow. Hence, the amount of energy released by the flow on the structure is a function of the characteristics of both flow and structure (for instance the relative stiffness between the two). Also the flow–structure interaction is likely to change over time, for instance due to the development of irrecoverable (plastic) deformations in the structure. So, it is not possible to predict the transfer of energy (and equally of momentum), unless a specific flow and a structure of interest are modelled. Here, however, we provide an analysis of the linear momentum and kinetic energy of the flow, measured at various distances from the initial position of the columns, in order to identify an upper bound on the maximum energy that may be imparted to structures knocked by granular flows, under the simplifying assumption of disregarding the effects of any structure–flow interactions.

Considering an imaginary vertical section perpendicular to the direction of horizontal propagation of the flow (see the vertical plane depicted in Figure 15) the flow mass transiting through such a section is a function of time. This is nil until the flow front reaches the position, then it gradually increases, then decreases and eventually becomes nil again once the flow has come to a stop (see Figure 15). In the following analysis, five locations along the flow path, shown in Fig. 16, are considered. Each location is identified by a letter (see Figure 16). A convenient measure of the maximum energy that can be transferred from the flow to the impacted structure is what we here call the flux of kinetic energy. This is evaluated as follows: for a given location of interest, the kinetic energy of the particles passing through the vertical plane oriented in the direction perpendicular to the flow is recorded during a specified time interval Δt and their total kinetic energy, ΔE, is evaluated. The flux of kinetic energy through the plane is given by the ratio, $\Delta E/\Delta t$, a quantity with the dimensions of power. It is convenient to normalise this quantity, so that a comparison of fluxes between columns of various sizes can be carried out. A simple way of doing so is by normalising both numerator, ΔE, and denominator, Δt. Thus, we define the normalised flux, P, as:

$$P \equiv \frac{\Delta E}{E_0} / \frac{\Delta t}{\sqrt{H_i / g}} \tag{13}$$

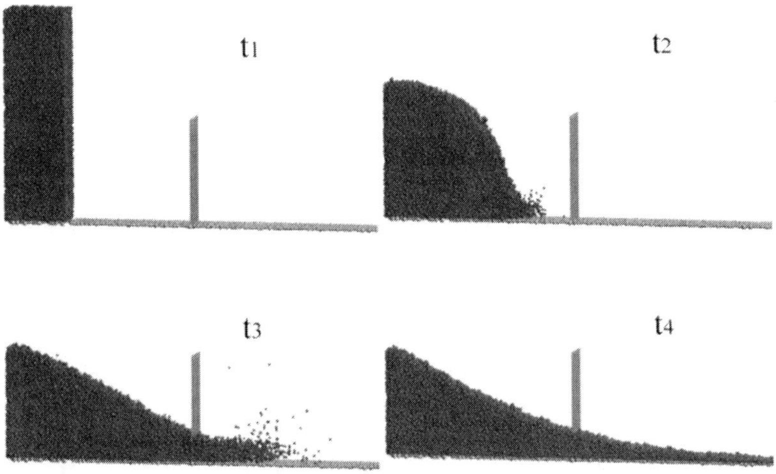

Fig. 15. Schematic view of granular flows past a structure.

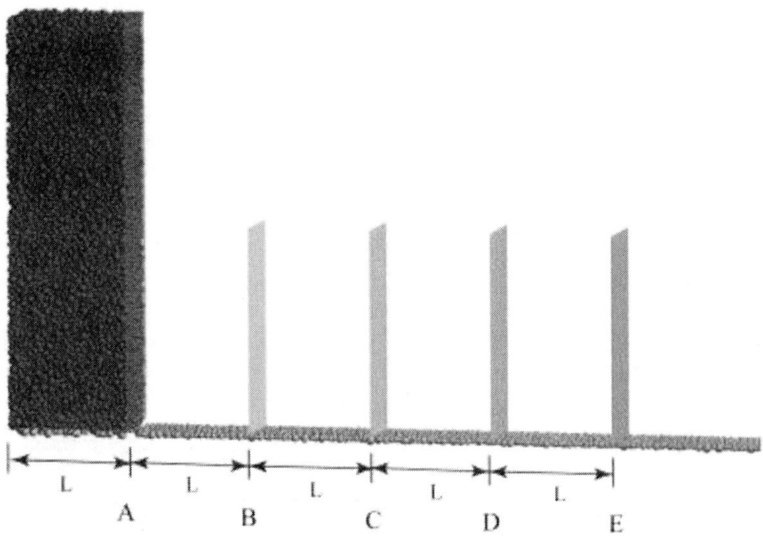

Fig. 16. Location of the sections where the granular flow is measured along the granular flow path.

Substituting E_0 from Eq. (11) into Eq. (13) and rearranging, Eq. (13) can be written as:

$$P = \frac{\Delta E}{\Delta t} \cdot \frac{2}{M\sqrt{g^3 H_i}} \tag{14}$$

Eq. (14) makes clear that $\frac{2}{M\sqrt{g^3 H_i}}$ is the multiplying factor to normalise the measured flux $\Delta E/\Delta t$. The flux, P, represents the maximum energy that can be transferred from the flow to the impacted structure. In fact, if all the energy were to be transferred away from the flow, the flow would be suddenly deprived of all its kinetic energy and therefore it would come to a stop, which is evidently an unrealistic scenario. In reality, only a portion of the energy of the flow is lost in the interaction with the structure that will cause the flow to slow down but not to stop. So, P can be thought of, as an upper bound on the maximum destructive power that the flow may impart on the impacted structure. In Fig. 17(a), the flux of kinetic energy at the selected locations is plotted versus dimensionless time. It can be noted that the section with highest flux is B. Obviously the fluxes in the case damping is present are smaller than the case of undamped flow. An interesting finding is the fact that the peak takes place at a different time, with the time of peak for the damped system shifting progressively ahead of the peak time for the undamped one. Also, the difference of value between the peaks for the damped and undamped systems increases with the distance of the section investigated from the column initial position reaching up to 50% of the peak value for the undamped system. Fig. 17(b) illustrates the evolution of the height of the granular mass at different locations. From the figure, it can be concluded that the further the location is away from the slope source region, the lower the height of the final granular mass is.

Fig. 18 illustrates the evolution of the flux of kinetic energy for different column aspect ratios. From the figure, it can be observed that the position of the section where the flux is highest depends on the aspect ratio. For instance, in the case of small aspect ratios (e.g. $a = 0.93$), the flux of destructive energy at location A is largest, with only a small amount of particles travelling to locations further down section B. For intermediate aspect ratios (e.g. $a = 3.26$, 5.91), the largest flux takes place at location B. For large aspect ratios (e.g. $a = 9.27$), the largest flux occurs again at section A.

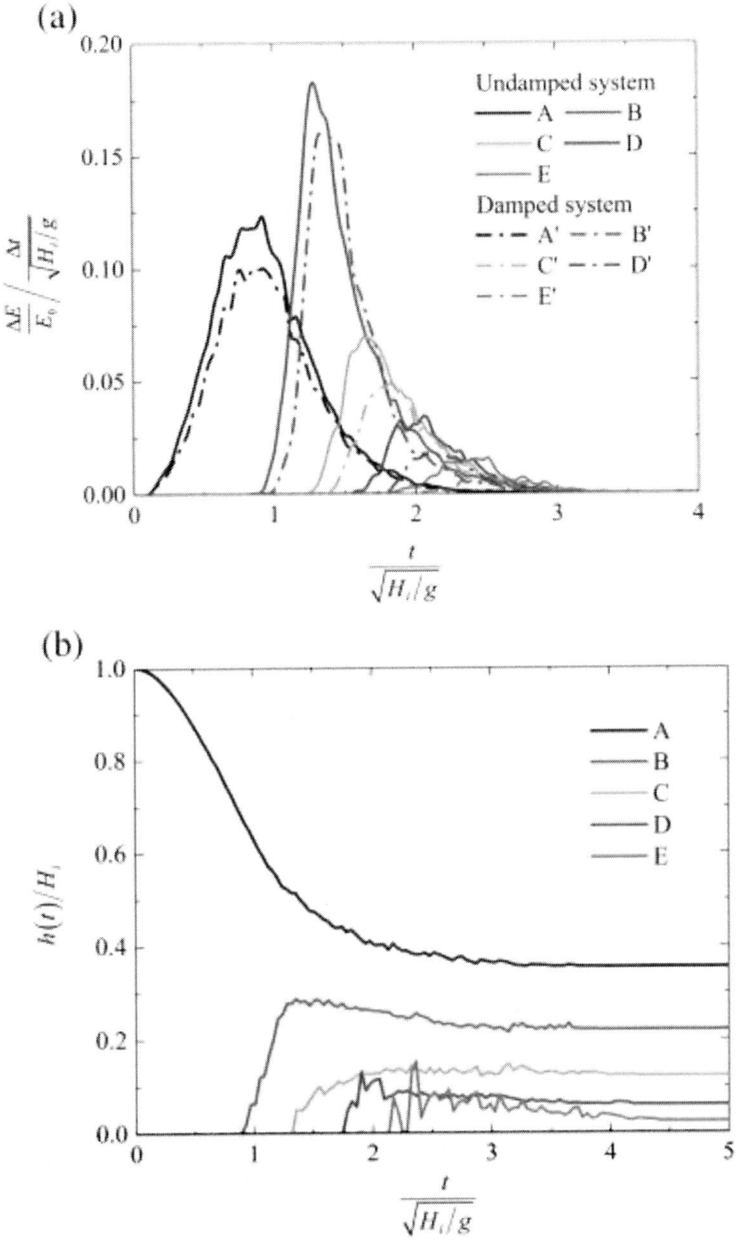

Fig. 17. Evolution of the flux of destructive energy and height of debris mass. (a) Flux of kinetic energy at different locations (a = 3.26). (b) Height of flow at different locations (a = 3.26).

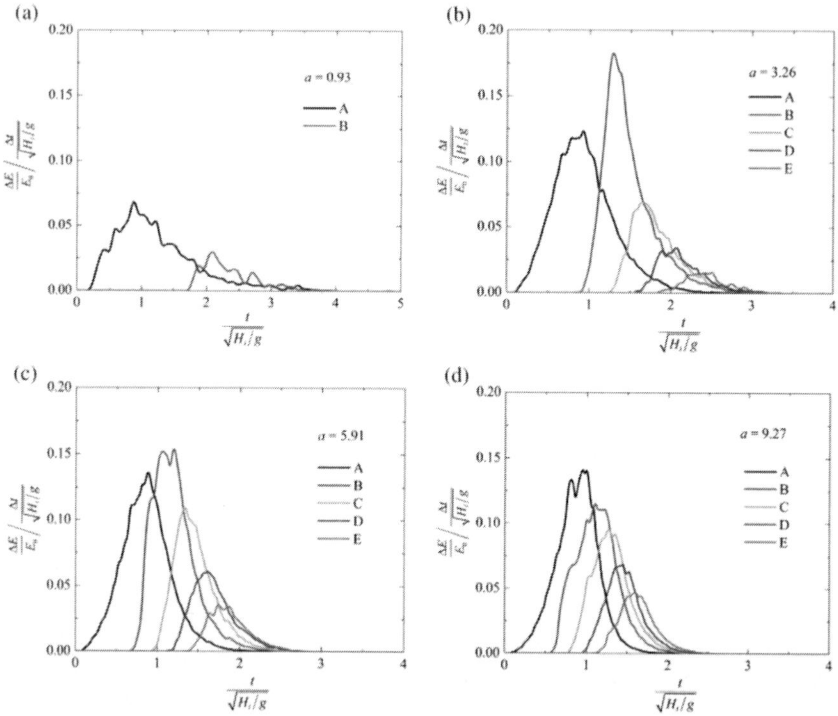

Fig. 18. Evolution of the flux of destructive energy for different granular columns.

4.7. Distribution of flux of kinetic energy and linear momentum over depth

To be able to design protective structures as effectively as possible, the spatial distributions of kinetic energy and momentum over the depth of the considered sections are also needed. Considering section B, we have split the depth of the flow, h(t), into 5 parts and calculated the flux of kinetic energy through the section for each part so as to obtain a vertical profile of the flux of kinetic energy (see Figure 19(a)). Looking at the figure, it emerges that the profile of the flux is initially bilinear, with the maximum flux at middle height of the column, then the flux evolves into a linear profile whose amplitude progressively reduces over time until becoming nil. The bilinear distribution points out to the presence of an uppermost layer of particles in an agitated loose state, which, after some time, consolidates so that a linear distribution is obtained. The presence of this layer of loose material is confirmed by the calculation of the profile of mass rate normalised by the total flow mass in the section (see Figure 19(c)).

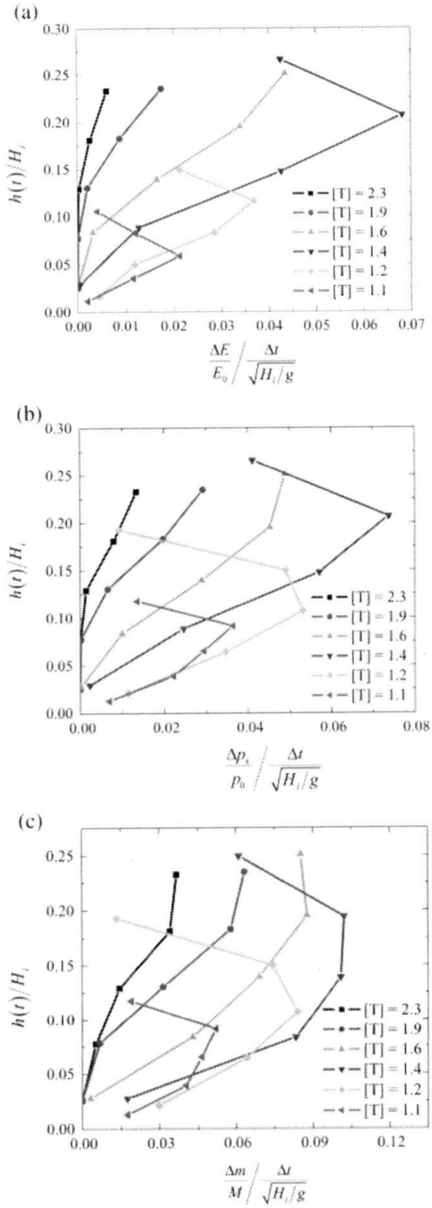

Fig. 19. Profile of kinetic energy, momentum and mass trough plane B at different times (a = 3.26). (a) Profile of kinetic energy flux (b) Profile of normalised momentum. (*Δpx* is the rate of momentum through the portion of section considered; (c) Profile of granular mass moving (*Δm* is the mass flowing through the portion of section considered.

A convenient measure of the maximum momentum that could be transferred by the flow to the impacted structure is the flux of linear momentum over time. Analogously to the flux of kinetic energy, we can evaluate the linear momentum of the particles passing through the vertical section of interest, Δp, during a specified time interval Δt. The flux of linear momentum through the whole section or parts of it, is given by $\Delta p/\Delta t$, a quantity with the dimensions of force (so we call it F). It is convenient to normalise this quantity so that comparison of fluxes between columns of various sizes can be carried out. A simple way of doing this is by normalising both numerator and denominator. So, we define the normalised linear momentum of the flow in the unit of time as:

$$F_{x,y,z} \equiv \frac{\Delta p_{x,y,z}}{p_0} / \frac{\Delta t}{\sqrt{H_i/g}} \tag{15}$$

with p_0 being an average linear momentum for the flow here defined in Eq. (12). Substituting p_0 from Eq. (12) and rearranging, Eq. (15) can also be written as:

$$F_{x,y,z} = \frac{\Delta p_{x,y,z}}{\Delta t} \cdot \frac{1}{Mg} \tag{16}$$

Eq. (16) makes clear that $1/Mg$ is the multiplying factor to normalise the momentum going through the plane in the unit of time. In terms of the value of the flux of the momentum at various sections in time, similar figures as those obtained for the flux of kinetic energy (Fig. 17 ; Fig. 18) are obtained which are not reported here for sake of brevity. To locate critical sections and times of interest one of the two fluxes, either the flux of kinetic energy or of momentum, is enough. However, it is of interest to practitioners appointed with designing engineering works for the mitigation of the flow hazard, to know the value of the flux of momentum over depth in order to have an idea of the distribution of the pressure that can act on the structure. In Fig. 19(b), the distribution of the horizontal component of the linear momentum along the direction of flow propagation, F_x, over depth is plotted against time. As it can be expected, the same shape of the profile as the profile of the flux of kinetic energy is found.

5. CONCLUSIONS

This paper presents a numerical investigation of the behaviour of dry granular flows generated by the collapse of prismatic columns via 3D Distinct Element Method (DEM) simulations under plane strain conditions. This type of analysis showcases the potential of the Distinct Element Method to investigate the phenomenology of dry granular flows and to gather unique information currently unachievable by experimentation. By means of dimensional analysis, the governing parameters of the problem were identified. Then, the influence of key variables of the problem was analysed. The main results are summarised as below:

(1)Different regimes of granular motion have been observed, depending on the initial aspect ratios. The DEM results qualitatively match the FEM analyses by Crosta et al. (2009) and the experimental results by Lube et al. (2005). The granular material slides along a plane which approximately corresponds to the active failure plane of the column in agreement with Rankine's theory.

(2)Quantitative relationships between the column aspect ratio and normalised runout distance and deposit height were established. Using the rolling resistance model, i.e. assigning a moment–relative rotation contact law, the DEM simulations give rise to runout distances and deposit height which match well the available numerical and experimental results.

(3)A detailed analysis of how energy is dissipated by granular flows was performed from which emerges that most of the energy of the columns is dissipated by inter-particle friction, with frictional dissipation increasing with the column aspect ratio. Also, the translational and rotational components of the kinetic energy of the flows, associated to particle rotational and translational motions respectively, were monitored during the simulations. It is found that the rotational component is negligible in comparison with the translational one; hence in order to calculate the destructive power of a granular flow slide, only the translational contribution of the kinetic energy is relevant.

(4)A methodology is presented to calculate the flux of kinetic energy over time carried by the granular flow through any vertical section of interest. This can be related to the energy released by landslide induced granular flows impacting against engineering structures under the simplifying assumption of neglecting all structure–flow interactions. This represents the first step towards achieving a computational tool quantitatively predicting the destructive power of a given flow at any location of interest along its path. This could be useful for the design of engineering works for natural hazard mitigation. To this end, also the distribution of the linear momentum of the flow over depth was calculated. It emerges that the distribution is initially bilinear, due to the presence of an uppermost layer of particles in an agitated loose state, but after some time becomes linear.

ACKNOWLEDGEMENTS

The second author has been supported by Marie Curie Actions-International Research Staff Exchange Scheme (IRSES): GEO-geohazards and geomechanics, Grant No. 294976.

REFERENCES

1. Balmforth, N.J., Kerswell, R.R., 2005. Granular collapse in two dimensions. J. Fluid Mech. 538, 399–428.
2. Belheine, N., Plassiard, J.P., Donzé, F.V., Darve, F., Seridi, A., 2009. Numerical simulation of drained triaxial test using 3D discrete element modeling. Comput. Geotech. 36, 320–331.
3. Calvetti, F., Nova, R., 2004. Micromechanical approach to slope stability analysis. In: Darve, F., Vardoulakis, I. (Eds.), Degradations and Instabilities in Geomaterials. Springer, Vienna, pp. 235–254.
4. Carrara, A., Crosta, G., Frattini, P., 2008. Comparing models of debris-flow susceptibility in the alpine environment. Geomorphology 94, 353–378.
5. Carrigy, M.A., 1970. Experiments on the angles of repose of granular materials. Sedimentology 14, 147–158.
6. Casagli, N., Ermini, L., Rosati, G., 2003. Determining grain size distribution of the material composing landslide dams in the Northern Apennines: sampling and processing methods. Eng. Geol. 69, 83–97.
7. Cleary, P.W., Campbell, C.S., 1993. Self-lubrication for long runout landslides: examination by computer simulation. J. Geophys. Res. Solid Earth 98, 21911–21924.
8. Collins, G.S., Melosh, H.J., 2003. Acoustic fluidization and the extraordinary mobility of sturzstroms. J. Geophys. Res. Solid Earth 108, 1–14.
9. Craig, R.F., 1997. Soil mechanics. E & FN SPON, London and New York. Crosta, G.B., Imposimato, S., Roddeman, D., Chiesa, S., Moia, F., 2005. Small fast-moving flowlike landslides in volcanic deposits: the 2001 Las Colinas Landslide (El Salvador). Eng. Geol. 79, 185–214. http://dx.doi.org/10.1016/j.enggeo.2005.01.014. Crosta, G.B., Frattini, P., Fusi, N., 2007. Fragmentation in the Val Pola rock avalanche, Italian Alps. J. Geophys. Res. Earth Surf. 112.
10. Crosta, G.B., Imposimato, S., Roddeman, D., 2009. Numerical modeling of 2-D granular step collapse on erodible and nonerodible surface. J. Geophys. Res. Earth Surf. 114, 1–19.
11. Cundall, P.A., Strack, O.D.L., 1979. A discrete numerical model for granular assemblies. Geotechnique 29, 47–65 (doi: citeulike-article-id:8653064).
12. De Blasio, F.V., Elverhøi, A., 2008. A model for frictional melt production beneath large rock avalanches. J. Geophys. Res. Earth Surf. 113 F02014.
13. Goren, L., Aharonov, E., 2007. Long runout landslides: the role of frictional heating and hydraulic diffusivity. Geophys. Res. Lett. 34. Hungr, O., Evans, S.G., 2004. Entrainment of debris in rock avalanches: an analysis of a long run-out mechanism. Geol. Soc. Am. Bull. 116, 1240–1252.
14. Iverson, R.M., 1997. The physics of debris flows. Rev. Geophys. 35, 245–296.

15. Iverson, R.M., Reid, M.E., Logan, M., LaHusen, R.G., Godt, J.W., Griswold, J.P., 2011. Positive feedback and momentum growth during debris-flow entrainment of wet bed sediment. Nat. Geosci. 4, 116–121.
16. Johnson, K.L., 1985. Contact Mechanics. Cambridge University Press, Cambridge. Kerswell, R.R., 2005. Dam break with Coulomb friction: a model for granular slumping? Phys. Fluids 17, 1–16.
17. Lacaze, L., Phillips, J.C., Kerswell, R.R., 2008. Planar collapse of a granular column: experiments and discrete element simulations. Phys. Fluids 20, 1–12.
18. Lajeunesse, E., Monnier, J.B., Homsy, G.M., 2005. Granular slumping on a horizontal surface. Phys. Fluids 17, 1–15.
19. Linares-Guerrero, E., Goujon, C., Zenit, R., 2007. Increased mobility of bidisperse granular avalanches. J. Fluid Mech. 593, 475–504.
20. Lube, G., Huppert, H.E., Sparks, R.S., Freundt, A., 2005. Collapses of two-dimensional granular columns. Phys. Rev. E. 72, 1–10.
21. Mangeney, A., Bouchut, F., Thomas, N., Vilotte, J.P., Bristeau, M.O., 2007. Numerical modeling of self-channeling granular flows and of their levee-channel deposits. J. Geophys. Res. 112.
22. Mangeney, A., Roche, O., Hungr, O., Mangold, N., Faccanoni, G., Lucas, A., 2010. Erosion and mobility in granular collapse over sloping beds. J. Geophys. Res. Earth Surf. 115.
23. Melosh, H.J., 1979. Acoustic fluidization: a new geologic process? J. Geophys. Res. Solid Earth 84, 7513–7520.
24. Modenese, C., 2013. Numerical Study of the Mechanical Properties of Lunar Soil by the Discrete Element Method. DPhil, University of Oxford (D.Phil Thesis).
25. Palmer, A.C., 2008. Dimensional Analysis and Intelligent Experimentation. World Scientific. Phillips, J.C., Hogg, A.J., Kerswell, R.R., Thomas, N.H., 2006. Enhanced mobility of granular mixtures of fine and coarse particles. Earth Planet. Sci. Lett. 246, 466–480.
26. Pohlman, N.A., Severson, B.L., Ottino, J.M., Lueptow, R.M., 2006. Surface roughness effects in granular matter: influence on angle of repose and the absence of segregation. Phys. Rev. E. 73, 031304.
27. Pöschel, T., Buchholtz, V.V., 1993. Static friction phenomena in granular materials: Coulomb law versus particle geometry. Phys. Rev. Lett. 71, 3963–3966.
28. Rankine, W.J.M., 1857. On the stability of loose earth. Philos. Trans. R. Soc. Lond. 147, 9–27.
29. Roche, O., Attali, M., Mangeney, A., Lucas, A., 2011. On the run-out distance of geophysical gravitational flows: insight from fluidized granular collapse experiments. Earth Planet. Sci. Lett. 311, 375–385.
30. Rothenburg, L., Bathurst, R.J., 1992. Micromechanical features of granular assemblies with planar elliptical particles. Geotechnique 42, 79–95.
31. Saucedo, R., Macías, J.L., Sheridan, M.F., Bursik, M.I., Komorowski, J.C., 2005. Modeling of pyroclastic flows of Colima Volcano, Mexico: implications for hazard assessment. J. Volcanol. Geotherm. Res. 139, 103–115.
32. Shreve, R.L., 1968. Leakage and fluidization in air-layer lubricated avalanches. Geol. Soc. Am. Bull. 79, 653–658.
33. Staron, L., Hinch, E.J., 2005. Study of the collapse of granular columns using DEM numerical simulation. J. Fluid Mech. 545, 1–27.
34. Staron, L., Hinch, E., 2007. The spreading of a granular mass: role of grain properties and initial conditions. Granul. Matter 9, 205–217.
35. Tang, C.L., Hu, J.C., Lin, M.L., Angelier, J., Lu, C.Y., Chan, Y.C., Chu, H.T., 2009. The Tsaoling landslide triggered by the Chi-Chi earthquake, Taiwan: insights from a discrete element simulation. Eng. Geol. 106, 1–19.
36. Tsuji, Y., Tanaka, T., Ishida, T., 1992. Lagrangian numerical-simulation of plug flow of cohesionless particles in a horizontal pipe. Powder Technol. 71, 239–250.
37. Utili, S., Crosta, G.B., 2011a. Modelling the evolution of natural slopes subject to weathering: Part I. Limit analysis approach. J. Geophys. Res. Earth Surf. 116, F01016.

38. Utili, S., Crosta, G.B., 2011b. Modelling the evolution of natural slopes subject to weathering: Part II. Discrete element approach. J. Geophys. Res. Earth Surf. 116, F01017.
39. Utili, S., Nova, R., 2008. DEM analysis of bonded granular geomaterials. Int. J. Numer. Anal. Methods Geomech. 32 (17), 1997–2031.
40. Weatherley, D., Boros, V., Hancock, W., 2011. ESyS-Particle Tutorial and User's Guide Version 2.1. Earth Systems Science Computational Centre, The University of Queensland.
41. Zenit, R., 2005. Computer simulations of the collapse of a granular column. Phys. Fluids 17, 1–4.
42. Zhao, T., 2014. Investigation of Landslide-Induced Debris Flows by the DEM and CFD. DPhil, University of Oxford (D.Phil Thesis).
43. Zhao, T., Houlsby, G., Utili, S., 2012. Numerical simulation of the collapse of granular columns using the DEM. International Symposium on Discrete Element Modelling of Particulate Media. RSC Publishing, Birmingham, pp. 133–140.

CHAPTER 4

Analysis of Engineering Geology Indices in Three Units of Atamir Formation in Kope Dagh Zone

Bizhan Shirdel

Department of Geology, Karaj Branch, Islamic Azad University, Karaj, Iran

ABSTRACT

This is a field report on a comprehensive study of the Atamir Formation from the engineering geology perspective using the related indices. The Atamir Formation of the Cretaceous Period, which has outcropped in the form of thick frequencies of grey-knotted sandstone and black shales, is situated in the Kope Dagh zone. A survey of discontinuities together with bedding was carried out to study slope stability. The layers have a general east-west trend with a gentle slope towards the south. Because of the tectonic and stratigraphic differences, and with the purpose of facilitating surveys related to joint study of the outcrop, the formation in the study region was divided into three units. The lower unit is made of shale, the middle of sandstone, and the upper of marlstone. All three units were studied from the perspective of geomechanical classification, rock mass indices, geological strength, geomechanical indices, and wedge instability analysis under dry and wet conditions, and the results were investigated in the form of various images and figures. The Dips software was used to display the rose diagram and stereographic projection of each unit, the Swedge software to analyze instability of the wedges, and the Roctab software to analyze the geomechanical parameters and present the outputs along with the description of each unit.

Keywords: Atamir Formation, Kope Dagh Zone, Engineering Geology Indices, Joint Study, Software Analysis, Wedge Stability Analysis

1. INTRODUCTION

Based on field surveys and geological surveys such as outcrop studies, the Atamir Formation was divided into three separate units to study the characteristics of the discontinuities and of the rock mass in each one. This study intended to review results of research carried out on the Atamir Formation in detail and present a field report. Therefore, the units were studied and the results were compared.

2. RESULTS

2.1. The Atamir Formation Unit One K (at) 1

This lithology included black shales. In the study region, there were three dominant discontinuity sets (two joint sets together with bedding) and random discontinuities. Table 1 lists the characteristics of these discontinuities. Table 2 lists geotechnical characteristics of the various discontinuities in the rock masses. Figures 1-3 show the Stereographic characteristics of Figures 4-7 show statistical analysis.

Rock masses were classified based on the methods proposed by Bieniawski (RMR), Barton (Q), and Hoek (GSI), all of which are discussed below.

2.1.1. Geomechanical Classification (RMR)

1) Compressive strength of intact rock

The value of this parameter was determined based on the strength characteristic of outcropped masses and on performing tests.

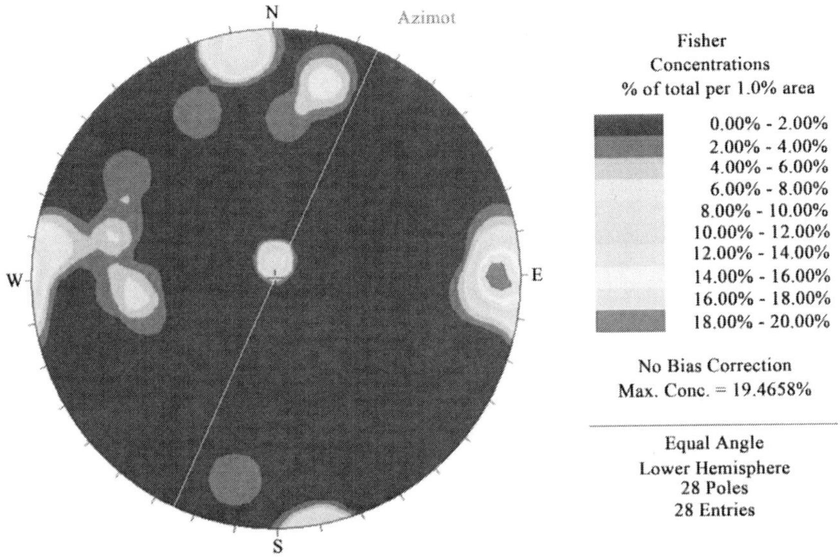

Figure 1. Stereographic image of the discontinuities in unit one of the Atamir Formation K (at) 1.

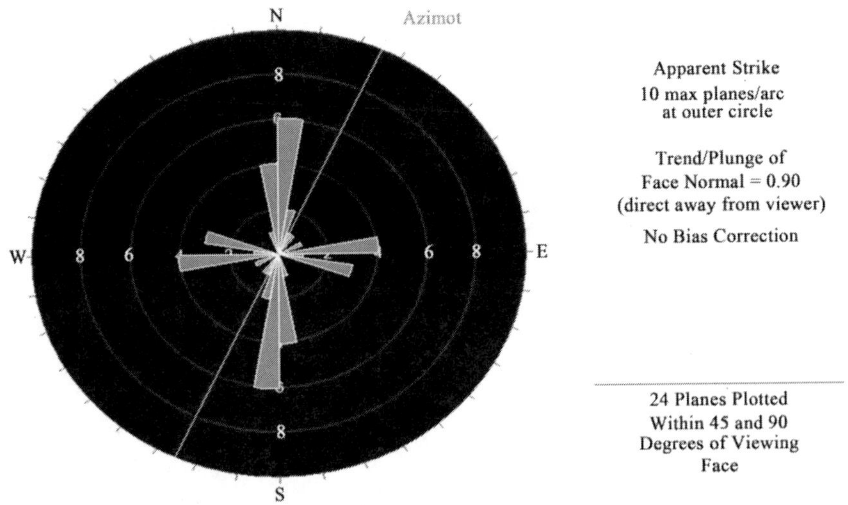

Figure 2. Rose diagram of the discontinuities in unit one of the Atamir Formation K (at) 1.

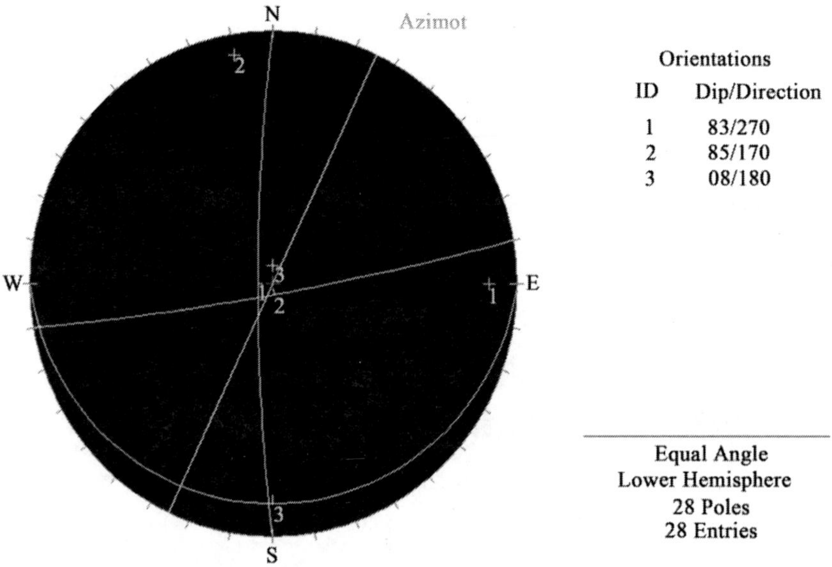

Figure 3. Stereography of discontinuities in unit one of the Atamir Formation K (at) 1.

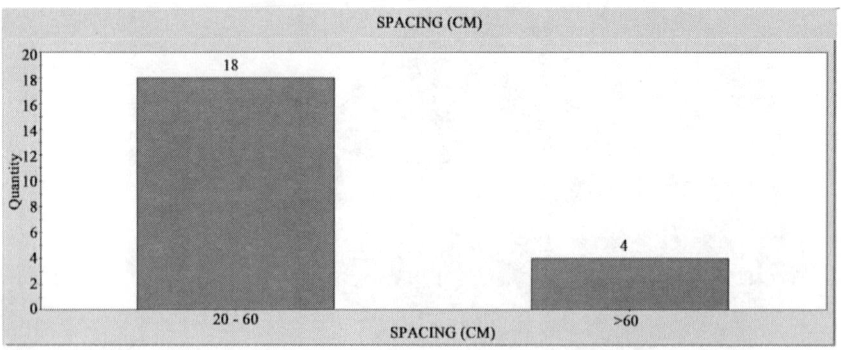

Figure 4. Statistical analysis of discontinuity spacings in unit one of the Atamir Formation K (at) 1.

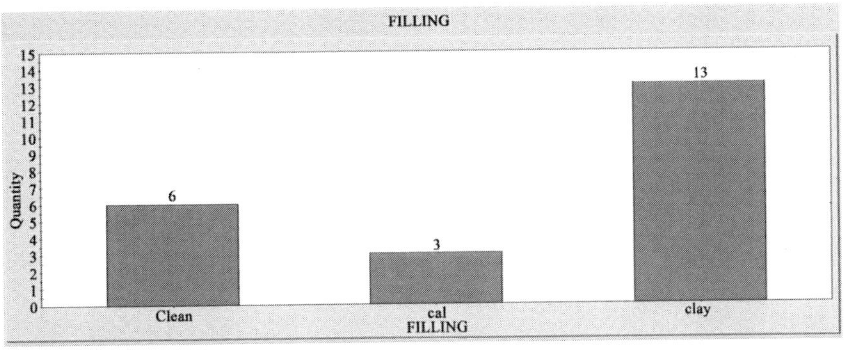

Figure 5. Statistical analysis of discontinuity fillings in unit one of the Atamir Formation K (at) 1.

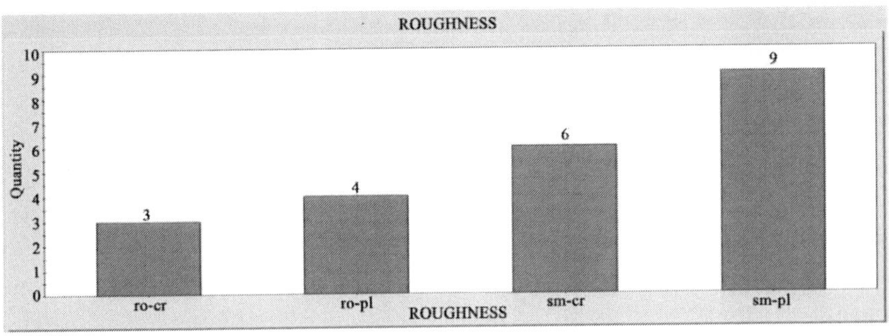

Figure 6. Statistical analysis of discontinuity roughness in unit one of the Atamir Formation K (at) 1.

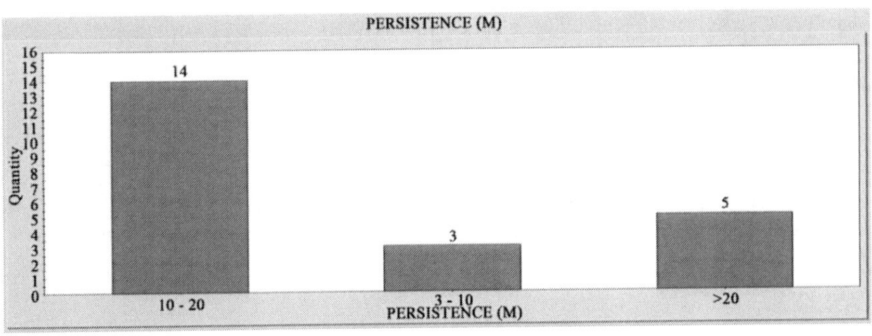

Figure 7. Statistical analysis of discontinuity orientation in unit one of the Atamir Formation K (at) 1.

Table 1. Characteristics of the discontinuities in unit one of the Atamir Formation K (at) 1.

Discontinuity	ID	Dip	Dip Direction
Js1	1	83	270
Js2	2	85	170
Bedding	3	8	180

Table 2. Geotechnical characteristics of the various discontinuities in the rock masses of unit one in the Atamir Formation K (at) 1.

Joint Sets	Dip Direction	Joint Frequency	Persistence (m) >20%	10% - 20%	3% - 10%	1% - 3%	<1%	Joint Roughness (%) Very Rough	Rough	Rough-Smooth	Smooth	Slickenside
Js1	270\83	50	-	80	20	-	-	-	-	20	80	-
Js2	170\85	30	-	70	30	-	-	-	30	70	-	-
Bedding	180\8	20	100	-	-	-	-	-	-	100	-	-

Joint Alteration	Joint Water	Aperture (mm) >5 mm %	1 - 5 mm %	0.1 - 1 mm %	<0.1 mm %	Close	Joint filling Soft filling mm 5 < %	Soft filling mm 5 > %	Hard filling mm 5 > %	Hard filling mm 5 < %	Clean %	Joint spacing (m-mm) <60 mm %	60 - 200 mm %	200 - 600 Mm %	0.6 - 2 m %	>2 m %
		-	100	-	-	-	-	100	-	-	-	-	-	100	-	-
W2	Dry	40	60	-	-	-	-	-	40	-	60	-	-	50	50	-
		-	100	-	60	-	100	-	-	-	-	-	-	-	-	100

2) Rock quality designation

The rating of 8 may be given depending on the state of the discontinuity system.

3) Joint spacing

The rating for this parameter was the least value of spacing. In the study region, there were three sets of discontinuities (two sets of dominant discontinuities together with bedding), and the minimum discontinuity spacing was less than 10 centimeters. Considering the geomechanical classification parameters, the rating of 8 - 10 was given to this parameter.

4) Condition of discontinuities

The joints in the rock masses in unit one of the Atamir Formation, K (at) 1, were 10 - 20 meters long, clay, calcite, and few traces of iron oxide were the main filling, the joints were smooth to a little rough and, with respect to euhidrality, the rock masses were planar and moderately weathered. Considering the existing parameters, the rating given to this parameter was 10.

5) Groundwater

Presence of groundwater in a region is undoubtedly very important, especially if permeability is high. Considering the conditions of the rock masses in the region, the seepage state was determined and given the rating of 4.

6) Discontinuity orientations

This parameter is determined by the mode of transgression of the structure towards the jointed rock mass, by the dips of the layers (and of the discontinuities), and by the orientation of the structure (which has been used by Feenstra & Wickham in 1975 [1] and suggested by Bieniawski 1973) [2] .

The presented parameters 2 - 1 to 2 - 5 are used for determining the value of RMR that, for the rock masses in unit one of the Atamir Formation, K (at) 1, was 34 - 36. Based on this, these rock masses were placed in the poor class; pertaining data are shown in Table 3.

2.1.2. Qualitative Index of the Rock Mass (Q)

The index Q for the rock masses in unit one of the Atamir Formation, K9 (at) 1, was estimated based on the following 6 parameters.

1) Rock quality designation (RQD)

Considering the condition of the rock masses, the value of 43 - 48 was given to the RQD parameter of unit one in the Atamir Formation, K (at) 1.

Table 3. Geomechanical classification of the rock masses in unit one of the Atamir Formation K (at) 1.

Parameter	Value	Rating
UCS (Mpa)	25 - 50	4
RQD	25 - 50	8
Spacing (mm)	60 - 200	8 - 10
Condition of discontinuities		10
Groundwater	Flow	4
RMR	34 - 35	
Class no.	Poor rock	
Cohesion of the rock mass (KPa)	115	
Friction angle of the rock mass (deg)	15 - 25	

2) Number of joint sets (J_n)

Based on the data obtained from the geological field surveys, this parameter received the value of 12 for unit one of the Atamir Formation, K (at) 1.

3) Discontinuity surface roughness (J_r)

This parameter shows the degree of roughness of the discontinuities and, based on Barton's theory, the weakest set of critical discontinuities or discontinuities filled with clay materials in the related zone must be determined. Considering the conditions, the value of one was given to this parameter for unit one of the Atamir Formation, K (at) 1.

4) Joint alteration number (J_a)

Considering the joint filling and the apparent conditions of discontinuity surfaces, this parameter received the value of 4 for unit one of the Atamir Formation, K (at) 1.

5) Joint water reduction factor (J_w)

This parameter measures water pressure, which inversely affects shear strength of the discontinuity. Based on data collected in the study region, the value of 0.66 was considered for this parameter.

6) Stress reduction factor (SRF)

This parameter is determined based on the characteristics of the study region including the existing stresses, the tectonic conditions, the shear zones, etc. Since the necessary tests were not performed, the estimated value of one was used for this factor.

Considering the ratings of the six parameters, and using the presented relationship, the expected value determined for the Q index was 0.592 - 0.66 for the rock masses in unit one of the Atamir Formation, K (at) 1, which places them in the class of very poor rocks. Data are shown in Table 4.

2.1.3. Geological Strength Index (GSI) of the Rock Masses

Using the following three methods, the value of the geological strength index for the rock masses in unit one was determined.

1) Geological strength index of the rock mass (GSI) (Figure 8)

The GSI system is the only classification system that is directly related to engineering parameters such as Mohr-Coulomb and Hoek-Brown parameters. Conventional methods for determining strength parameters and rock mass modulus in Iran and other countries include plate-loading tests and block shear in situ. However, these tests can be performed only if exploration drilling has been carried out and, in addition to that, they are very costly. Therefore, many studies have been conducted and numerous attempts made for developing and introducing methods to indirectly determine engineering parameters of rock masses, one of which is the GIS system [3] . In the following sections, the GSI of rock masses in unit one will be discussed.

2) Determining GSI based on field observations

Based on field studies and observations, the value of geological strength index of rock masses in unit one of the Atamir Formation, K (at) 1 was 35 - 45.

2.1.4. Determination of Geomechanical Parameters of the Rock Masses in Unit One of the Atamir Formation K (at) 1

To determine these parameters, the GSI was considered 35 - 45. Since cohesive strength and angle of internal friction are among the input parameters in most methods of analysis, the Mohr-Coulomb and Hoek-Brown criteria had to be fitted to determine these two parameters. Results are listed in Table 5 and shown in Figures 9-11.

2.1.5. Analysis of Wedge Instability in Unit One of the Atamir Formation, K (at) 1, under Dry and Wet Conditions

Analysis of wedge instability was performed using SWEDGE. This software, which was developed by the ROCSCIENCE Company, analyzes wedge stability in rock slopes, degree of cohesion, geomechanical characteristics, etc., studies stability from various aspects, analyzes areas with rock fall potential using the StereoNet network, follows discontinuities and determines and analyzes intersection points of wedges, and plots the StereoNet using input data. Figure 12 shows stereographic images of the joint sets. Applying the primary parameters, the characteristics of the largest wedges that can be formed in the walls are determined. According to the performed analysis, wedge slide will happen along the interface between two discontinuities, and the factor of safety (FS) will be 0.07743. Figures 13-18 show other views.

Table 4. The Q index of the rock masses in unit one of the Atamir Formation, K (at) 1.

Parameter	Description	Value
RQD	43 - 48	43 - 48
J_n	Three joint sets with random	12
J_r	Smooth or rough to Craw or Planar	1
J_a	Softening or low friction clay mineral coatings	4
J_w	Inflow	0.66
SRF	Medium stress, favorable stress condition	1
Q	0.592 - 0.66	
Class No.	Very poor rock	

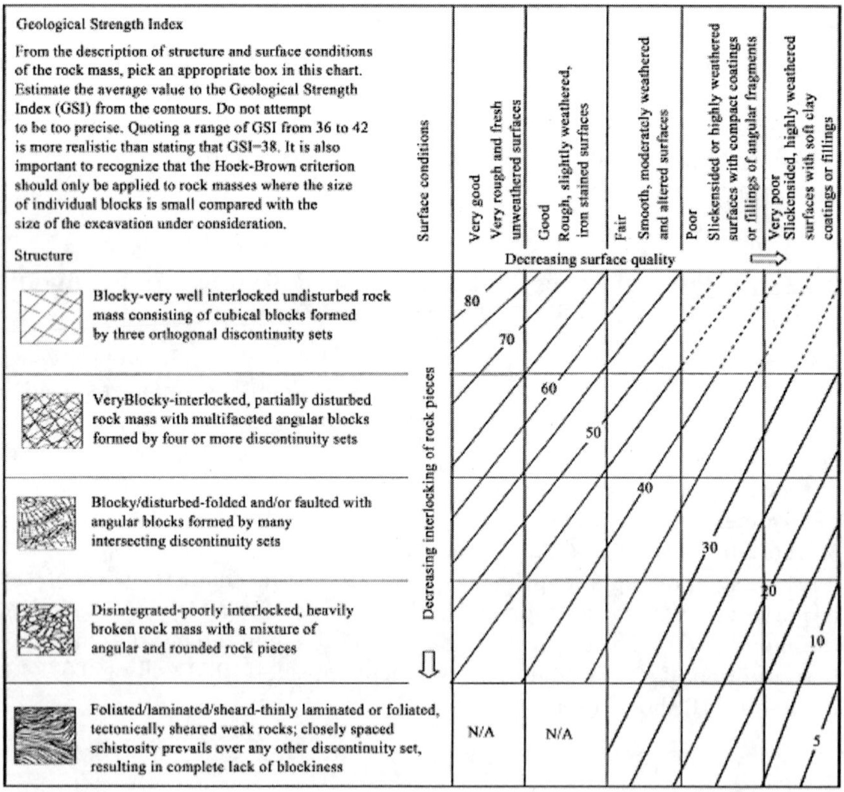

Figure 8. The GSI range of the rock masses in unit one of the Atamir Formation K (at) 1 in the GSI classification diagram.

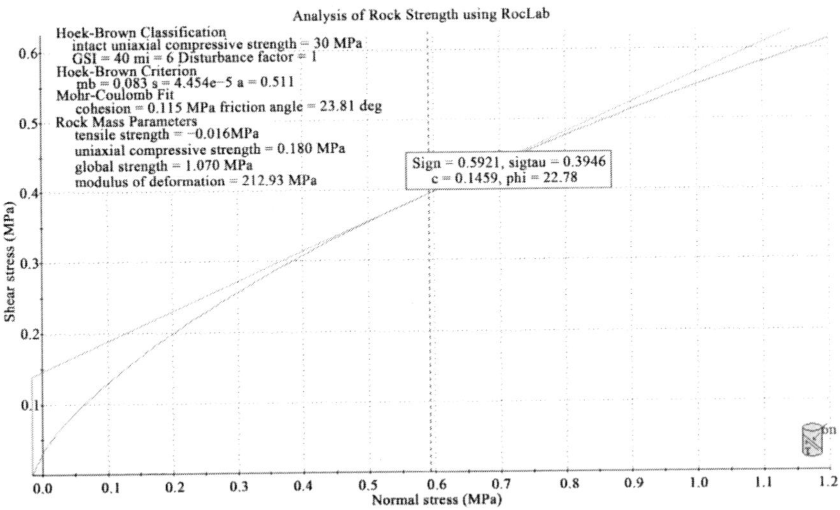

Figure 9. Fitting the data on the rock masses in unit one of the Atamir Formation, K (at) 2, to the Hoek-Brown and Mohr- Coulomb criteria based on GSI = 40 and normal shear stress.

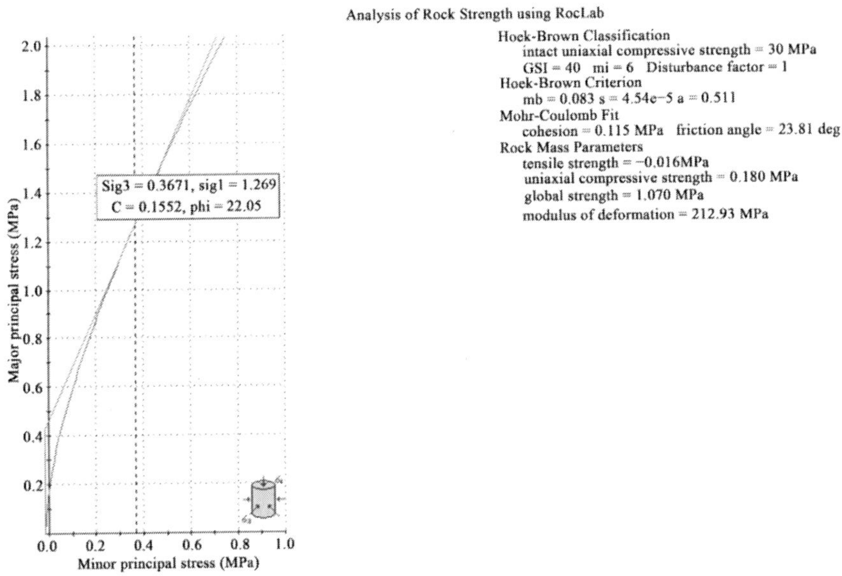

Figure 10. Fitting the data on the rock masses in unit one of the Atamir Formation, K (at) 2, to the Hoek-Brown and Mohr- Coulomb criteria based on GSI = 40 and major principal stress.

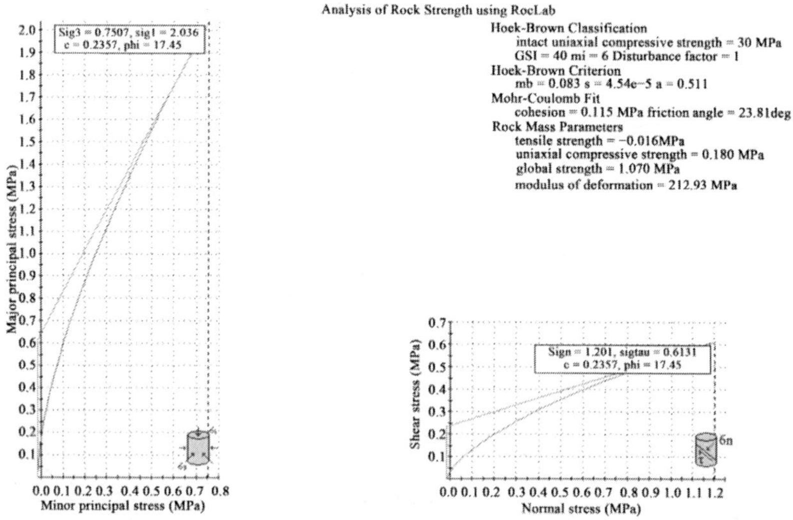

Figure 11. Analysis of output data for determining geomechanical parameters of the rock masses in unit one based on GSI = 40.

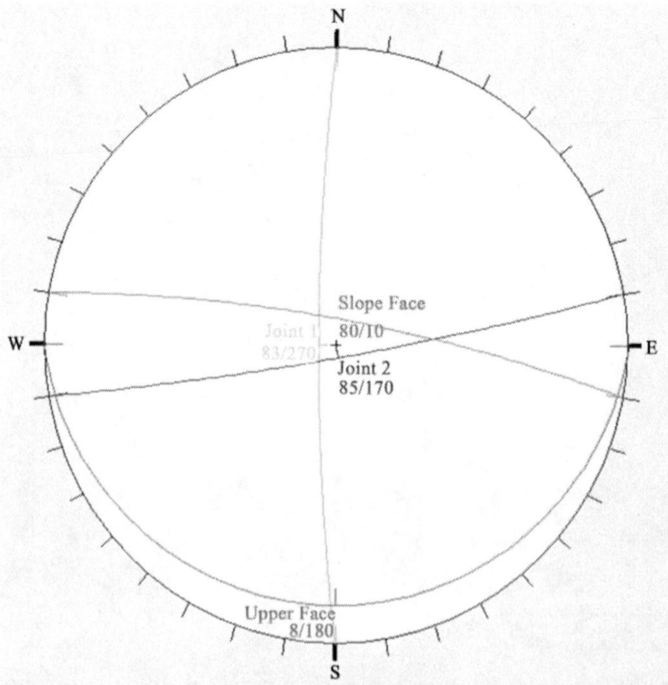

Figure 12. Stereographic images of joint sets.

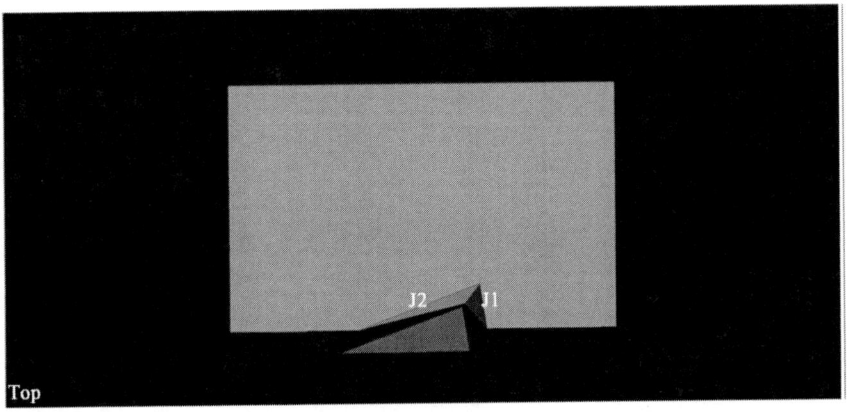

Figure 13. Top view of wedge slide analysis.

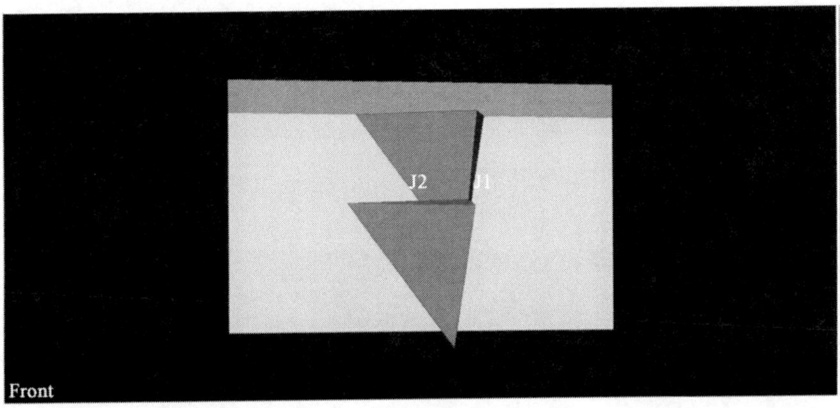

Figure 14. Front view of wedge slide analysis.

Figure 15. Wedge slide: front view.

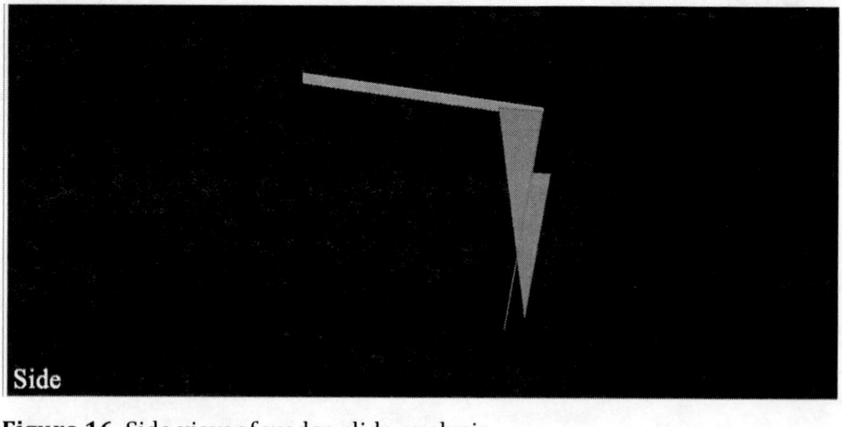

Figure 16. Side view of wedge slide analysis.

Figure 17. Perspective view of wedge slide analysis.

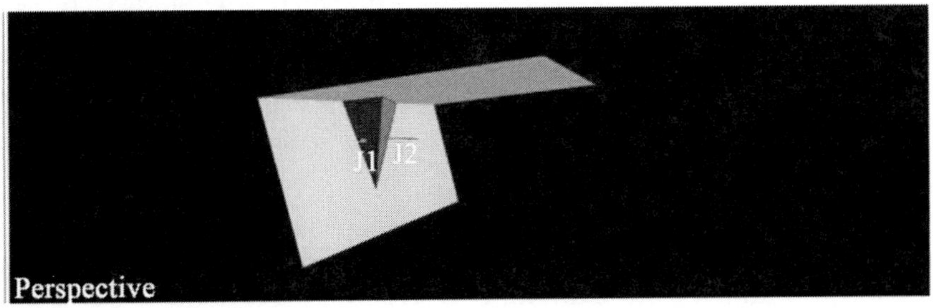

Figure 18. A view of wedge slide analysis obtained by applying directions of water penetration.

Table 5. Geotechnical parameters of the rock masses in unit one of the Atamir Formation K (at) 1.

Parameter		Min	Max	Ave
GSI		35	45	40
Intact uniaxial compressive strength (Mpa)		10	50	30
Mi		3	9	6
Material constants	m_b	0.0235	0.923	0.083
	s	2.65	6.73	4.54
	A	0.515	0.502	0.511
Cohesion (Mpa)		0.089	0.431	0.115
Friction angle (Deg)		12.76	62.76	23.81
Rock mass parameter (MPa)	Tensile strength	−0.010	−0.05	−0.016
	Uniaxial compressive strength	0.09	0.98	0.180
	Global strength	0.9	2.1	1.070
	Modulus of deformation	180.97	560.98	212.93

2.1.6. Analysis of Wedge Instability under Wet Conditions in Unit One of the Atamir Formation, K (at) 1

Although water is not directly involved in mass (slope) movement, it is an important factor because:

1) Increased water content of slopes caused by rain and snowmelt makes them heavy. Water can penetrate into pores and fractures and replace air and, since water is heavier than air, weights of rocks increase. Because weight is the same as force, and force applies stress on surfaces, increased stress can cause instability.

2) Water can alter slope angle (angle of the slope at which the slope is stable).

3) Water can be absorbed or repelled by soil mineral substances. Water absorption causes the electrical poles of water molecules to stick to the surfaces of mineral materials and penetrate into them. Therefore, increased water content makes rocks heavier, leading to reduced rock strength. In general, wet clay soils have less strength compared to dry clay soils. Therefore, water absorption reduces strength and, since clay is the material that fills the spaces between the joints in this unit, increased water content gains more importance.

4) Groundwater is present almost everywhere under the surface of the ground. This water fills the empty spaces between rock particles and even the cracks within rocks. Groundwater level changes due to rainfall and rises in wet seasons leading to greater water penetration. In dry seasons, groundwater level falls causing less water penetration. These changes in groundwater levels can be an important factor in slope stability.

If water penetrates through joint surfaces, the factor of safety will become zero (FS = 0) (Figure 19).

2.2. Unit Two of the Atamir Formation, K (at) 2 (from 300 + 3 to 500 + 5 Meters)

This lithology includes dark grey sandstones, there are three dominant discontinuity sets in the study region (two joint sets together with bedding), and the discontinuities are random. Table 6 and Table 7 shows the characteristics of the discontinuities. It is shown in Figures 20-22.

Figure 19. Water penetration into joint sets in unit one of the Atamir Formation.

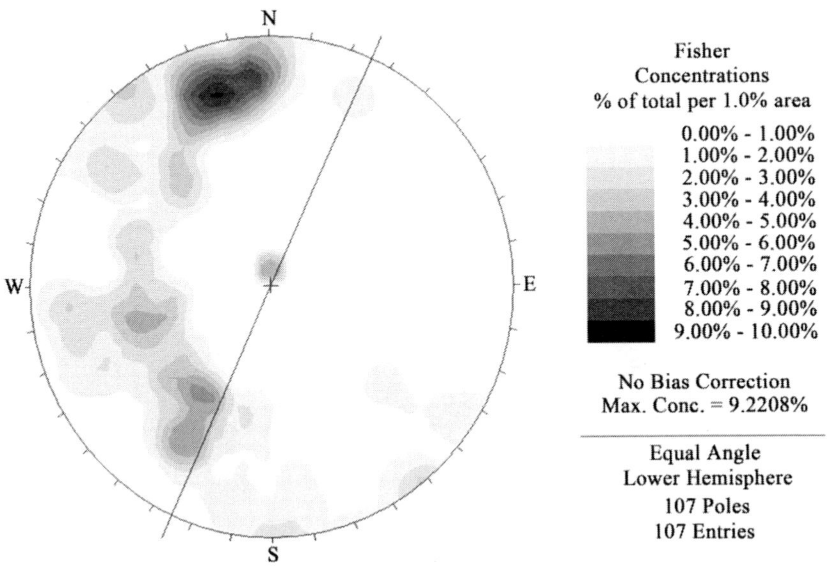

Figure 20. Stereographic image of the discontinuities in unit two of the Atamir Formation K (at) 2.

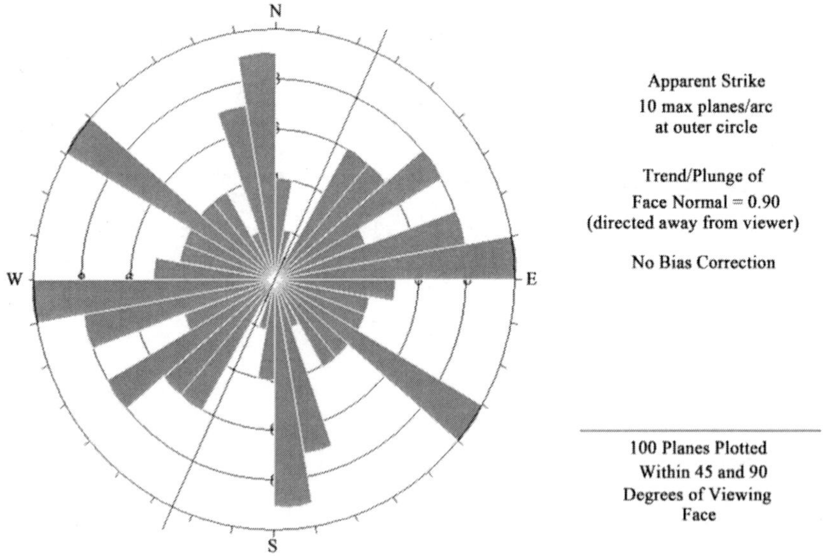

Figure 21. Rose diagram of the discontinuities in unit two of the Atamir Formation K (at) 2.

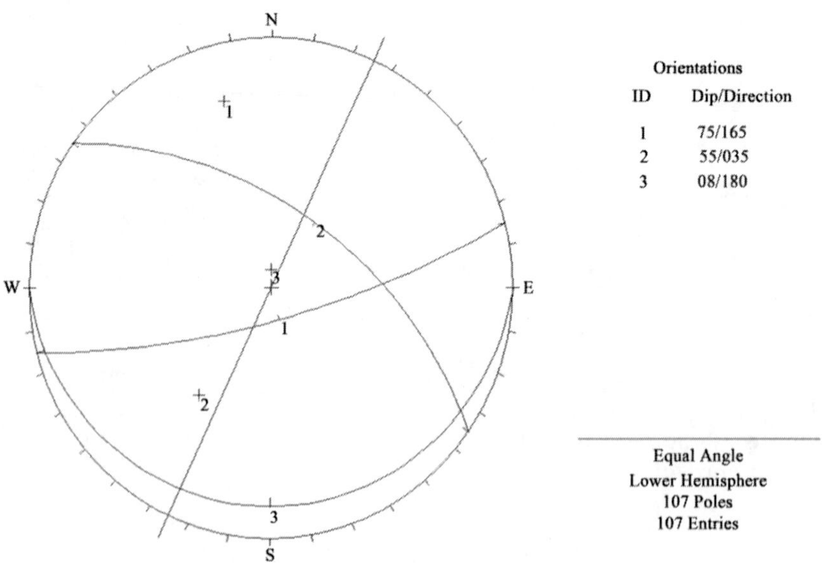

Figure 22. Stereographic image of the discontinuities in unit two of the Atamir Formation K (at) 2.

Table 6. Characteristics of the discontinuities in unit two of the Atamir Formation K (at) 2.

Discontinuity	ID	Dip	Dip direction
Js1	1	75	165
Js2	2	55	035
Bedding	3	8	180

Table 7. Geotechnical characteristics of the various discontinuities in the rock masses of unit two in the Atamir Formation K (at) 2.

Joint sets	Dip Dip direction		Persistence (m)						Joint roughness (%)			
			>20%	10% - 20%	3% - 10%	1% - 3%	<1%	Very rough	Rough	Rough-Smooth	Smooth	Slickenside
Js 1	165\75	60	80	20	-	-	-	-	80	20	-	-
Js 2	035\55	20	30	70	-	-	-	-	30	70	-	-
Bedding	180\8	20	100	-	-	-	-	-	100	100	-	-

Joint alteration	Joint water	Aperture (mm)					Joint filling					Joint spacing (m-mm)				
		>5 mm %	1 - 5 mm %	0.1 - 1 mm %	<0.1 mm %	Close	Soft Filling mm 5 < %	Soft Filling mm 5 > %	Hard Filling mm 5 > %	Hard Filling mm5< %	Clean %	<60 mm %	60 - 200 mm %	200 - 600 mm %	0.6 - 2 m %	>2 m %
		90	10	-	-	-	-	50	-	50	-	-	-	100	-	-
W1	Dry	60	40	-	-	-	-	-	70	-	30	-	-	70	30	-
		-	100	-	-	-	70	30	-	-	-	-	-	-	-	100

Figures 23-26 show statistical analysis for unit 2.

Tables 8-17 show information of data for the Atamir Formation K (at) 2 like the order of previous zone.

The GSI range of the rock masses in unit two of the Atamir Formation is shown in Figure 27.

Fitting data related to the rock masses in unit two of the Atamir Formation is shown in Figures 28-30.

Stereographic images of joint sets are shown in Figure 31. Wedge slide analysis is also shown in Figures 32-36.

2.3. Unit Three of the Atamir Formation, K (at) 3 (from 500 + 3 to 540 + 3)

This lithology consists of dark grey sandstones, and there are four dominant discontinuity classes (three joint classes together with bedding) and random discontinuities.

Stereography of the discontinuities is shown in Figure 37-39.

Statistical analysis of discontinuity from Figures 40-43.

Range of GSI in the rock masses of unit three in the Atamir Formation is shown in Figure 44.

Fitting the data on rock masses in unit three is shown in Figures 45-47.

Stereographic images of the joint sets for unit 3 is shown in Figure 48.

Views of wedge slide analysis are shown in Figures 49-53.

3. CONCLUSIONS

All three units had very poor to poor rocks that required rapid stabilization at most 10 hours after blasting operations. The following geomechanical and geotechnical characteristics were observed:

1) Layered sedimentary rocks sloping towards the outer domain with slopes close or equal to dip slope.
2) Joints and foliations that form weak extended surfaces and intersect the domain surface.
3) Intersecting joints that cause wedge slide failure.
4) Presence of soft, layered, strongly jointed shales having weak layers.
5) Saturation or wetting of joint surfaces having clay filling by water from rainfall, snowmelt, changes in groundwater levels, and reduced frictional force (considering the failure mechanism, and ground surveys, the effects of last spring's rain on the occurrence of this phenomenon cannot be denied).
6) Wet and semi-humid climates and successive snowfalls and, in addition, drought and wet periods.
7) Formations such as marl and shale ones, or intermittent marl and shale formations, are susceptible to mass movements, and the more unsaturated clay fills the joints, the more susceptible the formation will be to mass movement.
8) The tectonic condition (topography) of the dip slope (slopes exceeding 30%), and slope and orientation of the faults, layers, and fractures.
9) Failure to start stabilization in time according to the RMR table.

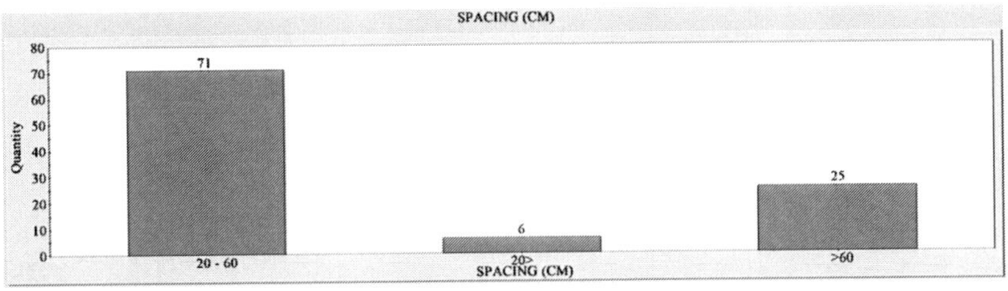

Figure 23. Statistical analysis of discontinuity spacing in unit two of the Atamir Formation K (at) 2.

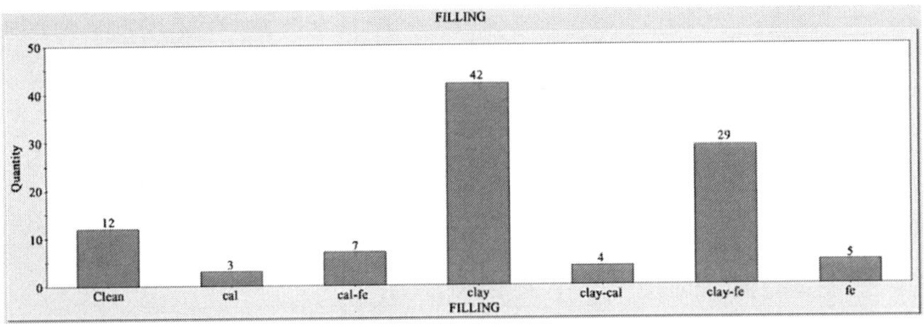

Figure 24. Statistical analysis of discontinuity filling in unit two of the Atamir Formation K (at) 2.

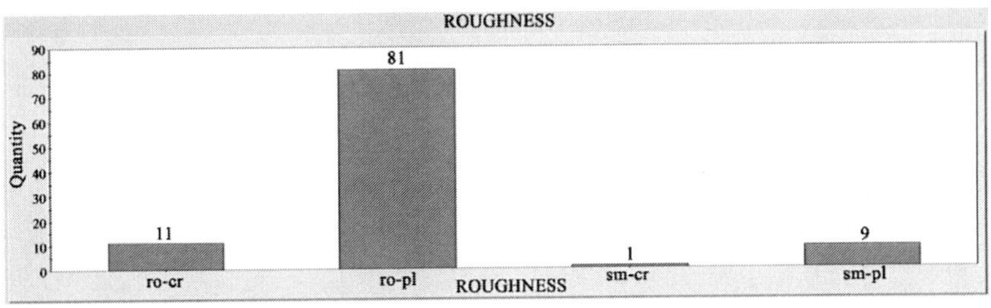

Figure 25. Statistical analysis of discontinuity roughness in unit two of the Atamir Formation K (at) 2.

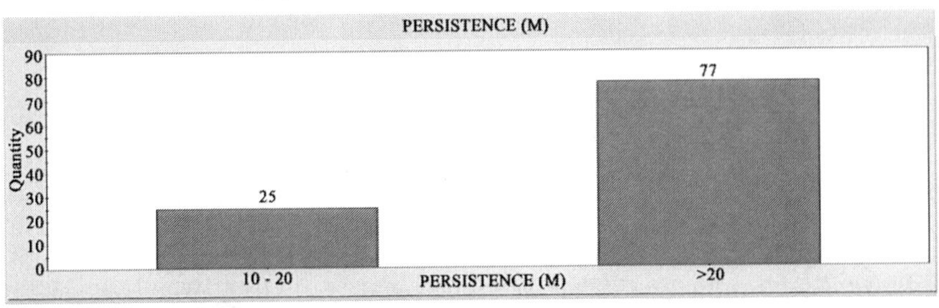

Figure 26. Statistical analysis of discontinuity orientations in unit two of the Atamir Formation K (at) 2.

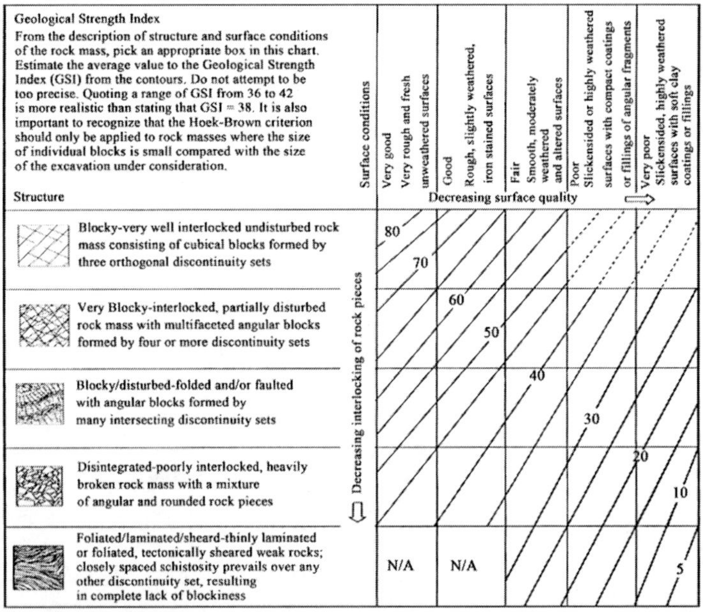

Figure 27. The GSI range of the rock masses in unit two of the Atamir Formation K (at) 2 in the GSI classification diagram.

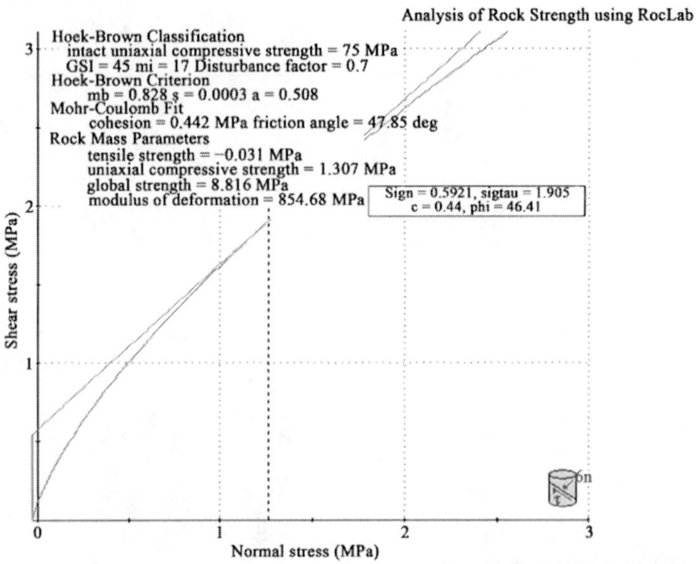

Figure 28. Fitting data related to the rock masses in unit two of the Atamir Formation, K (at) 2, to Hoek-Brown and Mohr- Coulomb criteria.

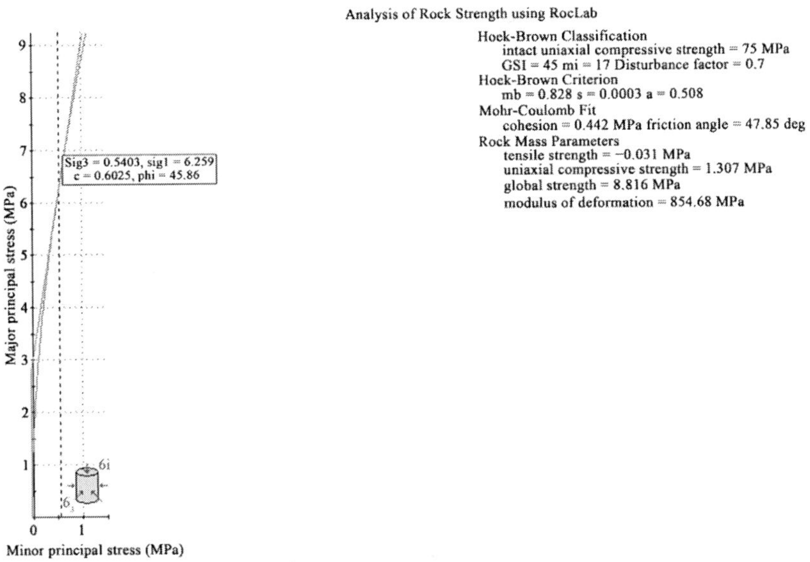

Figure 29. Fitting the data related to the rock masses of unit two to the Hoek-Brown and Mohr-Coulomb criteria based on GSI = 45 and on major principal stress.

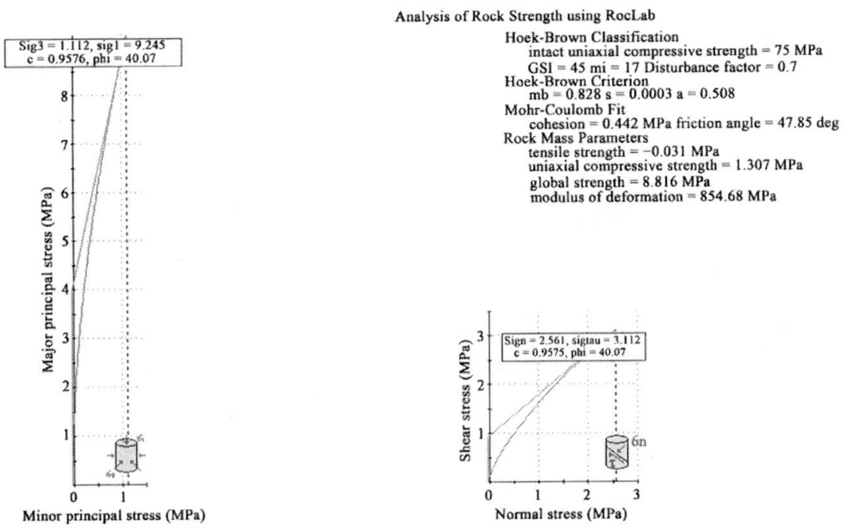

Figure 30. Analysis of output data determining geomechanical parameters of the rock masses in unit two based on GSI = 45.

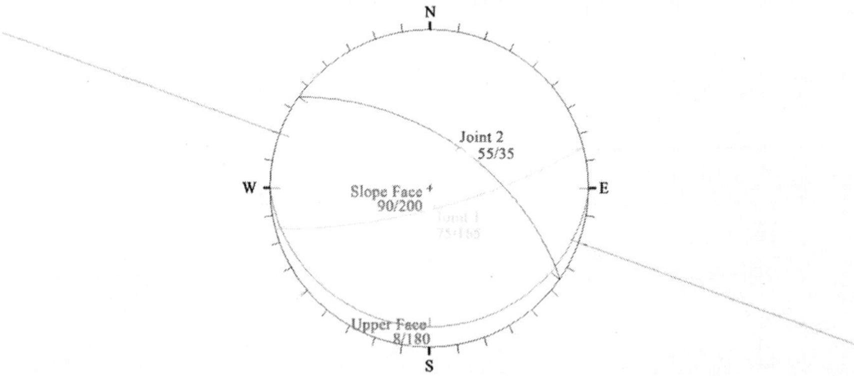

Figure 31. Stereographic images of joint sets.

Figure 32. Top view of wedge slide analysis.

Figure 33. Front view of wedge slide analysis.

Figure 34. Front view of wedge slide.

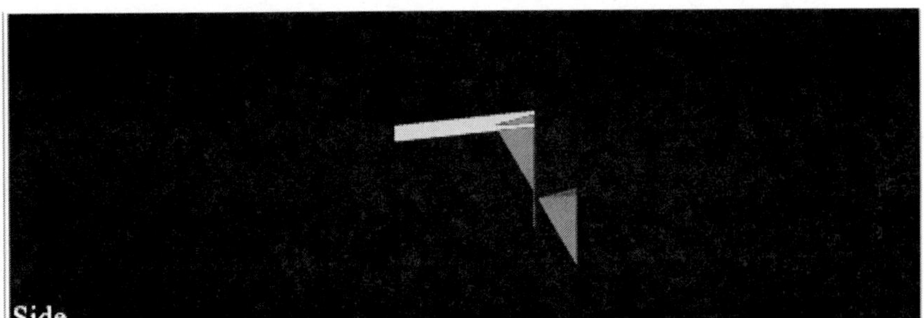

Figure 35. Side view of wedge slide analysis.

Figure 36. Perspective view of wedge slide analysis.

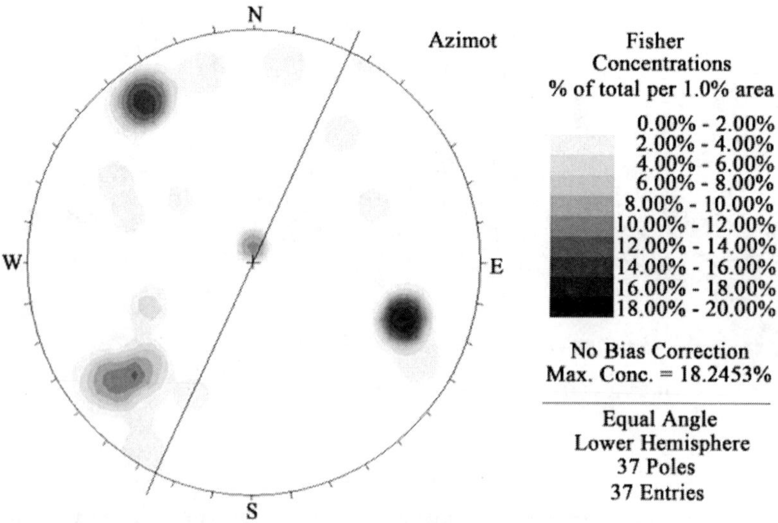

Figure 37. Stereographic image of the discontinuities in unit three of the Atamir Formation K (at) 3.

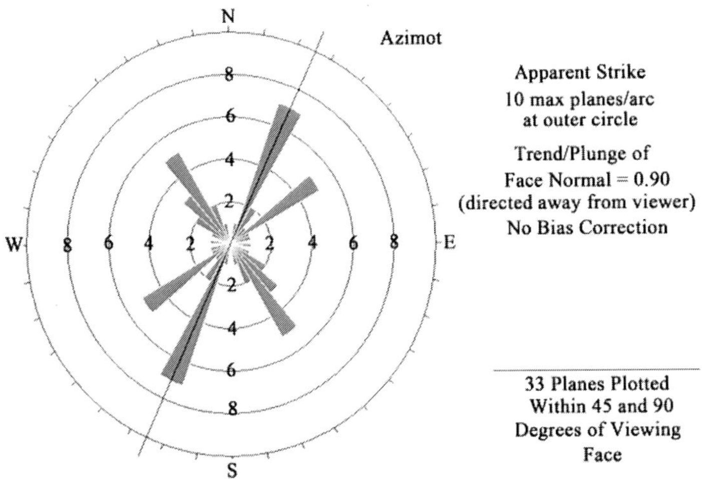

Figure 38. Rose diagram of the discontinuities in unit three of the Atamir Formation K (at) 3.

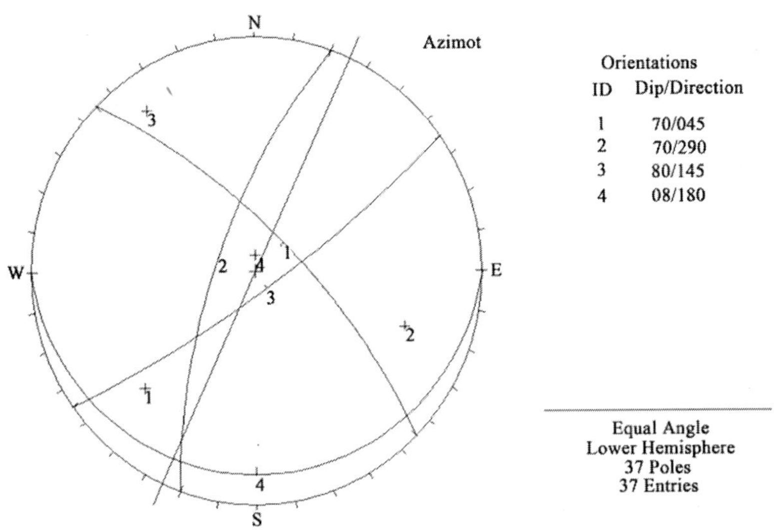

Figure 39. Stereography of the discontinuities in unit three of the Atamir Formation K (at) 3.

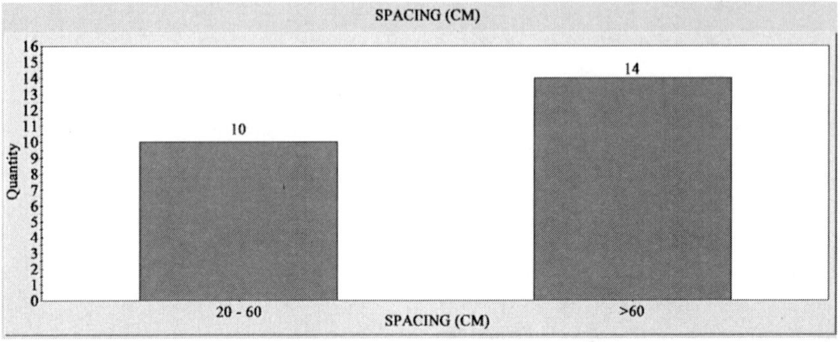

Figure 40. Statistical analysis of discontinuity spacings in unit three of the Atamir Formation K (at) 3.

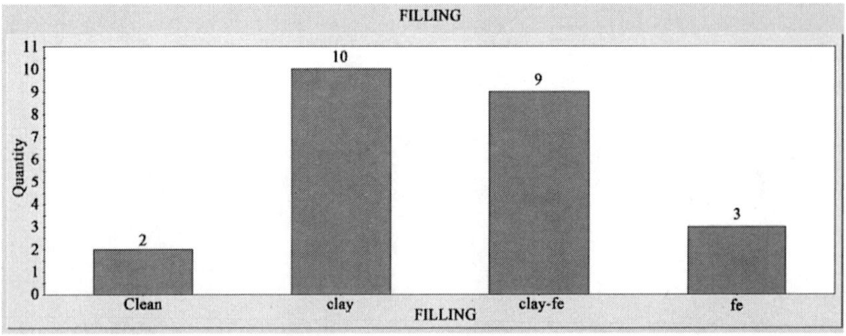

Figure 41. Statistical analysis of discontinuity filling in unit three of the Atamir Formation K (at) 3.

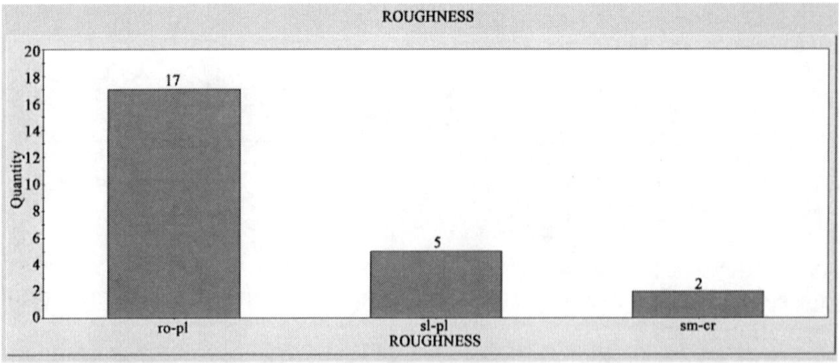

Figure 42. Statistical analysis of discontinuity roughness in unit three of the Atamir Formation K (at) 3.

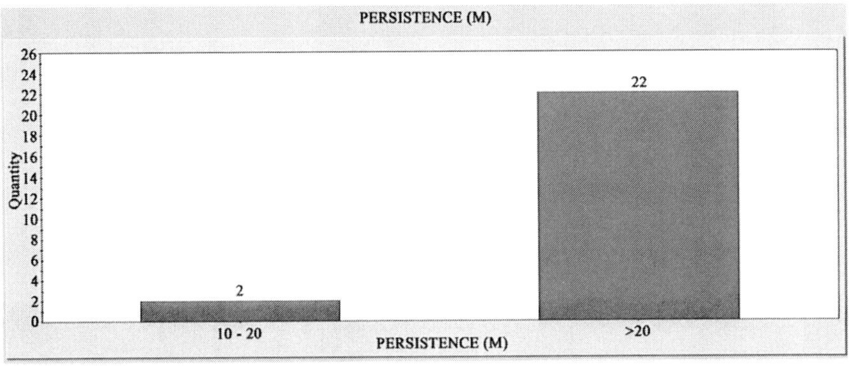

Figure 43. Statistical analysis of discontinuity orientation in unit three of the Atamir Formation K (at) 3.

Table 8. Geomechanical classification of the rock masses in unit two of the Atamir Formation K (at) 2.

Parameter	Value	Rating
UCS (MPa)	40 - 60	4 - 6
RQD	25 - 50	8
Spacing (MM)	60 - 200	8 - 10
Condition of discontinuities	-	10
Groundwater	Dam	7
RMR	37 - 41	
Class No.	Dam	
Cohesion of the rock mass (KPa)	442	
Friction angle of the rock mass (deg)	15 - 25	

Table 9. The Q index of the rock masses in unit two of the Atamir Formation K (at) 2.

Parameter	Description	Value
RQD	40 - 50	40 - 45
J_n	Three joint sets with random to four joint sets	12
J_r	Smooth to planar	1
J_a	Softening or low friction clay mineral coatings	4
J_w	Dry to medium inflow of pressure	1 - 0.66
SRF	Medium stress, favorable stress condition	1
Q	0.62 - 0.83	
Class No.	Very poor rock	

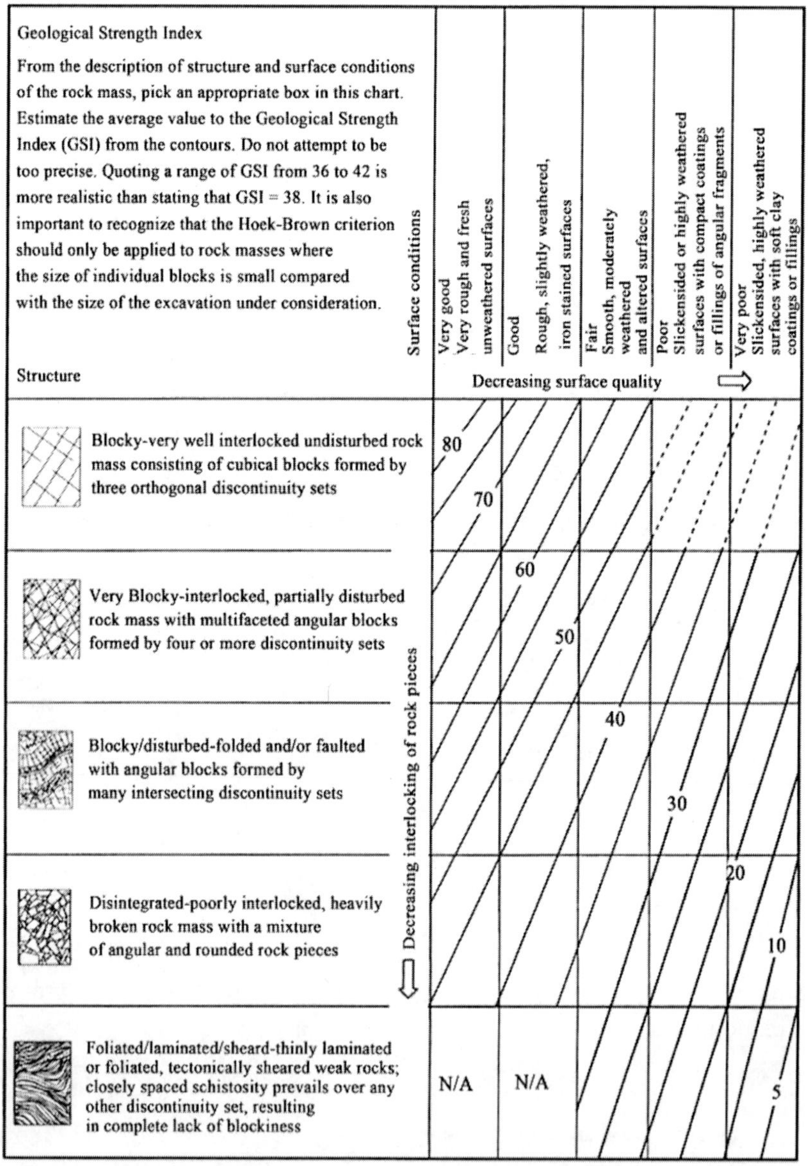

Figure 44. Range of GSI in the rock masses of unit three in the Atamir Formation K (at) 3.

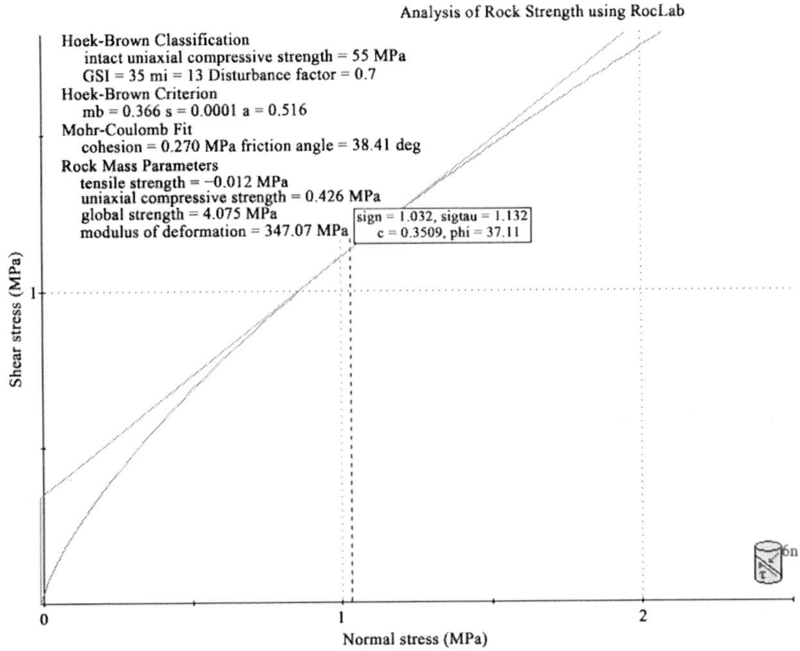

Figure 45. Fitting the data on rock masses in unit three to the Hoek-Brown and Mohr-Coulomb criteria, based on GSI = 35 and on normal shear stress.

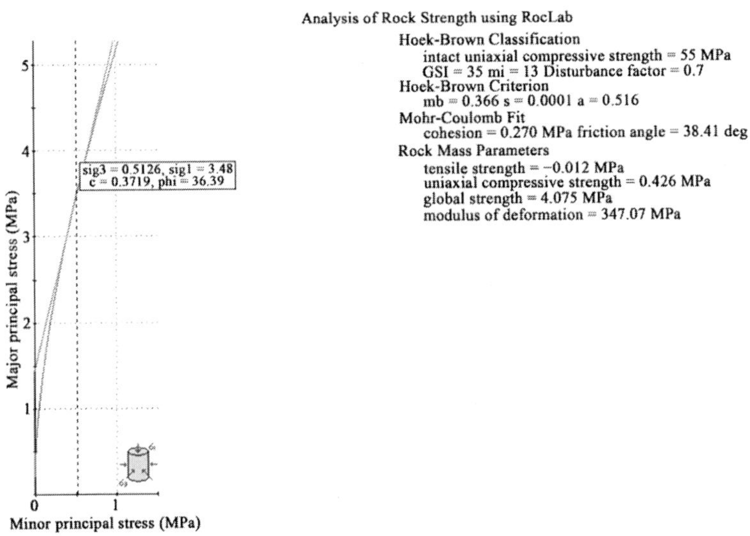

Figure 46. Fitting the data on rock masses in unit three to the Hoek-Brown and Mohr-Coulomb criteria, based on GSI = 35 and on major principal stress.

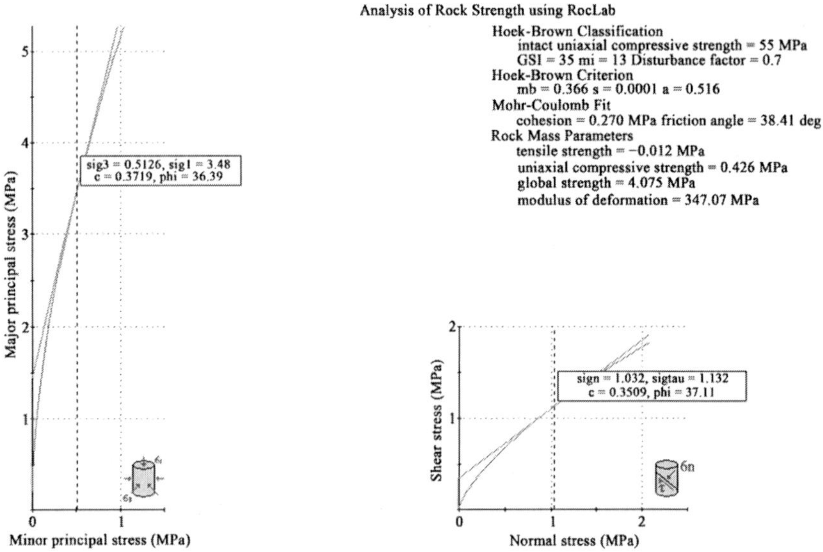

Figure 47. Analysis of output data determining the geomechanical parameters of rock masses in unit three, based on GSI = 35.

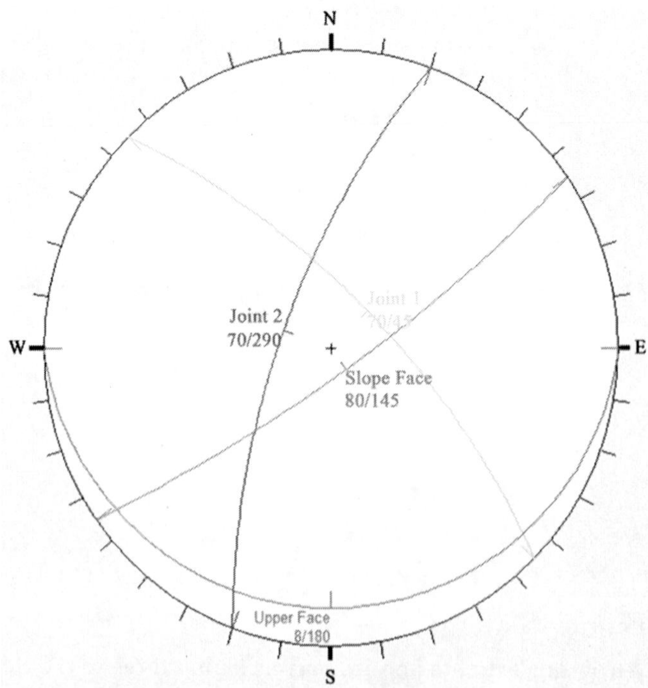

Figure 48. Stereographic images of the joint sets.

Figure 49. Top view of wedge slide analysis.

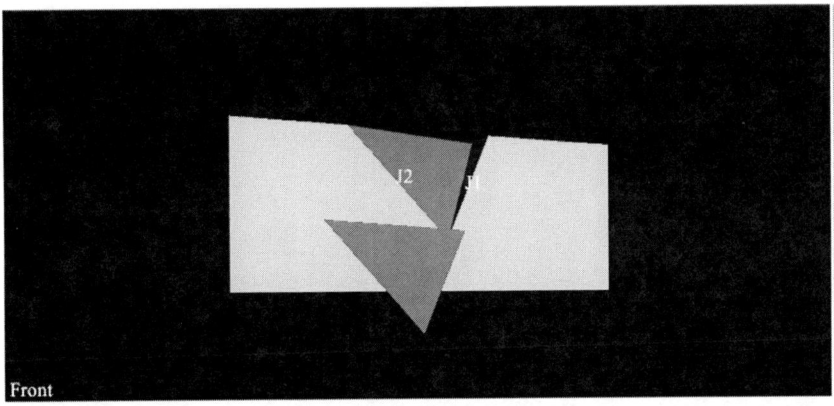

Figure 50. Front view of wedge slide analysis.

Figure 51. Front view of wedge slide.

Figure 52. Side view of wedge slide analysis.

Figure 53. Perspective view of wedge slide analysis.

Table 10. Geomechanical parameters of the rock masses in unit two of the Atamir Formation K (at) 2.

Parameter		Min	Max	Ave
GSI		40	50	45
Intact uniaxial compressive strength (Mpa)		50	100	75
Mi		11	23	17
Material constants	m_b	0.154	1.754	0.828
	S	0.0001	0.0014	0.0003
	A	0.478	0.612	0.508
Cohesion (Mpa)		0.210	0.501	0.442
Friction angle (deg)		62.87	90.63	47.85
Rock mass parameter (Mpa)	Tensile strength	−0.019	−0.08	−0.031
	Uniaxial compressive strength	0.096	2.68	1.307
	Global strength	6.98	10.67	8.816
	Modulus of deformation	600.75	1000.5	854.68

Table 11. Characteristics of the discontinuities in unit three of the Atamir Formation K (at) 3.

Discontinuity	ID	Dip	Dip direction
Js1	1	70	45
Js2	2	70	290
Js3	3	80	145
bedding	4	8	180

Table 12. Geotechnical characteristics of various discontinuities in rock masses of unit three in the Atamir Formation K (at) 3.

Joint sets	Dip Dip direction		Persistence (m)						Joint roughness (%)			
			>20%	10% - 20%	3% - 10%	1% - 3%	<1%	Very rough	Rough	Rough-Smooth	Smooth	Slickenside
Js1	45\70	50	100	-	-	-	-		100	-	-	-
Hs2	290\70	25	70	30	-	-	-	-	20	80	-	-
Js3	145\80	10	-	100	-	-	-	-	-	100	-	-
Bedding	180\8	15	100	-	-	-	-	-	100	-	-	-

Joint alteration	Joint water	Aperture (mm)					Joint filling					Joint spacing (m-mm)				
		>5 mm %	1 - 5 mm %	0.1 - 1 mm %	<0.1 mm %	Close	Soft filling mm 5 < %	Soft filling mm 5 > %	Hard filling mm 5 > %	Hard filling mm 5 < %	Clean %	<60 mm %	60 - 200 mm %	200 - 600 mm %	0.6 - 2 m %	>2 m %
		-	100	-	-	-	-	50	-	50	-	-	-	-	100	-
W1	Dry	100	-				100	-	-	-	-	-	-	100	-	-
		40	60	-	-	-	-	-	70	-	30	-	-	30	70	-
		-	100	-	-	-	70	30	-	-	-	-	-	-	-	100

Table 13. Geomechanical classification of rock masses in unit three of the Atamir Formation K (at) 2.

Parameter	Value	Rating
UCS (Mpa)	40 - 60	4 - 6
RQD	25 - 50	8
Spacing (mm)	>60 - 200	8
Condition of discontinuities		10
Groundwater	Dam	7
RMR	35 - 39	
Class No.	Dam	
Cohesion of the rock mass (KPa)	190	
Friction angle of the rock mass (Deg)	15 - 25	

Table 14. The Q index of rock masses in unit three of the Atamir Formation K (at) 3.

Parameter	Description	Value
RQD	37 - 43	37 - 43
J_n	Three joint sets with random to four joint sets	12
J_r	Smooth to planar	1
J_a	Softening or low friction clay mineral coatings	4
J_w	Dry to medium inflow or pressure	1 - 0.66
SRF	Medium stress, favorable stress condition	1
Q	0.51 - 0.89	
Class No.	Very poor rock	

Table 15. Geomechanical parameters of the rock masses in unit three of the Atamir formation K (at) 3.

Parameter		Min	Max	Ave
GSI		30	40	35
Intact uniaxial compressive strength (MPa)		30	80	55
Mi		9	17	13
Material constants	m_b	0.457	0.754	0.366
	S	0.0001	0.0009	0.0001
	A	0.489	0.654	0.516
Cohesion (Mpa)		0.190	0.451	0.270
Friction angle (Deg)		19.05	50.63	38.41
Rock mass parameter (MPa)	Tensile strength	−0.021	−0.09	−0.012
	Uniaxial compressive strength	0.023	1.98	0.426
	Global strength	1.65	9.76	4.075
	Modulus of deformation	600.75	1000.5	347.07

Table 16. Geomechanical parameters of the rock masses in the three units of the Atamir Formation.

Unit	RMR	Q	GSI	Factor of safety (FS) of the wedge
Unit one of Atamir Formation K (at) 1	34 - 36 Poor	0.592 - 0.66 Very poor	35 - 45 Poor	0.037
Unit two of the Atamir Formation K(at) 2	37 - 41 Poor	0.62 - 0.83 Very poor	40 - 50 Poor	0
Unit three of the Atamir Formation K (at) 3	35 - 39 poor	0.51 - 0.89 Very poor	30 - 40 Poor	0

Table 17. Geomechanical parameters of the rock masses in unit three of the Atamir Formation K (at) 3.

Set number	Characteristics and parameters of rock mass	Rating rock masses				
		81 - 100 (I)	61 - 80 (II)	41 - 60 (III)	21 - 40 (IV)	<20 (V)
1	Rock classification	Very good	Good	Relatively good	Poor	Very poor
2	Average time of self supporting capacity	10 years for active opening of 15 meters	6 months for active opening of 8 meters	1 week for active opening of 5 meters	10 hours for active opening of 2.5 meters	30 minutes for active opening of 1 meter
3	Cohesion of rock mass (Mpa)	>0.4	0.3 - 0.4	0.2 - 0.3	0.1 - 0.2	<0.1
4	Angle of internal friction	>45 degrees	35 - 45 degrees	25 - 35 degrees	15 - 25 degrees	<15 degrees
5	Load bearing capacity (T/M^2)	440 - 600	280 - 440	135 - 280	45 - 135	30 - 45

Therefore, rockslide has been unavoidable in rock masses of all three units. The point of great importance is the mass movement of the rocks towards the road and channel that has blocked them and will cause floods during future rainfalls. Based on observations made, the process of landslide under discussion has not ended, and it seems that a long time has to pass (from several months to several years) for stability to be established.

Condition of the rock masses (Table 16).

Considering the RMR stability table (Table 17).

REFERENCES

1. Feenstra, R. and Wickham, J. (1975) Evolution of Folds around Broken Bow Uplift, Ouachita Mountains, Southeastern Oklahoma. AAPG Bulletin, 59, 974-985.
2. Bieniawski, Z.T. (1973) Engineering Classification of Jointed Rock Masses. Civil Engineer in South Africa, 15, 353.
3. Cai, M., Kaiser, P.K., Uno, H., Tasaka, Y. and Minami, M. (2004) Estimation of Rock Mass Strength and Deformation Modulus of Jointed Hard Rock Masses Using the GSI System. International Journal of Rock Mechanics and Mining Sciences, 41, 3-19.

CHAPTER 5

Developing 3-D Mine-Scale Geomechanical Models in Complex Geological Environments, as Applied to the Kiirunavaara Mine

J. Vatcher [a], S.D. McKinnon[b], J. Sjöberg [c]

[a] *Division of Mining and Geotechnical Engineering, Luleå University of Technology, SE-971 87 Luleå, Sweden*

[b] *The Robert M. Buchan Department of Mining, Goodwin Hall, 25 Union St. Queen's University, Kingston, Ontario, K7L 3N6, Canada*

[c] *Itasca Consultants AB, Aurorum 2, SE-977 75 Luleå, Sweden*

ABSTRACT

An understanding of the relationship between the geological environment and rock mass behaviour induced by mining activities can lead to hazard reduction through knowledge-based design. However, characterisation of complex and heterogeneous rock masses that typify mining environments is difficult. A methodology to characterise these types of rock masses, based largely on classical statistics, geostatistics and an extension of previous quantitative structural domaining work, is presented and applied to the Kiirunavaara Mine, Sweden. In addition to a new perspective on intact rock strengths of geological units at the mine, a correlation was found between modelled volumes of clay, modelled RQD, newly identified structural domains and falls of ground. These relationships enabled development of a conceptual model of the role of geology in rock mass behaviour at the mine. The results demonstrate that the proposed methodology can be useful in characterisation of complex rock masses.

Keywords: Geomechanical model Geostatistics Structural domaining Rock mass characterisation

1. INTRODUCTION

Mining induced rock mass behaviour is a result of the interaction between the mining and geological environments. An understanding of the role of 3-D geomechanical characteristics in rock mass behaviour can lead to significant opportunities for risk-mitigating engineering design. Typically, geomechanical models of mining environments are developed by using classification methods such as Geological Strength Index (GSI) or Rock Mass Rating (RMR). These methods were originally developed and calibrated for civil engineering excavations, such as tunnels, and are well suited to homogeneously jointed rock masses. Alone, however, they are unsuited to heterogeneous rock masses in which the strength and stiffness is affected by characteristics in addition to discontinuities, which typify the geological environment of many mines. Adaptations of GSI for heterogeneous rock masses have been developed (Hoek and Karzulovic, 2001; Marinos and Hoek, 2001); however, they are still mostly based upon planes of weakness, and as identified by Mandrone (2006), are less applicable when weakness exists in other forms, such as alteration. Over the life of a mine, geomechanically relevant data is often collected for different purposes, and there can be difficulty synthesizing these multiple sources of data into a coherent geomechanical model. Also, data availability and coverage issues specific to mining caused by economical and physical constraints (such as shotcreting limiting available mapping faces) often compound the difficulties associated with this process.

This paper proposes a methodology to create a 3-D mine-scale geomechanical model for complex and heterogeneous rock masses using readily available data from mines, such as geologically logged core, Rock Quality Designation (RQD) and mapped discontinuities. Luossavaara-Kiirunavaara Aktiebolag's (LKAB) Kiirunavaara Mine, in northern Sweden, is used to illustrate the techniques. This geomechanical model created for the mine is then compared to fall of ground information, with the intention of developing a conceptual model of causes of rock mass behaviour and identifying indicators of problem areas. In 2008, the Kiirunavaara Mine became seismically active (refer to Dahnér et al. (2012) for a detailed description). Several studies were undertaken by LKAB to understand the underlying causes and nature of the behaviour (Sjöberg et al., 2011, 2012; Vatcher et al., 2014). Recent underground mapping, core logging and laboratory testing campaigns done in support of these projects have revealed that the rock mass is heterogeneous, with variable characteristics across both the 5 km length and > 1000 m depth of the orebody. An additional potential complication for geomechanical characterisation of the rock mass is that there are volumes of clay alteration, which are often meters to tens of meters in their longest dimension.

Previous studies at the Kiirunavaara Mine have been limited in scope with respect to development of a geomechanical model. The majority of these studies were limited in breadth of analysis, often focusing on portions of the available data at the time of each study (Henry and Marcotte, 2001; Lindgren, 2013; Mattsson et al., 2010; Rådberg et al., 1989). Much additional data collection and analysis has been done since these studies with the intention of increasing the geomechanical understanding of the rock mass, providing a good opportunity to enhance knowledge.

The data analysis techniques presented to create large-scale 3-D geomechanical models were intentionally selected for their compatibility with characteristics of the geology and typical limitations associated with data from mining environments, such as sparse coverage and few samples. The methodology utilises standard statistical tests, geostatistics, and an extension of previously published techniques related to quantifiable identification of structural domains.

2. KIIRUNAVAARA MINE

The Kiirunavaara Mine is a large sublevel caving mine that produces approximately 28 million tonnes per annum of iron ore. With the deepest production levels currently at approximately 800 m below surface (Level 1051 m), the rock mass is seismically active and rockburst prone. The newest main haulage level is Level 1365 m (see Fig. 1), and sublevels currently have a 28.5 m separation. Mine coordinates are approximately aligned with the cardinal directions, and the Y coordinates (positive axis aligned with south), which crosscut the orebody, are used to divide the orebody into blocks for production and ore handling (see the numbered ore pass groups along the orebody in Fig. 1).

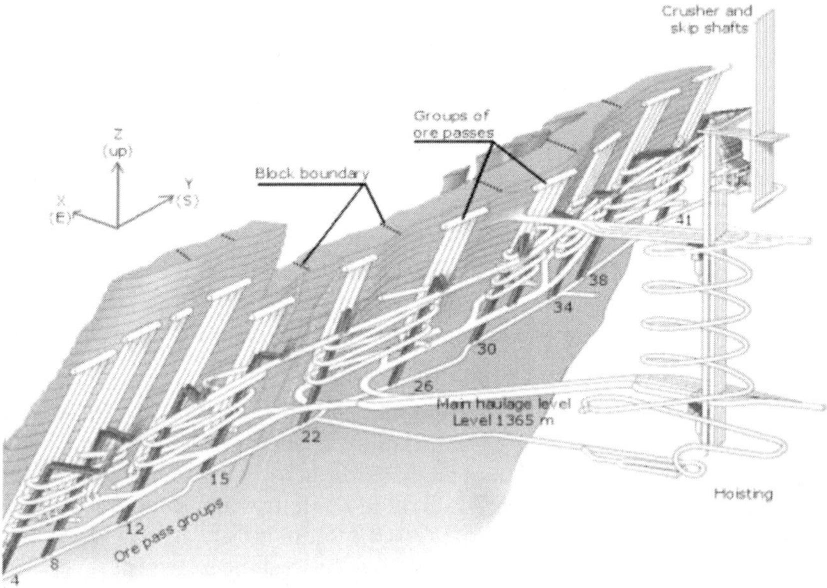

Fig. 1. Sketch of the orebody and mine layout. Production blocks and associated ore pass groups are numbered based on their Y coordinate (numbered at the draw points in this image). Modified image from LKAB.

The magnetite orebody extends approximately 5 km along strike, which is nearly north–south, with a varying width from meters to over 150 m and a dip of 50-70° towards the east (positive X-axis in mine coordinates). For a detailed description of the geology, refer to Geijer (1910). The most specific geological classifications at the mine are done during core logging, whereas underground geological mapping uses more generic geological classifications. The mine distinguishes ore rock types based on their grade and contaminant concentration. Footwall rock types are divided into three major classifications; granite, skarn or trachyte-trachyandesite (referred to as syenite porphyries at the mine). The trachyte-trachyandesite is further divided into 5 subgroups based on their mineralogy, texture and/or alteration, referred to as Sp1 through Sp5. Porphyry dykes are also common in the footwall region. Hangingwall material, mainly rhyodacite (referred to as quartz-bearing porphyries at the mine), is similarly divided into 5 subcategories (Qp1 through Qp5) based on their mineralogy, texture and/or alteration. Underground geological mapping identifies the ore units based on grade and contamination (the same units as mapped during core logging), however footwall and hangingwall materials are mapped in less detail underground than they are during core logging. Fewer hangingwall data are available by either mapping or core logging due to underground access and ore-targeted drilling.

Some of the rock mass at the Kiirunavaara Mine has undergone significant alteration. This alteration is in the form of both replacement of minerals by clay as well as leaching, leaving the rock porous (Berglund and Andersson, 2013). The clay-altered lens-shaped volumes are visible underground and in core from diamond drilling (as clay and possibly as core losses), and their extent ranges from centimeters to tens of meters throughout the mine.

3. DATA AND METHODOLOGY

The selection of data was based on potential importance to geomechanical characterisation, availability, and reliability. Data analyzed in the development of the geomechanical model of the mine was selected because of their possible relations to intact rock strength, zones of strength and stiffness contrasts, rock mass quality and structural domains. The forms of these data sources available at the Kiirunavaara Mine are described below. Some of this data is routinely collected at the mine, and some was collected during specifically targeted campaigns in support of this and other projects at the mine.

- The diamond drill core database consists of approximately 590 000 m of mostly 28 mm diameter core, from approximately 3000 boreholes (LKAB, 2014a). The extent of the data is shown by Volume A in Fig. 2. With few exceptions, all drilling is from underground, with fans of four holes on average at 50 m spacing along strike targeting orebody definition. The majority of drilling is done from the footwall due to access restrictions. On surface, geologists log these drill holes, identifying geological units, core losses and zones of poor material (such as clay), and RQD.
- A total of 56 unconfined compressive strength (UCS) tests were completed on 11 of the rock units, which are sourced from previous testing done by LKAB employees (such as Andersson and Israelsson, 2009), unpublished work (Westblom, 2014) and as a part of this study. Competent and representative samples were selected. Samples were extracted from many locations in the mine, mostly contained by Volume B in Fig. 2.
- Approximately 740 Point Load Strength tests were completed on 20 of the rock units on 28 mm core as a part of this study. The majority of these samples were from Volume C in Fig. 2. Weaker samples were preferentially eliminated due to sampling technique. An additional 141

Point Load Strength tests were completed by Lindgren (2013) on lump samples from varied locations throughout the mine. These 141 tests were not used in this study to reduce the introduction of potential sources of error caused by 1) the different sample type, and 2) the different source location.

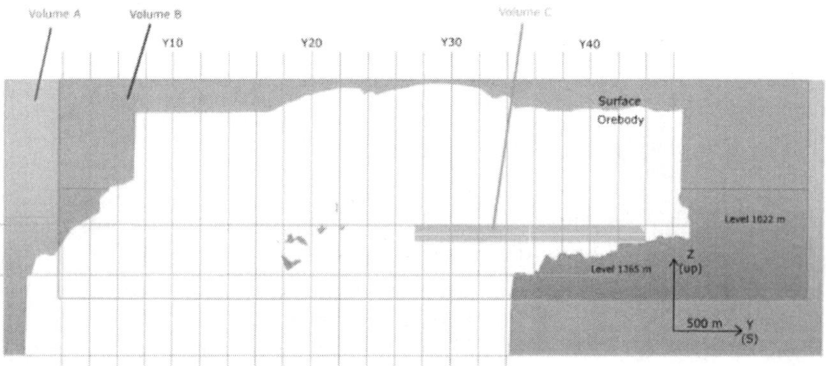

Fig. 2. Data extents (not coverage) represented by boxes overlaid on an outline of the orebody (longitudinal view from the footwall towards the hangingwall). All volumes extend into the hangingwall and footwall. Mine coordinates along orebody strike (Y coordinates) are labeled. Volume A represents the data extents of the drill cores. Volume B shows the data extents of the discontinuities mapped underground. Volume C illustrates the location of extensive underground mapping campaigns to better identify geological features, such as structures and clay altered volumes.

- Geological data exists from routine underground drift mapping, which focuses on definition of the orebody and the location of different types of ore rock units. The majority of mapping data is in the footwall and in the orebody. Extensive shotcreting limits mapping anywhere except working faces. In addition, a recent underground mapping campaign over the span of one year has provided a more detailed definition of clay zones in Volume C in Fig. 2.
- Over 4000 discontinuities exist in the database from underground drift mapping (LKAB, 2014b). The extents of these data are represented by Volume B in Fig. 2. All mapped joints have information about their orientation.
- Geological Strength Index (GSI) has been mapped during underground campaigns (LKAB, 2014c). Data is available between Level 1051 m and

Level 1543 m within Volume B (Fig. 2), but data coverage is limited due to shotcreting.

- A recent campaign acquired and analyzed data from underground mapping to provide an understanding of the kinematic regime of the rock mass (Berglund and Andersson, 2013). This is the first major underground campaign specifically related to faults at the mine, and was mostly conducted in Volume C (Fig. 2).
- Rock falls occurring during production are individually documented in reports and stored in a database at the mine. All rock falls have been systematically recorded since circa 2010, with selected events recorded in 2008 and 2009.
- The Kiirunavaara Mine has an extensive seismic network with related information stored in the database. Dahnér et al. (2012) provides a description of the network and mining-induced seismicity.

In complicated rock masses such as at the Kiirunavaara Mine, the importance of using statistics as a basis for analysis is heightened. With such large complexity and quantity of available data, there is a likelihood that geomechanical patterns become too convoluted to analyze without the assistance of statistics. As such, the methodology to build the geomechanical models is based on statistical analysis of physical data, using operational experience and separate data sources for verification where possible.

Geostatistical techniques have been used with success to assess rock masses from a geomechanical perspective for a variety of purposes (Ayalew et al., 2002; Coli et al., 2012; Ozturk and Simdi, 2014; Stavropoulou et al., 2007). These techniques are well suited to data-rich environments that have large-scale changes in characteristics. Due to its robust and quantifiable nature and suitability with complex and heterogeneous geological environments, geostatistics was selected as one of the main spatial analysis techniques to develop the geomechanical model at the Kiirunavaara Mine.

The proposed methodology employs the use of standard geostatistical techniques with additional statistical analysis to understand the large-scale 3-D patterns of rock mass characteristics, such as structural domains, zones of high or low stiffness material and rock mass quality. These areas are first analyzed separately and then subsequently synthesised in 3-D space to create one geomechanical model. Geostatistical analysis is limited to identification of volumetric characteristics of the rock mass, and specifically excludes identification of discrete features such as faults, for which targeted underground mapping is better suited.

As with all modelling, control data for calibration and verification purposes is essential. Good candidates include data related to rock mass behaviour as well as seismicity, rock falls and drift mapping. Data sets which are too sparse for statistical analysis or which are in an unsuitable format (such as maps or photos) can be used as control data.

4. GEOMECHANICAL CHARACTERISATION

4.1. Intact rock strength

A complete understanding of the intact strength differences between geological units does not yet exist at the Kiirunavaara Mine. In the literature, petrographic data has been used to evaluate intact compressive rock strength through regression, expert systems, and artificial intelligence (Alvarez Grima and Babuška, 1999; Gokceoglu, 2002; Gokceoglu et al., 2009; Singh et al., 2001), however, insufficient data was available for this study to use such an approach. As an alternative, UCS and Point Load Strength tests were selected to evaluate intact strength. Typically, the results from Point Load Strength tests are used to gain additional strength information by scaling the Point Load Strength Indices ($Is_{(50)}$) to UCS values. Numerous empirical relationships exist in literature for the scaling of $Is_{(50)}$ data to UCS values (Bieniawski, 1975; D'Andrea et al., 1965; ISRM, 1985; Kahraman, 2001; Kaya and Karaman, 2015; Kohno and Maeda, 2012). Most of these relationships use linear scaling as suggested by ISRM (1985):

$$UCS = kIs_{(50)} \tag{1}$$

where k is the scaling factor. The scaling factor suggested in the testing standards ranges from 20–25, however it is acknowledged in the standards and confirmed by others (previously listed) that this value may vary between 15 and 50 and is highly dependent upon rock type (ISRM, 1985). Some authors, such as Kaya and Karaman (2015), have found even lower scaling factors.

Since this scaling is dependent upon rock type, it is critical to first approach grouping of rock units by intact rock strength using statistical tests. The purpose of this analysis is to identify 1) the underlying distribution of strength parameters, 2) groups of geological units that share common strengths, and 3) the estimated mean strengths and standard deviations of these groups. There is insufficient data availability at the mine to determine if intact rock strength is dependent upon location, not only geological units.

Due to the large number of samples, the Kolmogorov-Smirnov test was used on the $Is_{(50)}$ data to evaluate all possible groupings of geological units. This method uses the cumulative distribution functions of two samples to evaluate if differences exist between the location and shape of the distribution functions, and was selected due to its robust, nonparametric nature. Each sample was represented by one $Is_{(50)}$ value; an average value for the sample was used if there was more than one acceptable data point. $Is_{(50)}$ data were then scaled to UCS data for each group of geological units individually to estimate the strength of each group according to the following:

$$k = \left(\frac{\mu_{UCS}}{\mu_{Is_{(50)}}}\right) \tag{2}$$

where k is the scaling factor, μ_{UCS} is the average UCS from the UCS tests and μIs50 is the average $Is_{(50)}$.

Based on the results from the Kolmogorov-Smirnov tests and considerations of sample size, it is evident that some mapped geological units likely share common underlying distributions. The most likely grouping of geological units into strength groups based on the statistical analysis are shown in Fig. 3, where the distribution of the grouped and scaled $Is_{(50)}$ data is shown by box plots, providing information about the distributions of each group through the visualisation of their quartiles. No significant difference between the distributions of 1) the 5 types of hangingwall units or 2) the remaining ore types were found, however not all units were well represented. Units that have limited sample numbers are starred (*). Additional point load tests would improve the precision and accuracy of this analysis. Based on Kolmogrov-Smirnov tests and χ^2 tests, all data before and after grouping by multiple geological units indicated normal distributions, with the exception of the ore material with few samples (ore, higher contaminant levels). This may be a result of few samples or at least two unique distributions of strength within these geological units.

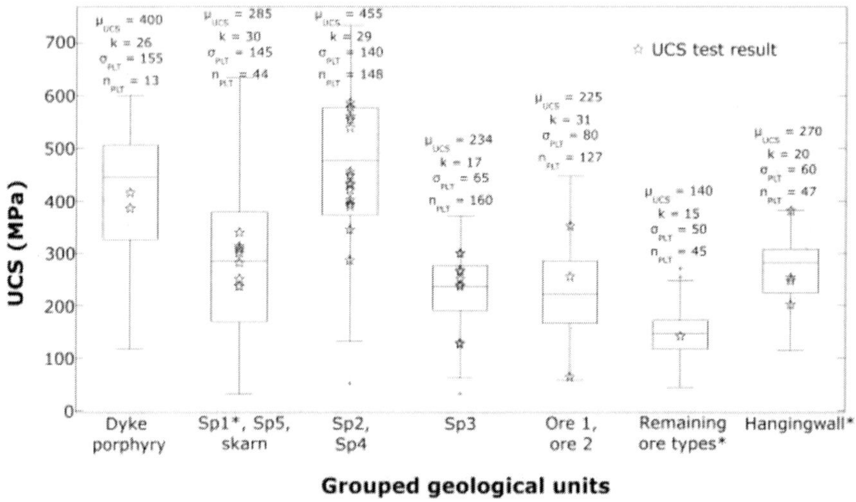

Fig. 3. Scaled Point Load Strength test results (box plots) and UCS test results (stars) by grouped geological units. Boxplots of the point load tests for each grouped geological unit show outliers (considered as +/− 2.7 * standard deviation) as crosses. Average UCS from the UCS tests (μ_{UCS}), the scaling factor (k), the scaled standard deviation from the Point Load Strength tests (σ_{PLT}), and the number of Point Load Strength tests (n_{PLT}) for each group are presented over the box plots.

The UCS data for each group of geological units are overlaid on the scaled $Is_{(50)}$ data in Fig. 3 as stars. Additional UCS tests would significantly improve the accuracy of the estimated strengths, in particular for groups with few UCS samples. Both the UCS data and the Point Load Test data had a relatively wide spread, in particular in the majority of the footwall material. The porphyritic dyke and Sp2/Sp4 groups showed higher strengths than the other groups. Kolmogorov-Smirnov tests on the scaled point load test data and the UCS data sets indicated that the distributions between the two data sets (Point Load Test data and UCS test data) for individual grouped geological units were consistent.

There is a significant spread of estimated intact rock strengths for the geological units present at the Kiirunavaara Mine. Lindgren (2013) showed that increasing the pervasive alteration into softer minerals, such as chlorite, carbonates and clay, lowers the intact rock strength through analysis of Point Load Strength tests conducted on lump samples; however, the current study does not have sufficient data to evaluate the role of alteration on intact strength at the mine. Although very clear during core logging and underground site investigations, the intact rock strength tests do not illustrate the significant strength and stiffness difference of some of the weaker units found at the mine, such as clay, due to preferential sampling.

4.2. Volumes with clay alteration

Due to the significantly different material properties of clay compared to the surrounding rock, and due to the total volume of clay present, the locations of volumes of clay alteration are an important consideration for the geomechanical model at the Kiirunavaara Mine. 3-D volumes of clay based on isotropic search parameters from mapped data were developed by Mattsson et al. (2010). However, this work has not been updated to reflect newly acquired data and knowledge.

Geostatistical techniques were applied to understand the underlying spatial distribution of these clay volumes. This includes an analysis of appropriate search parameters via semi-variograms and then interpolation of the data. The interpolated model from the drill core data was compared to drift mapping to evaluate accuracy of the technique for this rock mass.

A common difficulty in identifying clay material in drill cores is that clay is often washed away by the drilling water and does not remain as part of the core. To account for this, both mapped clay and core loss in the core logging data (LKAB, 2014a) were considered to be indications of clay altered zones.

Indicator semi-variograms were created to determine 1) if directional anisotropy exists, and 2) the appropriate search parameters for the composited clay data from core mapping. Sample compositing was used to reduce bias caused by variable sample length. The clay and core loss data were downhole composited on 0.5 m intervals, as the majority of sample lengths were shorter than this value. The semivariograms indicated that the major and minor search ellipse axes are horizontal (Fig. 4). The major axis is semi-perpendicular to orebody formation. The semi-major axis was steeply dipped towards the north. The major and semi-major axes had approximately the same range, however the range was significantly reduced in the minor axes direction. As evident in Fig. 4, the minor axis, approximately aligned along orebody strike, has a significantly smaller range than the borehole spacing (circa 50 m). This indicates that the data from the drilling has insufficient density for the use of geostatistical techniques to estimate clay volumes at the Kiirunavaara Mine. This could be related the possible genesis method of the hydrothermal clay alteration at the mine; hydrothermal fluid travels through existing discontinuities until it reaches a volume of the rock mass which is susceptible to larger amounts of alteration. The preferential travel along existing discontinuities can result in locations of clay that have a discrete nature, rather than a volumetric one.

Due to this discrete nature, this data a difficult candidate to consider for interpolation methods.

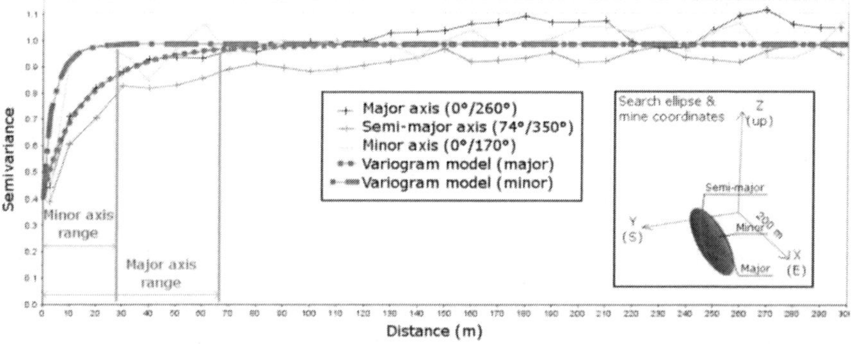

Fig. 4. Semi-variograms in the major, semi-major and minor search ellipse directions. Search ellipsoid is visualised with mine coordinates.

However, a simple spatial analysis of the density of clay indicators from the core data in 3-D space using search radii identified from the semi-variograms yielded a model of the clay volumes (not clay alteration found in joints) that could be calibrated to clay mapped underground where data was available, see Fig. 5a) and b). Fig. 5c) shows the calibrated clay model with the underground mapping data. Calibration was done visually to determine which isosurface value in the clay model corresponded to the mapped data. Within the calibration volume, this isosurface represents the density at which the clay indicators represent a volume of the rock mass that has more clay. The only mapped clay volumes that were not represented in the model were volumes which had no associated drill core data. The calibrated model does not necessarily represent volumes that have entirely been altered to clay, rather volumes that have a higher concentration of clay than their neighbours. At the Kiirunavaara Mine, this simplistic form of spatial analysis of core data has the potential to assist with prediction of location and size of clay volumes prior to drifting. Tighter exploration drill spacing and additional underground mapping in other areas of the mine have the potential to improve future models.

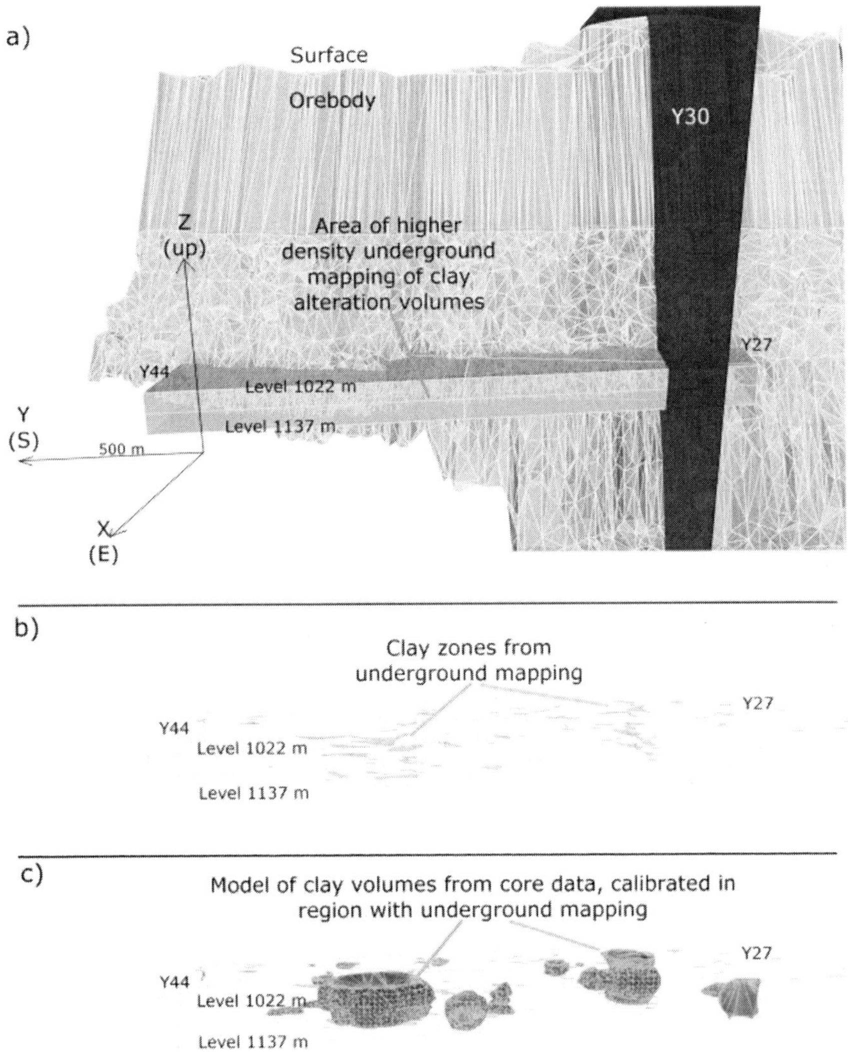

Fig. 5. Calibration of model of clay volumes from core data with underground mapping data (view from the hangingwall towards the footwall). Area of higher density underground mapping shown in a), mapped clay volumes are presented in b) with the overlaid clay model from the core data in c).

4.3. Rock mass quality

When considering the data available over the entire mine, neither GSI (LKAB, 2014c) nor RMR data sets are good candidates for geostatistical analysis. The RMR data set at the mine is very sparse in both number of sample points as well as their physical locations. GSI has more data points than RMR, but still suffers from poor coverage. However, the RQD data set (LKAB, 2014a) is much more complete, and is amenable to geostatistical analysis.

Two previous works exist using geostatistics to estimate RQD at the Kiirunavaara Mine. The first was limited to a section of the rock mass located in the north (Henry and Marcotte, 2001), and the second used isotropic search parameters to develop a mine-scale model (Cotesta, 2011). To update the model with newly mapped data and to provide additional analysis, a new mine-scale model that used anisotropic search parameters was created.

Using similar techniques as previously discussed, anisotropy was found to exist through the analysis of semi-variograms. Despite results from Henry and Marcotte (2001) illustrating that the range of the semi-variograms indicated insufficient drill spacing for geostatistical analysis of the analyzed portion of the rock mass (only the northern portion), the semi-variograms of the current database for the entire mine indicated otherwise. Based on the ranges from the semi-variograms search radii were evaluated to be 100 m along strike, 100 m across strike, and 80 m vertically, which are greater than the maximum drill spacing of 50 m.

Models of RQD were created using inverse distance, as a preliminary evaluation technique. Kriging methods have the potential to improve these models. An example vertical slice through the mine-scale RQD model is illustrated in Fig. 6 in combination with a longitudinal projection of the orebody. Distinct zones of lower RQD crosscut the orebody, in particular in the southern end (Fig. 6). These zones are physically associated with the calibrated clay volume model; low RQD is found both inside and in the immediate vicinity of the volumes that are estimated to have more clay alteration (Fig. 6). Consistent with the mine-scale results found by Cotesta (2011), RQD increases with depth in the northern portion of the rock mass, with a sharp contrast at approximately the Level 950 m.

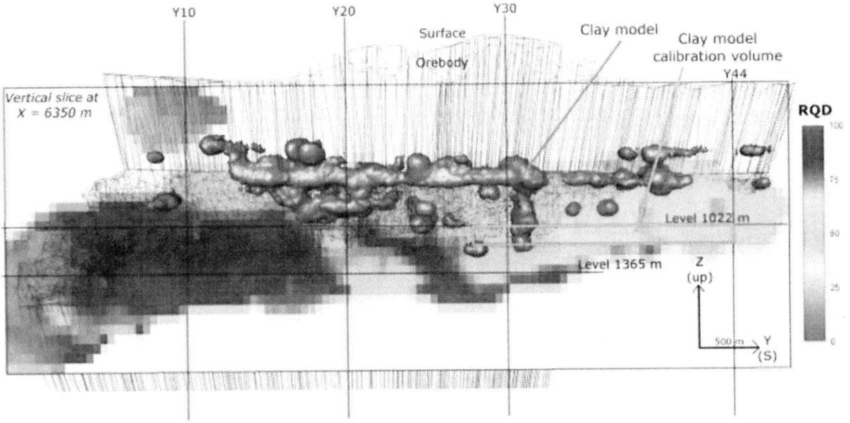

Fig. 6. Example vertical slice through the mine-scale RQD model through X = 6350 m viewed from the footwall overlaid on a longitudinal projection of the orebody. The RQD model is represented by the block model, where values are scaled by colour. The mine-scale clay model (from Section 4.2) is superimposed on the RQD model, in addition to its calibration volume.

4.4. Structural domains

The last structural domaining done at the mine scale was completed in 1995 (Holmstedt, 1994, 1995), however the structural domains found by Rådberg et al. (1989) are more commonly used. Since that time, some additional work on identifying domains has been done sporadically for specific volumes of the mine. The current study updates the structural domain model by incorporating historical and newly acquired data (LKAB, 2014b).

Due to the importance of the identification of structural domains, multiple techniques have been developed to assess their boundaries (Escuder Viruete et al., 2001; Jimenez-Rodriguez and Sitar, 2006; Martin and Tannant, 2004; Piteau and Russell, 1971; Zhou and Maerz, 2002). Most techniques have limitations, however, such as their reliance upon a priori geological knowledge, lack of evaluation in 3-D space, appropriate distance metrics for clustering, etc., which may misrepresent the underlying spatial distribution of joint sets.

Available discontinuity data at the mine scale in the Kiirunavaara Mine is limited to drift mapping of joint orientations at active faces (shotcrete limits mapping of completed excavation). The nature of this data eliminates the possibility of multivariate clustering as suggested by Zhou and Maerz (2002), as only dip and dip direction is readily available at the mine scale.

Piteau and Russell (1971) developed an effective and simple technique to identify the locations of structural domains. Their technique is based upon the mathematical definition of structural domains; within each domain, the slope of the cumulative sum of dip or dip direction is constant. The major limitation of their domain location technique is that it is based upon evaluating data along a 1-D ray or in a 2-D plane. Identifying the correct locations of domains using this technique assumes that the domains can be defined upon the user selected ray or plane, and that there are no other domains in other directions. The success of this approach is highly dependent upon the underlying spatial distribution of the joint sets.

An extension of Piteau and Russell (1971) has been developed to quantitatively identify domains in 3-D. Identified domains are then statistically evaluated using similar strategies as presented by Mahtab and Yegulalp (1984) and Martin and Tannant (2004). The proposed technique for quantitative identification of structural domains is detailed below.

1. Create a block model representing the span and extent of the original, cleaned data. Block dimensions are user specified; the intended scale of the domains, measurement spacing and geological environment are evaluated to determine this.

2. Within each block, for every joint measured, calculate the slope of the cumulative sum of 1) strike and 2) dip. Filtering of blocks based on user-defined minimum number of data points and linear fit parameters is recommended.

3. Using the slope of the cumulative sum and/or dip direction, apply geostatistical interpolation methods to supplement gaps in the data. The decision of which data to use should be based upon statistical analysis of the underlying data. With most data sets, the slope of the cumulative sum of dip is the most appropriate source to use, as when strike is used, there is a numerical discrepancy between 0° and 359° that does not represent a large physical direction difference. The methodology could be adapted for other mapped joint characteristics, such as joint roughness, however due to data availability at the mine this was not explored.

4. Using isosurfaces (3-D analog of an isoline or contour) of the block model, identify continuous volumes in 3-D space representing volumes of distinct structural domains. The selection of isosurfaces requires user input. Particular attention should be paid to large changes in shape/location of isosurfaces. Distinct and individual

isosurfaces represent domains; however, the numeric values of the isosurfaces are irrelevant.

5. Statistically and visually compare the stereonets for the determined structural domains.

This technique was tested using the underground discontinuity mapping data at the Kiirunavaara Mine to evaluate the location and shape of the structural domains. Domains were identified using cubic blocks of 30 m edge length, that had greater than five data points and an r^2 value from linear regression greater than 0.95. A significant number of the structures measured at the mine had strikes that were orientated around 0° N, so the cumulative sum of dip was the most appropriate metric. Isotropic inverse distance was applied to the block model to interpolate the data. Alternative methods of interpolation have the potential to improve this model.

Example slices through the interpolated block model are shown in Fig. 7 in conjunction with a wireframe of the orebody, where each block is coloured based on its value of the slope of cumulative sum of dip. Changes in colour represent distinct 3-D volumes that share similar slopes of cumulative sum of dip, indicating distinct domains.

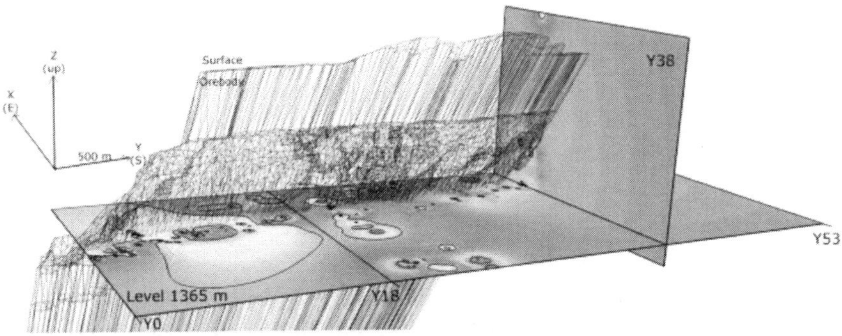

Fig. 7. Example slices through the slope of the cumulative sum of dip block model (30 m cubes) used to identify structural domains (seen from the footwall towards the hangingwall). The colour of each block represents the slope of the cumulative sum of dip for all joint sets mapped inside of that block. The values themselves are inconsequential; it is the location of changes in the values that represents possible structural domain boundaries.

The 3-D isosurfaces from the domain block model were used to group and test the original data points. Mahtab and Yegulalp (1984) suggested a method of statistically comparing stereonets based on pole count in equal area windows, and then using the χ^2 test to test if the distribution of the pole count for two stereonets is different, where a difference represents a new domain. Due to sample size limitations associated with the χ^2 test, Martin and Tannant (2004) adopted the window technique, however used the Pearson product–moment correlation coefficient (Pearson's r) as an indication of similarity between two stereonets. For the Kiirunavaara Mine, Pearson's r, Spearman's rank correlation coefficient (Spearman's rho (ρ)) as well as the Kendall rank correlation coefficient (Kendall's tau (τ) coefficient) were tested to evaluate statistical similarity between stereonets. For the joint sets in the Kiirunavaara Mine, the use of Spearman's ρ correlation coefficient and Kendall's τ were found to be most appropriate since the joint set data series after the count of poles in the windows are discrete and non-normal.

The results from the domaining analysis at the Kiirunavaara Mine highlight that the domains have 3-D shapes at the mine-scale, which cannot always be accurately represented in 2-D. The methodology is useful not only to identify domains and their shape, but also to explore rotations of joint sets within one domain. The magnitude of a joint set's rotation and intended use of the structural domains form the basis of the decision about if it constitutes a domain boundary. Statistical comparison of the stereonets can assist with this process via quantification of how different two stereonets are. At the Kiirunavaara Mine, one of the identified domains consists of a joint set that rotates within the volume. The three volumes presented in Fig. 8 are statistically from the same domain (see domain stereonet in the upper left corner of Fig. 8) using both Spearman's ρ correlation coefficient and Kendall's τ. However, as shown by the other stereonets in Fig. 8, the orientation of the main joint set varies slightly between the individual volumes.

At the Kiirunavaara Mine, transitions between domains at the mine-scale were rarely sharp. One example to the contrary, however, corresponds with a known, mapped geological feature. Despite not being used as a priori boundary during the analysis, where the rock mass transitions from north to south into a more clay altered volume (from approximately Y 28 and southwards), there is a sharp domain change. In this southern portion of the orebody, only two domains have been identified at the mine-scale (Fig. 9). They consist of many joints, and joint set identification is difficult (Fig. 9). The joint sets have a more random component in the heavily clay altered volumes of the rock mass.

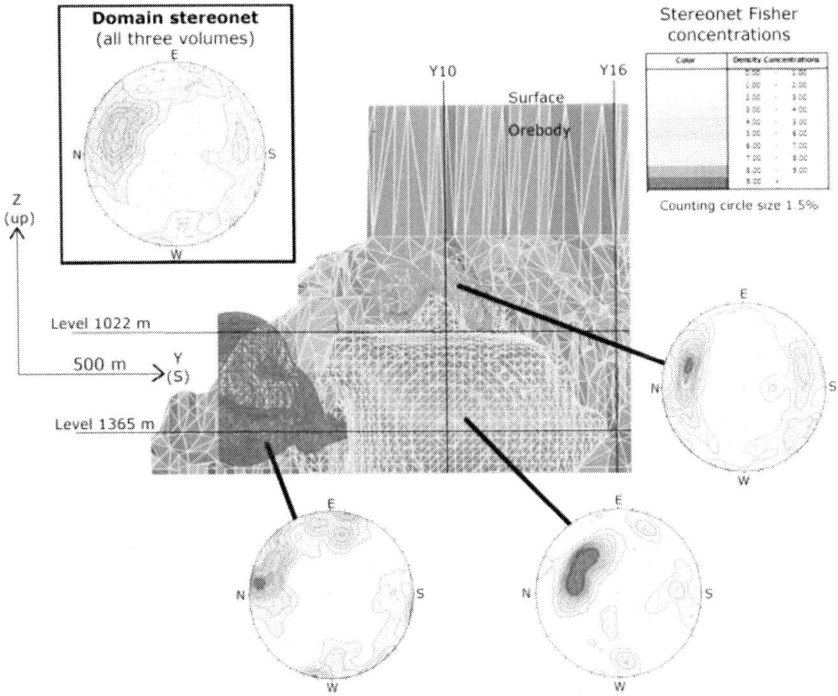

Fig. 8. Example of one domain (domain stereonet in upper left) at the Kiirunavaara Mine with slight variation of orientation of the major joint set for different volumes within the domain. Fisher concentrations for the poles are shown on equal area lower hemisphere stereonets. Stereonets are rotated so they are aligned with the mine coordinate system.

The 3-D domains identified for the rock mass using the suggested technique offered a significant improvement when compared to vertical extension of domains previously analyzed as suggested by Rådberg et al. (1989). More unique domains were identified, in different locations, with clearer definition of joint sets and transition zones between domains. Statistical comparison of the old and new domains showed that the new domains were more distinct than the domains identified by Rådberg et al. (1989).

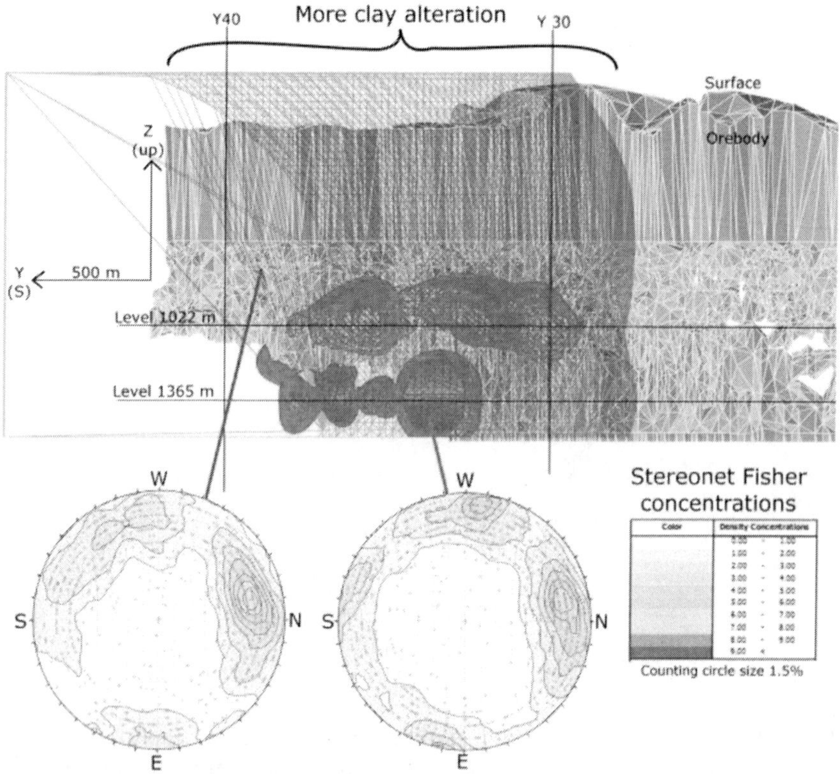

Fig. 9. Domains in the southern portion of the rock mass overlaid on the orebody as seen from the hangingwall. Domains are represented by the shaded wireframes, and their associated lower hemisphere equal area stereonets are provided. Stereonets are rotated so they are aligned with the mine coordinate system.

4.5. Faults

Numerous faults (discontinuities with identifiable shear displacement) are present at the mine. Berglund and Andersson's (2013) underground mapping work around Y 34 on the Level 1051 m, Level 1060 m, and Level 1079 m identified many faults. The majority of these faults are steeply dipping and follow the strike of the orebody (Berglund and Andersson, 2013). Currently, their length, spacing and persistence is unknown. Evidence of faults can also be seen during core logging in the form of slickenslides along discontinuity surfaces. Visible underground, many of these ore parallel structures, which exhibit slickenslides, are hematite coated with a mirror like surface and undulate on the scale of meters. Few orebody perpendicular faults have been mapped underground. Seismic clustering of events by the mine has identified possible fault structures, however only one structure has been confirmed by underground mapping, which was subsequently used in numerical stress analysis by Sjöberg et al. (2011). The currently available data does not enable a mine-scale characterisation to include faults, since mapped data is mainly limited to that done by Berglund and Andersson (2013).

5. CORRELATION TO FALLS OF GROUND

Based on the geomechanical characterisation, the portion of the rock mass which contains significant volumes of clay has different characteristics than the unaltered portion. The different characteristics are not limited to the clay volumes; rather the entire rock mass surrounding the clay. This is apparent from information related to jointing. Surrounding the clay volumes there are lower RQD values, few distinguishable structural domains, and a more random component to the discontinuities. Because of the seemingly controlling nature the clay zones have over the geomechanical characteristics of the rock mass, it is logical to evaluate if there is a correlation between the calibrated clay model and falls of ground.

The falls of ground at the Kiirunavaara Mine are often associated with seismic events, in the form of shakedown and/or bursting. With available data, it is difficult to accurately distinguish between these causes of the falls of ground, so all of these events were considered in this analysis. All fall of ground events at the mine are comparable since the support system is the same for almost the entire extent of the rock fall database. The direction of the major principal in situ (pre-mining) stress at the Kiirunavaara Mine is horizontal and approximately perpendicular to the strike of the orebody, whilst the intermediate and minor principal in situ stresses are nearly equal (Sandström, 2003). When viewing the data from the approximate direction of principal in situ stress, a correlation between the modelled clay zones and the falls of ground is evident (Fig. 10). The majority of the falls of ground are concentrated in the rock mass surrounding the clay zones.

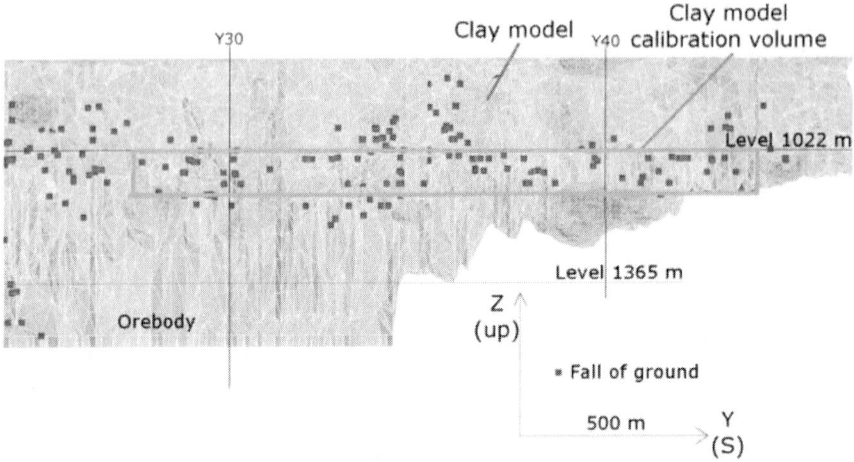

Fig. 10. Falls of ground (squares) versus the calibrated clay volume model (shown from the footwall towards the hangingwall). Viewed from the direction of major principal in situ stress (pre-mining).

There were significantly fewer falls of ground from approximately Y 23 and northwards. This could be related to the lack of clay volumes in the area, although additional calibration of the clay model in that volume is required.

6. DISCUSSION

Analysis of intact rock strength at the Kiirunavaara mine showed large variations not only between the groupings of geological units, but within them. This spread has implications for the geomechanical model and future rock engineering analyses. Representing the rock mass with constant strengths and stiffnesses is a simplification that cannot fully capture the rock mass response related to intact strength variation. The causes of this variation are unclear; the role of alteration in intact rock strength needs to be further evaluated. Additional UCS values (from UCS tests or triaxial tests) for geological groups with few of these data points may improve the scaling of the Point Load Strength tests. At this stage, how to scale the intact rock strength to rock mass strength for this complex geological environment remains unresolved.

At the Kiirunavaara Mine, a link exists between modelled clay volumes and falls of ground. Considering how the in situ stresses in this region likely behave, a possible explanation of this mine-scale behaviour is suggested. Since the volumes with clay alteration have a lower stiffness than the surrounding rock mass, the stresses will flow around the clay zones and concentrate in the more competent rock. This increase in stress magnitude in the surrounding rock mass has the potential to initiate new failure not necessarily induced by mining, such as those possibly indicated by the RQD model and the random component of the joint sets in the region. Stress concentrations caused by the clay volumes combined with mining induced stress changes have the potential to explain some of the behaviour. Numerical stress analysis in combination with more detailed study of the failure modes of the ground falls will be carried out in a future stage of this investigation to further explore this hypothesis and its implications on rock mass behaviour at the mine.

The methodology presented, as applied to the Kiirunavaara Mine, resulted in an increased understanding of the geomechanical characteristics of the rock mass. It successfully improved upon many aspects of geomechanical characterisation, such as the identification of new structural domains and a new understanding of the distributions of intact rock strengths. The methodology worked particularly well with the large quantity of borehole data combined with separate data sets for calibration. As an additional benefit, the development of a geomechanical model using these techniques and comparison of that model to rock mass behaviour highlighted where supplementary data collection could add the most value.

The methodology is robust; it has the potential to be adapted to different types of data in other mining environments. However, a priori knowledge of the rock mass was used to select the analysis techniques. Due to the smoothing nature of geostatistics, consideration must be given to if the parameters of interest are good candidates for interpolation. Data sets for interpolation must show spatial correlation over a greater distance than the data spacing, which is evaluated through semi-variograms. The highest resolution of the interpolation is a function of the input data set; interpolation requires supporting data points and accuracy of the results cannot be improved without well-positioned data. Provided the data supports interpolation, the presented methodology is well suited to geological environments with heterogeneity and alteration, and/or environments that have domain boundaries with irregular shapes in 3-D.

7. CONCLUSION

Through enhancement of the understanding of the geomechanical characterisation of the Kiirunavaara Mine, a conceptual model of some aspects of the rock mass behaviour was developed. Clay volumes (represented by a model based on borehole data calibrated to underground mapping) correlated to the geomechanical characteristics and behaviour of the rock mass. The rock mass in the immediate vicinity of the volumes of clay alteration had lower RQD values, more random jointing, and a higher concentration of falls of ground than the surrounding rock mass.

Consideration of multiple sources of data in 3-D space through the application of geostatistics was a key component to the development of this new conceptual model. The methodology presented, based on statistics, geostatistics and an extension of previous quantitative domaining work, is of interest to other mines that have complex and heterogeneous geological environments, specifically those which have volumes of alteration, structural domains which possibly have complex shapes in 3-D, or other characteristics which are well suited to interpolation methods. The methodology is well suited to geological environments that are appropriate data for interpolation; data types which can logically be interpolated between data points and that are spatially correlated over a greater distance than the data spacing. The use of geostatistics on multiple data sets that are often available at mines (such as borehole data, structural mapping data, etc.) has the potential to give indications of problem areas, provided the interpolation is well calibrated against data that is distinct from the input data.

ACKNOWLEDGEMENTS

This work would not have been possible without the generous financial contributions from the Hjalmar Lundbohm Research Centre (HLRC) and LKAB. The importance of the resources provided by LKAB in the form of personnel, time, equipment, and data cannot be understated. The authors gratefully acknowledge Christina Dahnér (LKAB) for her continued support as our mine contact and insightful contributions to discussions. We thank Jimmy Töyrä (LKAB) for his assistance with yet another short-notice review. Ulf B. Andersson is thanked for his review of the manuscript. We are also immensely grateful to the following people, who have aided the formulation of portions of this manuscript via countless in-depth discussions and much assistance with data acquisition (employed by LKAB unless otherwise stated): Lars Malmgren, Carlos Quinteiro, Stig Fjellborg, Henrikki Rutanen, Karin Lindgren, Stina Hallinder, Anousheh Abdollahpour, Mirjana Bošković, Helena Sturk, Johan Berglund (Vattenfall), and Tomas Björnell (Norconsult).

REFERENCES

1. Alvarez Grima and Babuška, 1999 M. Alvarez Grima, R. Babuška**Fuzzy model for the prediction of unconfined compressive strength of rock samples**Int. J. Rock Mech. Min. Sci., 36 (3) (1999), pp. 339-349,
2. Andersson and Israelsson, 2009 U.B. Andersson, J. Israelsson**Resultat av bermekaniska materialtester**(in Swedish) (LKAB Internal Report)(2009)
3. Ayalew et al., 2002 L. Ayalew, G. Reik, W. Busch**Characterizing weathered rock masses—a geostatistical approach**Int. J. Rock Mech. Min. Sci., 39 (1) (2002), pp. 105-114,
4. Berglund and Andersson, 2013 J. Berglund, U.B. Andersson**Kinematic analysis of geological structures in Block 34, Kiirunavaara**(LKAB Internal Report)(2013)(Ref no. 13–746, 46 pp.)
5. Bieniawski, 1975 Z.T. Bieniawski**The point-load test in geotechnical practice**Eng. Geol., 9 (1) (1975), pp. 1-11,
6. Coli et al., 2012 N. Coli, P. Berry, D. Boldini, R. Bruno**The contribution of geostatistics to the characterisation of some bimrock properties**Eng. Geol., 137–138 (2012), pp. 53-63,
7. L. Cotesta**Development of an RQD Voxet Model for LKAB's Kiruna Mine**(LKAB Internal Report)(2011)
8. D'Andrea et al., 1965 D.V. D'Andrea, D.E. Fogelson, R.L. Fischer**Prediction of compressive strength from other rock properties**Report of investigations (Ref no. 6702), U.S. Dept. of the Interior, Bureau of Mines, Washington, D.C. (1965)
9. Dahnér et al., 2012 C. Dahnér, L. Malmgren, M. Bošković**Transition From Non-seismic Mine to a Seismically Active Mine: Kiirunavaara Mine**The 2012 ISRM European Rock Mechanics Symposium (Eurock 2012), Stockholm, Sweden (2012)
10. Escuder Viruete et al., 2001 J. Escuder Viruete, R. Carbonell, M.J. Jurado, D. Martí, A. Pérez-Estaún**Two-dimensional geostatistical modeling and prediction of the fracture system in the Albala Granitic Pluton, SW Iberian Massif, Spain**J. Struct. Geol., 23 (12) (2001), pp. 2011-2023,
11. Geijer, 1910 P. Geijer**Geology of the Kiruna district, Vetenskapliga och praktiska undersökningar i Lappland**P.A.Norstedt, Stockholm (1910)
12. Gokceoglu, 2002 C. Gokceoglu**A fuzzy triangular chart to predict the uniaxial compressive strength of the Ankara agglomerates from their petrographic composition**Eng. Geol., 66 (1–2) (2002), pp. 39-51
13. Gokceoglu et al., 2009 C. Gokceoglu, H. Sönmez, K. Zorlu**Estimating the uniaxial compressive strength of some clay bearing rocks selected from Turkey by nonlinear multivariable regression and rule-based fuzzy models**Expert. Syst., 26 (2) (2009), pp. 176-190
14. Henry and Marcotte, 2001 E. Henry, D. Marcotte**Developments in ground control planning enhanced by RQD geostatistical analysis at LKAB's Kiruna mine**The 2001 ISRM European Rock Mechanics Symposium (Eurock 2001), Espoo, Finland (2001)
15. Hoek and Karzulovic, 2001 E. Hoek, A. Karzulovic**Rock Mass Properties for Surface Mines**W.A. Hustrulid, M.K. McCarter, D.J.A. Van Zyl (Eds.), Slope Stability in Surface Mining, Society for Mining, Metallurgy, and Exploraction, Inc. (SME), Littleton (2001), pp. 59-69
16. Holmstedt, 1994 A. Holmstedt**Bergmasseklassificering av Kiirunavaara malmkropp**(in Swedish) (LKAB Internal Report)(1994)(Ref no. 94–788)
17. Holmstedt, 1995 A. Holmstedt**Bergmasseklassificering av liggvägen I KUJ ovanför nivå 740 m**(in Swedish) (LKAB Internal Report)(1995)(Ref no. 95–782)
18. ISRM Commission on Standardization of Laboratory and Field Tests, 1985 ISRM Commission on Standardization of Laboratory and Field Tests**Suggested method**

for determining point load strength Int. J. Rock Mech. Min. Sci. Geomech. Abstr., 22 (1985), pp. 51-60,

19. Jimenez-Rodriguez and Sitar, 2006R. Jimenez-Rodriguez, N. Sitar**A spectral method for clustering of rock discontinuity sets**Int. J. Rock Mech. Min. Sci., 43 (2006), pp. 1052-1061, 10.1016/j.ijrmms.2006.02.003
20. Kahraman, 2001 S. Kahraman**Evaluation of simple methods for assessing the uniaxial compressive strength of rock**Int. J. Rock Mech. Min. Sci., 38 (2001), pp. 981-994,
21. Kaya and Karaman, 2015 A. Kaya, K. Karaman**Utilizing the strength conversion factor in estimation of the uniaxial compressive strength from the point load index**Bull. Eng. Geol. Environ. (2015), 10.1007/s10064-015-0721-1
22. Kohno and Maeda, 2012 M. Kohno, H. Maeda**Relationship between point load strength index and uniaxial compressive strength of hydrothermally altered soft rocks**Int. J. Rock Mech. Min. Sci., 50 (2012), pp. 147-157,
23. Lindgren, 2013 K. Lindgren**Punktlasttestning av olika bergartstyper i Kiirunavaaras liggvägg**(in Swedish) (Bachelor's Thesis 41921476)Luleå University of Technology, Luleå, Sweden (2013)
24. LKAB, 2014a LKAB**Core logging database**LKAB Internal Database (2014)(Accessed 2014-12-18)
25. LKAB, 2014b LKAB**Discontinuities database**LKAB Internal Database (2014)(Accessed 2014-12-18)
26. LKAB, 2014c LKAB**GSI database**LKAB Internal Database (2014)(Accessed 2014-12-18)
27. Mahtab and Yegulalp, 1984 M.A. Mahtab, T.M. Yegulalp**Similarity Test For Grouping Orientation Data In Rock Mechanics**The 25th U.S. Symposium on Rock Mechanics (USRMS), American Rock Mechanics Association (1984)
28. Mandrone, 2006 G. Mandrone**Engineering geological mapping of heterogeneous rock masses in the Northern Apennines: an example from the Parma Valley (Italy)**Bull. Eng. Geol. Environ., 65 (3) (2006), pp. 245-252,
29. P. Marinos, E. Hoek**Estimating the geotechnical properties of heterogeneous rock masses such as flysch**Bull. Eng. Geol. Environ., 60 (2) (2001), pp. 85-92,
30. M.W. Martin, D.D. Tannant**A technique for identifying structural domain boundaries at the EKATI Diamond Mine**Eng. Geol., 74 (3–4) (2004), pp. 247-264,
31. Mattsson et al., 2010 H. Mattsson, J. Berglund, B. Magnor**Strukturgeologisk modell över Kiruna gruvområde**(in Swedish) (LKAB Internal Report)(2010)(Ref no. 10–874)
32. Ozturk and Simdi, 2014 C.A. Ozturk, E. Simdi**Geostatistical investigation of geotechnical and constructional properties in Kadikoy–Kartal subway, Turkey**Tunn. Undergr. Space Technol., 41 (2014), pp. 35-45,
33. Piteau and Russell, 1971 D.R. Piteau, L. Russell**Cumulative sums technique: A new approach to analyzing joints in rock**Rock Mech., 13 (1971), pp. 1-29
34. Rådberg et al., 1989 G. Rådberg, S. Liedberg, A. Holmstedt**Domänindelning av Kirunavaara efter sprickors orientering**(in Swedish) (LKAB Internal Report)(1989)(Ref no. 89–737)
35. Sandström, 2003 D. Sandström**State of stress at the Kiirunavaara mine: its significance for rock mechanics analysis**(Licentiate Thesis (2003:02))Luleå University of Technology, Luleå, Sweden (2003)
36. Singh et al., 2001 V.K. Singh, D. Singh, T.N. Singh**Prediction of strength properties of some schistose rocks from petrographic properties using artificial neural networks**Int. J. Rock Mech. Min. Sci., 38 (2) (2001), pp. 269-284,
37. Sjöberg et al., 2011 J. Sjöberg, C. Dahnér, L. Malmgren**Forensic analysis of a rock burst event at the Kiirunavaara Mine – results and implications for the future**D. Sainsbury, R. Hart, C. Detournay, M. Nelson (Eds.), Proceedings of the 2nd International FLAC/DEM Symposium. Continuum and Distinct Element Numerical Modeling in Geomechanics--2011, Itasca International Inc., Melboune, Australia (2011), pp. 67-74

38. Sjöberg et al., 2012 J. Sjöberg, F. Perman, C. Quinteiro, L. Malmgren, C. Dahnér-Lindkvist, M. Boskovic**Numerical analysis of alternative mining sequences to minimise potential for fault slip rockbursting**Min. Technol., 121 (4) (2012), pp. 226-235

39. Stavropoulou et al., 2007 M. Stavropoulou, G. Exadaktylos, G. Saratsis**A Combined Three-Dimensional Geological-Geostatistical-Numerical Model of Underground Excavations in Rock**Rock Mech. Rock. Eng., 40 (3) (2007), pp. 213-243,

40. Vatcher et al., 2014 J. Vatcher, S.D. McKinnon, J. Sjöberg, C. Dahnér**Modelling methodology: Structural geology and rock mass behaviour at Kiirunavaara Mine**Rock Engineering and Rock Mechanics : Structures in and on Rock Masses, The 2014 ISRM European Rock Mechanics Symposium (Eurock 2014), CRC Press, Vigo, Spain (2014), pp. 643-648

41. Westblom, 2014 M. Westblom**Bergmekaniska tester av Syenitporfyrer, Leptiter och Granit**(in Swedish). (Draft Master's Thesis)Luleå University of Technology, Luleå, Sweden (2014)

42. Zhou and Maerz, 2002 W. Zhou, N.H. Maerz**Implementation of multivariate clustering methods for characterizing discontinuities data from scanlines and oriented boreholes**Comput. Geosci., 28 (7) (2002), pp. 827-839,

CHAPTER 6

Application of 2D Resistivity Imaging and Seismic Refraction Tomography to Identify Sungai Batu Sediment Depositional Origin

Saad R[1*], Muztaza Mn[1], Zakaria MT[1] and Saidin MM[2]

[1]*Geophysics Section, School of Physics, Universiti Sains Malaysia, Penang, Malaysia*

[2]*Centre for Global Archeological Research Malaysia, Universiti Sains Malaysia, Penang, Malaysia*

ABSTRACT

Geophysical survey such as 2D resistivity imaging and seismic refraction tomography are non-destructive methods that widely used in subsurface exploration including archaeological study. The purpose of this survey is to identify the sediment deposition types of Sungai Batu area for Ancient River. Two study sites were chosen to conduct 2D resistivity imaging and seismic refraction tomography surveys. The 2D resistivity imaging survey was conducted using Poledipole array with 2.5 m minimum electrode spacing while seismic refraction tomography was performed using 5 kg sledgehammer as seismic source with 5 m geophone spacing. Roll along techniques are apply for the two methods in the study site 1 and 2. The study concludes that the subsurface of the study area comprise of 3 major soil types. The top soil (1st type) consists of loose and dry alluvium which indicated with resistivity value of >100 Ohm.m. The second type was saturated alluvium (clay and sand) with resistivity and velocity values of 10-50 Ohm.m and <1400 m/s respectively. When this type of soil change to moist condition, the resistivity value increase to <100 Ohm.m but velocity value remain. The 2D resistivity imaging and seismic refraction tomography detected the river bed with the value of >300 Ohm.m and >3600 m/s respectively. The correlation of 2D resistivity imaging and seismic refraction tomography show that the depositional environment for this survey is causes by land sediments deposit.

Keywords: 2D resistivity imaging, seismic refraction tomography, sediments deposit

INTRODUCTION

Sungai Batu was located in Sungai Merbok district, which is situated between Sungai Petani and Gunung Jerai. Archaeological evidences such as metal artifacts, remains of furnaces, iron ores, tuyeres and slags has been used since 3rd-5th CE as a famous international entrepot [1]. The evidence of ancient structure and iron **smelting** with supporting data from geophysical methods show that Sungai Batu is an oldest civilization that has been recorded in Southeast Asia [2].

Mid-south of Kedah was recorded as marine area in the 1st-2nd century. The changes of **geomorphology** due to rise in sea level in year 1400 changes the area become landed region [3]. **Sediments** including rocks, soil, and mineral from weathering process was transported from river and ocean to the lower area, and strongly influenced by the regional geology. The sea level fluctuations are about 2-3 m in 200 years or less based on the historical and geological evidence [4]. Regression of sea level during Late Holocene until present day changes the geomorphology for this area. The mechanism of **sea level** changes is complex and related to the earth's rheology, eustasy and isostasy [5,6]. The sea level fall was recorded with the value of 0.7 and 0.5 mm/year for Peninsular Malaysia West and East coasts respectively in mid-Holecene to recent sea level trend [7]. Sungai Batu sedimentation process controls by the depositional process from Sungai Merbok and Malacca Strait. Sediments from Gunung Jerai was transported by Sungai Merbok to lower region with high of turbidity current until it settled down due to gravity factor. The hydrogenous sediments from Malacca Strait are transported due to long shore drift from the tidal current to formed range of landforms. There are several factors controlling the depositional environment in this area which is fluvial discharge including water volume, time variation, river mouth process and also tidal process. The sediments from the near shore or marine are well sorted with fine grained distribution due to current **velocity** and particle diameter [8].

RESISTIVITY value of unconsolidated sediments from Bering Sea was identify with value of 29.1-71.0 Ohm.cm at 25°C and irregularly increase 11 Ohm.cm in the first meter below the sediment surface [9]. A combination of laboratory analysis and multi-electrode resistivity field experiment for marine sediments status confirms that the multielectrode resistivity method is relevant to estimate the degree and state of sediment consolidation in real-time and in-situ [10]. A geophysical surveys, 2D resistivity imaging and **seismic refraction** tomography were carried out to identify the sediment depositional origin of Sungai Batu.

STUDY AREA

The study location was situated near river and residential area. The geomorphology of this area was flat landed fill with oil palm and rubber trees including streams, rivers and swamps at the eastern part. Sungai Batu area is located at northwestern of Peninsular Malaysia and near to Gunung Jerai and Sungai Merbok. Sungai Merbok is underlain by Mahang formation especially in eastern and southern part [11]. The formation formed in Middle Ordovician to Early Devonian age [12]. The rock unit comprises shale, red slate, grey slate and black slate. Hot and humid climate during a year leads the high chemical weathering to occurs, which changes the shale to soil and laterite [13]. The main **mineralogy** of rocks is mica and chloride. Black slate contains an abundance of carbon and red slate, consists of high iron oxide [14]. Granitic intrusion causes the rocks of surroundings area become metamorphose to produce schist rocks [12]. Gunung Jerai located at the northern part of study area which recorded as highest mountain and consist of rock cockerel, made of schist rocks with a few layers of quartzite (**FIGURE 1**).

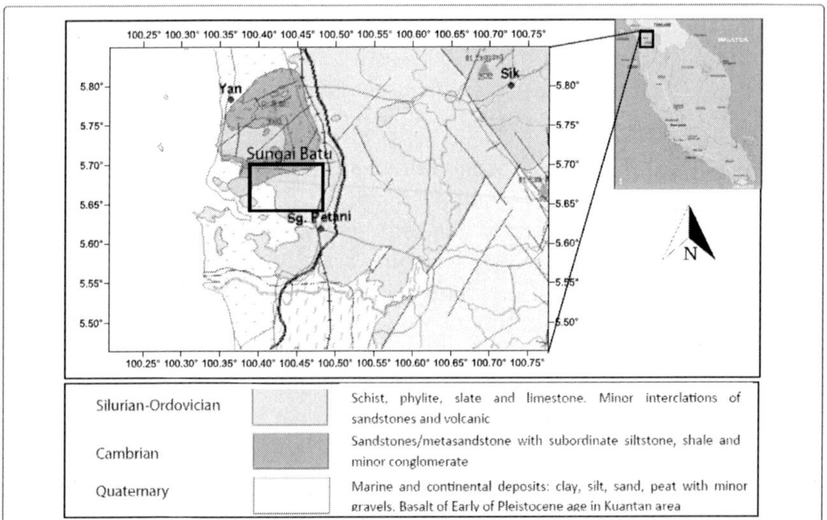

Figure 1: Geological map of Sungai Batu, Kedah [22].

FIGURE 1: Geological map of Sungai Batu, Kedah [22].

METHODOLOGY

2D resistivity imaging method utilizes direct current to investigate electrical properties of **subsurface** by measuring resistivity distribution of the materials. Electrical resistivity of the materials depends on a combination of ohmic and dielectric effects related to **lithology** of subsurface [15]. The apparent resistivity, ρ_a are calculated based on Equation 1.

$$\rho_a = k\frac{V}{I} \quad (1)$$

Where;

V=voltage, I=current, k=geometric factor

The depth of penetration is also depending on k value. For this survey, Pole-dipole array was used based on the deepest depth penetration beside provides a good resolution. Pole-dipole array is an asymmetrical arrangement which influence the **anomalies** by giving asymmetrical apparent resistivity. To eliminate the asymmetry affect, reverse measurement also taken (**FIGURE 2**).

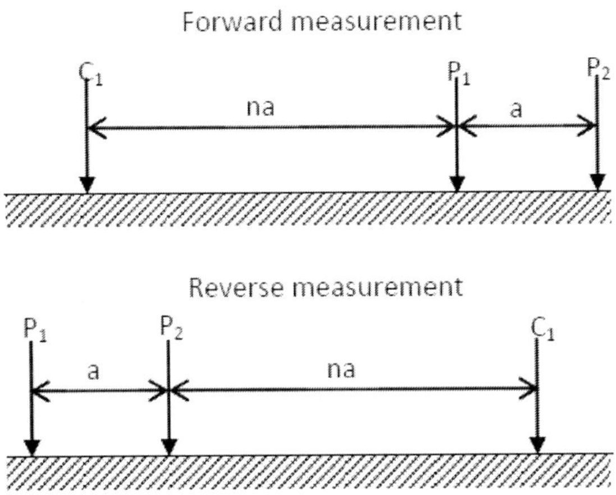

Figure 2: Forward and reverse measurement of pole-dipole array [23].

FIGURE 2: Forward and reverse measurement of pole-dipole array [23].

Rock types such as igneous and metamorphic; typically have high resistivity values depending on degree of fractures. Since water table in Malaysia is generally shallow, the fractures are commonly filled with **ground water** [16]. High fracturing of rocks will result a lower resistivity values. **TABLE 1** shows the resistivity values of rocks and soil types [17].

TABLE 1: Resistivity value of rocks and soil [17].

Material	Resistivity (Ωm)
Alluvium	10-800
Sand	60-1000
Clay	1-100
Groundwater (fresh)	10-100
Sandstone	$8\text{-}4 \times 10^3$
Shale	$20\text{-}2 \times 10^3$
Limestone	$50\text{-}4 \times 10^3$
Granite	$5 \times 10^3\text{-}1 \times 10^6$

Seismic refraction **tomography** is based on determination of time interval that elapses between an initiation of a seismic waves at a certain shot point and the arrival of refracted waves at one or more seismic detector (**FIGURE**

3). Seismic refraction tomography uses a wave's propagation in **ground surface** which is dependent on the velocity variation in difference medium [18]. The wave return to the surface as refracted waves which are sometimes called head waves [19].

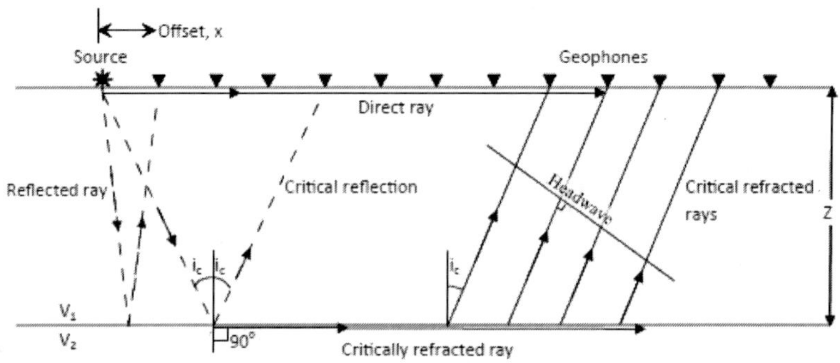

Figure 3: Seismic ray path of direct wave, reflection, and refraction ray [21].

FIGURE 3: Seismic ray path of direct wave, reflection, and refraction ray [21].

Seismic waves propagate with different velocity when travelling in different geological layer and refracted when they crosses a boundary between different type of geological layer. The thickness and velocity of material above and below an interface can be calculate by determining the arrival times for direct and refracted waves from seismic section. **TABLE 2** shows seismic velocity value varies with mineral content, lithology, **porosity**, density and degree of compaction [20].

TABLE 2: The P-wave velocities of various earth materials [21].

Material	P-wave velocity (m/s)
Air	332
Water	1400-1500
Granite	5500-5900
Sandstone	1400-4300
Limestone	5900-6100
Sand (Unsaturated)	200-1000
Sand (Saturated)	800-2200
Clay	1000-2500

Geophysical survey conducted in this study was 2D resistivity imaging and seismic refraction tomography methods which apply in two study sites (**FIGURE 4**).

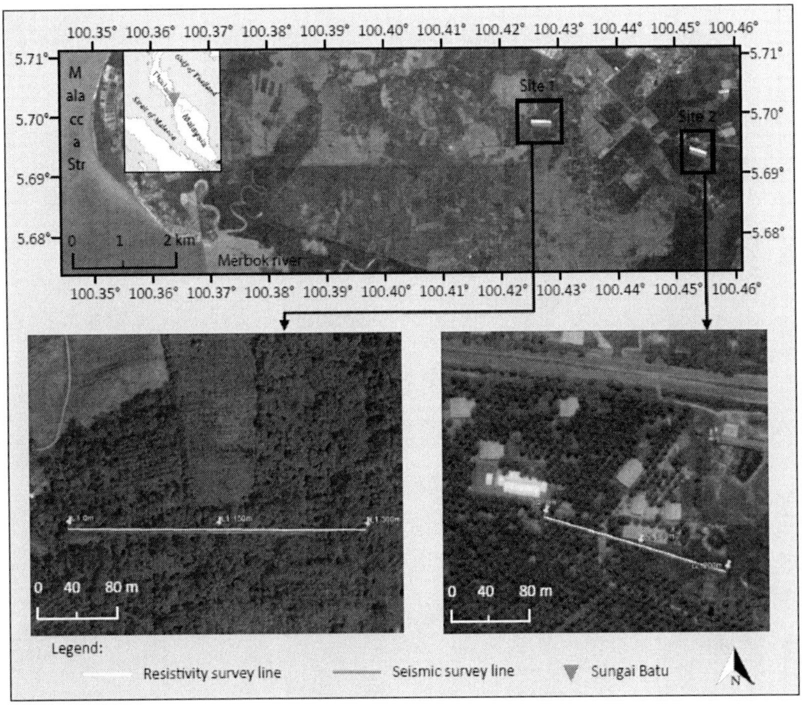

Figure 4: Two study sites at Sungai Batu, Kedah [24].

FIGURE 4: Two study sites at Sungai Batu, Kedah [24].

The 2D resistivity survey used ABEM SAS4000 Terrameter including ES10-64C electrode selector, smart cables, 41 stainless steel **electrodes** and jumpers. The survey lines are designed in such a way to fulfill study objectives. Line L1 at study site 1 was design with 2.5 m electrode spacing which make the total length of the spread is 100 m. Roll-up techniques was apply on line L1 with 4 times rolling-up with 50 m roll-up interval to make the total length of line L1 is 300 m. Line L2 at study site 2 was design using the same as line L1 of site 1 but applying 2 times rolling-up with 50 m roll-up interval to make the total length of line L2 is 200 m (**FIGURE 5**). The survey was performed using Pole-dipole array with remote electrode at 500 m away from each survey line and perpendicular to it. **TABLE 3** shows ABEM SAS4000 Terrameter data **acquisition** setting used for all survey lines.

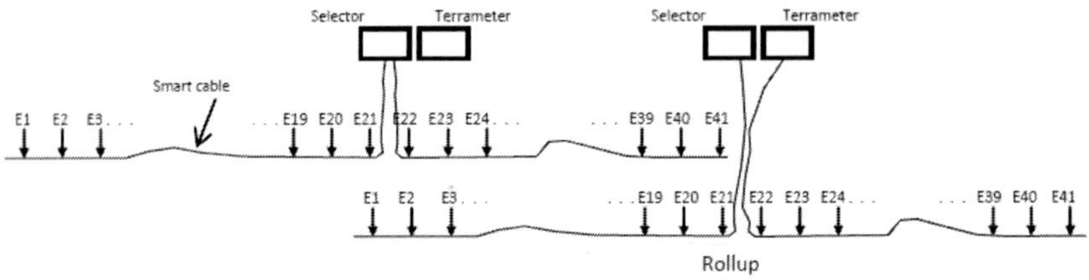

Figure 5: Rollup techniques.

FIGURE 5: Rollup techniques.

TABLE 3: ABEM SAS4000 Terrameter data acquisition setting.

Function	Setting	Value
Current	Maximum	50 mA
	Minimum	1 mA
Mode	Auto	-
Acquisition delay	-	0.3 sec
Acquisition time	-	0.3 sec
Total cycle time	-	3 sec
Stack	Maximum	3
	Minimum	1
Power line frequency	-	50 Hz

Seismic refraction tomography survey used one unit of ABEM MK8 seismograph, seismic cable with 24 take-out and 24 units of 28 Hz vertical geophones which planted on the ground at 5 m constant geophone spacing. Seismograph setting for data acquisition for line L1 and L2 on site 1 and 2 are the same (**TABLE 4**).

TABLE 4: ABEM MK8 seismograph data acquisition setting.

Function	Setting	Value
View mode	Auto	
Scale factor	-	1
Trace clip	-	1
Remove DC	On	-
Sampling interval (μs)	-	100
No. of samples	-	2048
Stack mode	Preview	-
Re-arm mode	Manual	-

Line L1 of site 1 is 230 m long and consists of two spreads. The first geophones of spread 1 located at 0 m and the 24th geophone at 115 m; while the first geophone of spread 2 located at 115 m and the 24th geophone at 230 m. Line L2 of site 2 is 200 m long and consists of two spreads. First geophone of spread 1 located at 0 m and 24th geophone at 115 m; while first geophone of spread 2 located at 85 m and 24th geophone at 200 m. Spread 2 was 30 m overlapped with spread 1. Seismic source used for both site was 5 kg sledgehammer with 7 shot points for each spread except spread 2 of line 2 for site 2 (**TABLE 5**).

TABLE 5: Seismic shot point location of line L1 and L2.

Site 1			Site 2		
Line	**Spread**	**Shot location (m)**	**Line**	**Spread**	**Shot location (m)**
L1	1	-20	L2	1	-30
-	-	2.5	-	-	2.5
-	-	27.5	-	-	27.5
-	-	57.5	-	-	57.5
-	-	87.5	-	-	87.5
-	-	112.5	-	-	112.5
-	-	135	-	-	145
-	2	95	-	2	55
-	-	117.5	-	-	87.5
-	-	142.5	-	-	112.5
-	-	172.5	-	-	142.5
-	-	202.5	-	-	172.5
-	-	227.5	-	-	197.5
-	-	250	-	-	-

2D resistivity data were processed using Res2Dinv. ver. 3.56.70 and Surfer8 software while seismic data was processed using FirstPix and Gremix15 (Interpex Limited) software for picking first arrival time, plotting travel time curve and velocity analysis. SeisOpt@2D. ver. 3.5 was used to produce 2D seismic profile tomography cross section. The processed results were plotted in Surfer8 software for correlation and interpretation.

RESULTS AND DISCUSSIONS

FIGURE 6 shows 2D resistivity section and seismic refraction profile of line L1 at study site 1. The resistivity profile shows that exploration depth is 25 m with resistivity value of 1-2000 Ohm.m (**FIGURE 6A**). A region with resistivity value of >100 Ohm.m indicate a top soil with depth of <5 m. A river bed identified with resistivity value of >300 Ohm.m at depth of 7-25 m. Low resistivity values of 10-50 Ohm.m is interpreted as saturated alluvium (clay and sand) which transported from land. Seismic refraction profile shows the investigation depth is 25 m with velocity value of 300-5600 m/s (**FIGURE 6B**). A region with velocity of >3600 m/s at depth of >15 m is interpreted as river bed. Low velocity value of <1400 m/s interpreted as moist/saturated alluvium zone which consist of clay and sand.

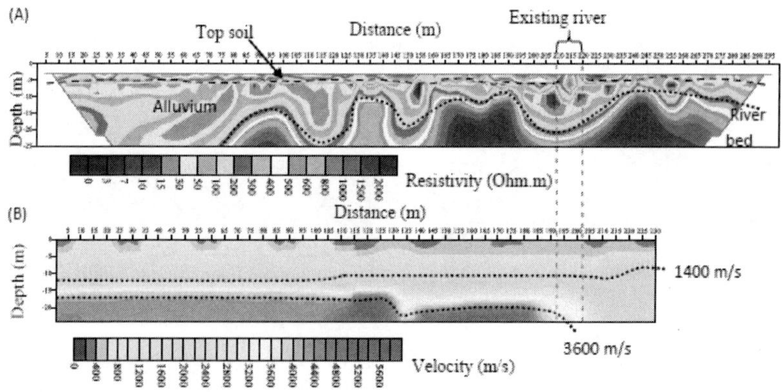

Figure 6: Cross-section of line L1 at study site 1. (A) Resistivity pseudo-section and (B) seismic refraction tomography profile.

FIGURE 6: Cross-section of line L1 at study site 1. (A) Resistivity pseudosection and (B) seismic refraction tomography profile.

FIGURE 7 shows 2D resistivity section and seismic refraction profile of line L2 at study site 2. The resistivity profile (**FIGURE 7A**) with resistivity value >300 Ohm.m indicated a river bed at depth of 7-25 m. A region with low resistivity value of 10-50 Ohm.m indicated the presence of saturated alluvium (clay and sand) which transported from land while moist zone was identified with resistivity value of <100 Ohm.m. From seismic profile

(**FIGURE 7B**), the river bed was identified at velocity of >3600 m/s at depth >10 m while velocity value of <1400 m/s interpreted as moist/saturated alluvium zone which consist of clay and sand.

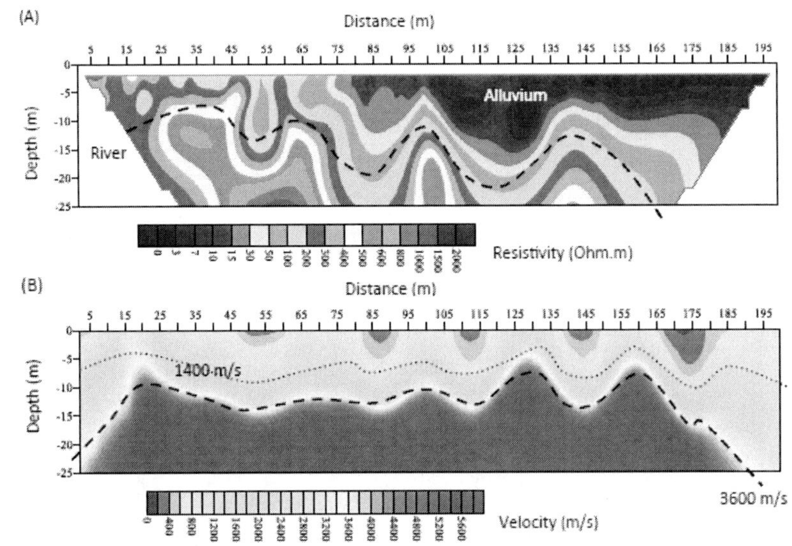

Figure 7: Cross-section of line L2 at study site 2. (A) Resistivity pseudo-section and (B) seismic refraction tomography profile.

FIGURE 7: Cross-section of line L2 at study site 2. (A) Resistivity pseudosection and (B) seismic refraction tomography profile.

CONCLUSION

Top soil of the study areas indicated by resistivity value of >100 Ohm.m is interpreted as loose and dry alluvium. The resistivity value of 10-50 Ohm.m and velocity value of <1400 m/s is interpreted as saturated alluvium (clay and sand) which transported from land and part of it is moist which indicated with resistivity value of <100 Ohm.m. The river bed identified by 2D resistivity imaging and seismic refraction tomography with the value of >300 Ohm.m and >3600 m/s respectively. The correlation of 2D resistivity imaging and seismic refraction tomography show that the **depositional environment** for this survey is causes by land sediments deposit.

ACKNOWLEDGEMENTS

The authors thank the technical staffs of the geophysics laboratory and all geophysics postgraduate students, School of Physics, Universiti Sains Malaysia for their assistance during the data acquisition, and the Centre of Global Archaeological Research (CGAR) Universiti Sains Malaysia for sponsoring the project (1002/ PARKEO/910328).

REFERENCES

1. Naizatul AM, Saidin M, Abdullah J (2011) The Ancient Iron Smelting in Sg. Batu, Bujang Valley, Kedah. Postgraduated Forum, Department of Anthropology, The Chinese University of Hong Kong.
2. Ramzul SI, Nordiana MM, Saad R, Saidin M, Maslinda U, et al. (2015) Investigation of Ancient River at Lembah Bujang , Kedah, Malaysia. Elect J Geotech Eng 20: 4385-4392.
3. Wheatley P (1961)The Golden Khersonese: Studies in the historical geography of the Malay Peninsula before A.D. 1500. Kuala Lumpur, University of Malaya Press.
4. Khoo TT (1996) Geomorphological Evolution of the Merbok Estuary Area and its Impact on the Early State of Kedah, Northwest Peninsular Malaysia. J Southeast Asian Ear Sci 13: 347-371.
5. Peltier WR (1987)In: Devoy RJN (ed.)Mechanisms of Relative Sea-Level Change and the Geophysical Responses to Ice-Water Loading. Sea Surface Studies: A Global View, London: Croom Helm57-94.
6. Pirazzoli PA (1991)World Atlas of Holocene Sea-Level Changes. Elsevier Oceanography Series, Amsterdam 58: 300.
7. Kamaludin H (2003) Mid-Holocene to recent sea level changes in Peninsular Malaysia: a tectonic implication. Geol Soc Malay Bull 46: 313-318.
8. Dominik J, Stanica A, Thomas R (2010) Soft Aediment Observation and Characterization in Field Courses. NEAR Curric Nat Environ Sci Terre Environ 88: 231-242.
9. Boyce RE (1968)Electrical Resistivity of Modern Marine Sediments from the Bering Sea.J Geophys Res73: 4759-4766.
10. Liu X, Jia Y, Zheng J, Shan H, Li H (2013) Field and laboratory Resistivity Monitoring of Sediment Consolidation in China's Yellow River Estuary. Eng Geol 164: 77-85.
11. Habibah J, Wan Fuad WH, Mohamad MT (2004) Pengaruh Jenis Batuan Sekitar Terhadap Taburan Pb dalam Sedimen Muara Sungai Merbok, Kedah. Geol Soc Malay Bull 48: 7-11.
12. Bradford EF (1970) Geology and Mineral Resources of the Gunung Jerai Area, Kedah. Geological Survey Malaysia District Memoir 13: 242.
13. Habibah J, Wan Fuad WH (2002) Komposisi Unsur Surih dan Major di Sepanjang Profil Luluhawa Syal Formasi Mahang di Sungai Merbok, Kedah. Geol Soc Malay Bull45: 59-64.
14. Almashoor SS (1974) Geology of Gunung Jerai, Kedah: M.Sc. thesis, Universiti Kebangsaan Malaysia.
15. Edwin SR, Cahit C (1988) Basic Exploration Geophysics: John Wiley and Sons, New York.
16. Nordiana MM, Saad R, Saidin M (2014) 2-D Resistivity Imaging of Buried Furnace at Sik, Kedah (Malaysia). Electron J Geotech Eng 19: 17635-17642.

17. Keller GV, Frischknecht FC (1996) Electrical Methods in Geophysical Prospecting. Pergamon Press, New York.
18. Telford WM, Geldart LP, Sheriff RE (1990) Applied geophysics.Cambridge Univ. Press, New York.
19. Sjögren B (1984) Shallow Refraction Seismics. Chapman and Hall Ltd., London.
20. Burger HR (1992) Exploration Geophysics of the Shallow Subsurface. Norton and Company, New York.
21. Ali N, Saad R, Saidin MM, Nordiana MM (2012) Applying Seismic Refraction Method in Depicting Geological Contact. Int Conf Geol Environ Sci 36: 59-63.
22. Geological Map of Sungai Batu, Kedah (1985) Minerals and Geoscience Department Malaysia. 8th Edition.
23. Loke MH (1999) Electrical Imaging Surveys for Environment and Engineering Studies. A practical guide to 2-D and 3-D surveys.
24. Google Earth (2016).

CHAPTER 7

Comparison of Magnetic Basement Depth Values from Spectral Technique, SPITH and Slope Techniques using HRAM

Nwosu OB[1],*, Umego MN[1] and Onuba LN[2]

[1]*Department of Physics and Industrial Physics NAU Awka, Nigeria*

[2]*Department of Geology, Anambra State University Uli, Nigeria*

ABSTRACT

The quantitative results obtained from the use of **spectral analysis** technique, Source parameter Imaging technique and empirical depth rule method (maximum and half slope techniques) in interpreting High Resolution AeroMagnetic of middle Benue trough data have been presented and compared. Before employing each technique, the regional residual seperation was done with least square method using polifit program to get the residual data for onward processing. The slope technique usually require that the residual values be contoured into map and then profiled along the most prominent anomaly to obtain the depth curves from where depth to magnetic sources could be calculated. The Spectral requires that the residual values be sub-divided into spectral blocks and radial energy spectrum be generated from where depth would be calculated from the slopes of the straight line plots. The SPI requires the griding of the residual values and the conversion of this to image map from where depth could be read off easily. The deep magnetic anomally source gotten ranges from 2.00 km to 6.29 km with an average depth of 3.25 km for SPI technique; 2.33 km to less than 5.66 km with an average depth of 3.65 km for Spectral analysis technique; and average of 3.74 km and 3.66 km for Half-slope and maximum slope technique respectively. These average depth could be taken as the magnetic basement depth and has the significant that if other conditions are met, this study area of marine sedimentary layer of the Albian Age is favourable for hydrocation accumulation. The shallow source depth gotten ranges from 0.02 km to 2.00 km with an average depth of 1.08 km for SPI; 0.05 km to less than 0.42 km with an average depth of 0.21 km for spectral analysis; and an average depth 0.80 km for Half slope technique. This can be viewed as magmatic intrussion into the sediment and could be responsible for the lead-Zinc **mineralization** found in the area. These methods thus compares favourably well and moreso the slope technique though empirical has compared favourably with more modern automatic source depth methods.

Keywords: High resolutio aeromagnetic data(HRAM); Magnetic anomaly; Spectral analysis; Source parameter imaging; Slope techniques; Magnetic basement depth

INTRODUCTION

Aeromagnetic Data analysis is an important tool in maping the magnetic basement depth beneath the sedimentary cover. One of the important parameter in quantitative interpretation is the depth of the anomalous body. In oil exploration, basement depth is needed which we have obtained through the empirical method based on slope technique (maximum and half-slope), the Spectral technique, and source parameter imaging techniques (SPI) [1-3]. This work seeks to compare these methods and their results.

The spectral analysis and SPI technique used here are amongst the modern automatic source depth method of determining magnetic basement depth. The slope techniques (the maximum and half slope technique) is an empirical methods based on slope techniques using empirical constants. This work has employed these methods to quantitatively interprete the HRAM data obtained from the most recent survey in the country. This work seeks to obtain the magnetic basement depth of the studied area, the basement topography and contour orientations and appraise the hydrocarbon accumulation potential of this area. It also looked at the advantages of these methods over one another and how they compared amongst themselves and those of other researchers.

Geography of the study area

The aeromagnetic map of Nigeria is shown in Figure 1a. The geographical cordinates of the study area is between Latitude 7.00'N to 8.00'N and Longitude 8.50'E to 9.50'E. It is represented by sheet numbers 251, 252, 271, 272 shown by Figure 1b.

Each square block of Figure 1b represents a map on the scale of 1:100,000. Each square block is (55 × 55) km2 covering an area of 3,025 km2, hence the study area is 12,100 km2.

The geology of the study area

The study area (the hatched portion in Figure 1c) is located in the middle Benue trough sedimentary basin Nigeria. Nigeria is made up of about seven inland sedimentary basins which are shown in Figure 1c. The Benue Trough is divisible into upper, middle and lower Benue trough. It originated from the early Cretaceous rifling of the Central West African basement uplift.

As shown in Figure 1d, the study area is characterized by the presence of thick sedimentary cover of varied composition whose age ranges from Albian to Maastrichtian, with the earliest sedimentation being the marine Asu-River group of Abian age [4].

Stratigraphically, the Cretaceous sedimentary succession in the study area as shown on Figure 1d,e comprise of the Asu -River Group of marine origin which is the oldest deposited sediment in the Middle Benue Trough followed by Ezeaku Formation, keana/Awe Formation, Awgu Formation and Lafia Sandstone as the youngest sediment [5]. The works of [6-10] have more on the geology of the Benue Trough. The important of the knowledge of the geology of the study area is with regards to hydrocarbon formation, as oil are usually formed from the remains of dead marine animals. The studied geology have therefore revealed that the earliest sediment in the study area is oil accumulation favourable Asu-river group of marine origin as obtainable in oil rich Niger-Delta part of the country.

Data Source and Acquisition

The aeromagnetic map of the digitized HRAM data used for this work is shown on Figure 2a. It was purchased from the Nigerian Geological Survey agency Abuja Nigeria, from the most recent 2009 survey caried out by

Fugro Airborne service. This data is in digitized form eliminating possible errors usually introduced during digitization of the old 1970's map. It is of higher resolution than those of 1970's which are in map form being flown at 500 m line spacing and 80 m terrain clearance as against the 1970's flown at a flight line spacing of 2 km, average terrain clearance of 150 m, and a nominal tie line spacing of 20 km. The HRAM data of the four different blocks was assembled and imported into SUFER32 software and used to produced the composite anomaly map of Figure 2b.

Amongst the notable researchers who have worked in this area with the old data of 1970's include [11-13].

Reduction to the pole

The aim of reduction to pole is to produce magnetic map that would represent an area being surveyed at the magnetic pole from the observed total magnetic field map. Assuming that all the observed magnetic field of the study area is due to induced magnetic effects, pole reduction can be calculated in the frequency domain using the following operator [14].

$$L(\theta) = \frac{1}{\sin \sin(I)\cos(I)\cos(D-\theta)}$$

Where

θ *is the wave number*

I *is the magnetic inclination*

D *is the magnetic declination*

In reduction to the pole procedure, the measured total field anomaly is transformed into the vertical component of the field caused by the same source distribution magnetized in the vertical direction. The acquired anomaly is therefore the one that would be measured at the North magnetic pole, where induced magnetization and ambient field both are directed downwards [15].

The Preliminary Qualitative Analysis (Regional- Residual Separation)

For potential field data like magnetic and gravity to be interpreted, the residual anomaly must be separated from the regional background field. Usually the targets are the anomalies of the basement rock, and their magnetic field is superimposed in the regional field that comes from larger deeper sources and the magnetic effect of all the underground magnetic sources. The regional field was modeled with a first order polynomial and the residual field regarded as an error between the model and the data using Polifit program. The residual data obtained was contoured into 2-D and 3-D maps in Figure 3a,b giving us the first impression of the study area.

The major anomalies in the 2-D contour map have been numbered 1-10 and these are the magnetic signatures of the subsurface basement rocks. The anomalies are divisible into low frequency anomalies (1,3,4,7,9,10) coming from deep seated bodies and high frequency anomalies (2,5,6,8)from shallow seated bodies. Juxtaposing the 2-D and 3-D maps revealed more clearely the undulations and likely traps in the basement surface. The orientation of the magnetic contours could be observed to be predominantly in NE-SW direction with subordinate E-W direction. Markudi area is more featureless while Gboko has interesting geological features and should be the exploration targets in the studied area.

Quantitative Interpretation

The theory and the result of the slope method, spectral analysis and the SPI used have been presented here.

The slope technique (maximum and half-slope): [16] was probably the first to relate depth of magnetic source to the horizontal extent of the portion of sloping flanks of his profile curves as illustrated in Figure 4.

Usually to estimate magnetic depth from this method, the horizontal extent of the portion of the profile curve that is nearly linear at the maximum slope (S) is measured, or the distance between the two points of tangency (half slope, {P}) is measured. The depth (Z) beneath the portion of the curve is calculated using equation 1.0 and 2.0 [17-19].

$$Z = K_1 S \; ; \; 1.67 \leq K_1 \leq 1.82 \; (\text{empirical constant } K_1 \; 1.82) \tag{1.0}$$

$$Z = K_2 P \; (\text{empirical constant } K_2 = 0.63) \tag{2.0}$$

To generate curves and calculate magnetic source depth, the residual contour map of Figure 3a was profiled NNI, BBI, AAI, OOI and WWI along the most prominent anomaly zones within the study area as shown on the profile map of Figure 5a [2].

The ploted magnetic profile curves which is a replica of the theoretical curves have been presented in Figure 5b.

Employing the theoretical equations of 1.0 and 2.0, the summary of the estimated depths for the five profiles taken are shown on Table1 using the maximum and half slope techniques.

The average magnetic basement depth from these two method is 3.70 km. It is important to use both the maximum and half slope method for they provides a check on the depth estimate and the care with which the graphical analysis is done.

The spectral technique: The application of the power spectrum method to potential field data has been well establised, and the determination of the anomalous body depth has also been given [20,21]. This method has been used extensively by many researchers like [22,23].

In its complex form, the two dimensional Fourier transform pair may be written as in eqn 3.0 and 4.0 [20,24]:

$$G(u,v)=\iint_{-\infty}^{\infty} g(x,y)\, e^{i(Ux-Vy)}\,dxdy. \qquad (3.0)$$

$$G(x,y)=\frac{1}{4\pi^2}\iint_{-\infty}^{\infty} g(u,v)\, e^{i(Ux-Vy)}\,dudv. \qquad (4.0)$$

Where u and v are the angular frequencies in x and y directions respectively

For depth estimations for magnetic field data this was expressed as

$$E(u,v)=exp(-2hr)\ [21] \qquad (5.0)$$

The exp (-2hr) term is the dominant factor in the power spectrum. The average spectrum of the partial waves falling within this frequency range is calculated and the resulting values together constitute the redial spectrum of the anomalous field. If we replace h with Z and r with f; then

$$Log\ E = -2Zf. \qquad (6.0)$$

Where Z is the required anomalous depth; and f frequency. Therefore the linear graph of logE against f *leaves* slope *(m)* =-2*Z*. The mean depth (*Z*) of the magnetic source is thus given by

$$Z=m/2. \qquad (7.0)$$

Employing this technique involves

-Divisions into Spectral cells and Windowing: The four residual blocks of the study area was subdivided into 16 spectral cells of 12.5 km by 12.5 km for easy handling of the large data involved and each signal was widowed 15 minutes by 15 minutes.

-Generation of radial energy spectrum: Digital signal processing software the Oasis Montaj was used to transform the residual magnetic data into the radial energy spectrum for each block.

-Plots of Log of Energy and the frequency: [21] have shown that the Log-power spectrum of the source has a linear gradient whose magnitude is dependent upon the depth of the source. The spectral energy against frequencies for the 16 spectral cells was plotted and shown on Figure 6.

These series of points which fall on one or more straight line segments represent bodies occurring within a particular depth range. The line segment in the higher frequency range is from the shallow sources and the lower harmonics are indicative of sources from deep seated bodies.The slope of these provide a measure of the mean depth to the ensemble of the anomalous bodies.

- Estimation of the depth to magnetic sources: The gradients $m = \frac{\Delta LogE}{\Delta freq}$

Equation of each of the line segments were first evaluated. The mean depth (z) of burial of the ensemble given as $Z = -\frac{m}{2}$ Equation was then calculated [21]. The coordinates and the two depth estimates (z1 and z2) for each of the sixteen spectral blocks are given on Table 2.

Therefore the result of the spectral analysis of aeromagnetic data over the project area suggests the existence of two main source depths; the deeper sources (z1) (2.23 km- 5.66 km; and avg of 3.53 km) represented by the steeper line of figure 6 and the shallow sources (z2) (0.05 km -0.42 km, avg of 0.21 km) represented by the other line.

Source parameter imagin (SPITH)

The SPITH method based on [25] method utilizes the relationship between source depth and the local wavenumber (k) of the analytic signal of the observed field. This can be calculated for any point within a grid of data via horizontal and vertical gradients to estimated magnetic depth.

The analytic signal A1(x,z) is defined by [26] as

$$A(x,z) = \frac{\partial M(x,z)}{\partial x} - j\frac{\partial M(x,z)}{\partial z}, \tag{5.0}$$

Where M(x,z) is the magnitude of the anomalous total magnetic field, j is the imaginary number, and z and x are Cartesian coordinate for the vertical direction and the horizontal direction perpendicular to strike, respectively. The local wave number K1 and K2 is expressed as.

$$k_1 = \frac{(n_k + 1)h_k}{h_k^2 + x^2}. \tag{6.0}$$

$$k_2 = \frac{(n_k + 2)h_k}{h_k^2 + x^2} \tag{7.0}$$

Where nk is the SPI structural index (subscript k=c, t or h), and nc=1 and nh=2 for the contact, thin sheet and horizontal cylinder models and hk is the depth to the top of the contact, respectively. The result of this analysis is displayed on an image and the correct depth estimate for each anomaly can also be determined. OASIS MONTAJ software was employed to grid the data and compute the SPI image and depth. SPI method makes the task of interpreting magnetic data significantly easier as shown by the SP images generated from residual field data of the studied area on Figure 7 [3].

The ends of the peak could be traced to the depth axis (Z) and the values read. The longest spikes are blue showing deep lying bodies and the shortes spikes are orange and pink showing shallow seated bodies. The depth values obtained from this method is displayed on Table 3.

Result Discussion and Comparison

The preliminary qualitative interpretation of the aeromagnetic data of part of middle Benue trough Nigeria described above have revealed two magnetic anomaly sources; the low frequency anomaly source related to deep seated bodies and the the high frequency anomaly source related to shalow seated magnetic bodies. The areas of deep seated bodies in the map are possibly the magnetic basement depth; while the shallower are possibly magmatic/igneous intrusions into the sedimentary basins which are possibly responsible for the leadzinc mineralization found in the area. This findings is in conformity with those of previous researchers which used the 1970's old data. This qualitative interpretations of the residual map also revealed conjugate pair of NE-SW and NW-SE fracture in the Benue trough as have been predicted earlier by the works of [27] which was attributed to the event at possible opening up of African and South American plates. Several undulations and likely traps have equally been found on the basement surface which could likely form traps for hydrocarbon accumulation. The spectral analysis and Landsat Imagery of the adjacent lower Benue trough by [28] equally revealed a two layer depth model and predominant NE-SW lineament trend.

The results of the quantitative interpretation to determine the deep and shallow magnetic sources using Slope, spectral and SPI technique have been presented on Table 3.

The revelation from the spectral study of 3.65 km, Slope techniqe study of 3.70 km and SPI study of 3.25 km respectively represents the magnetic basement depth. These are synonymous to depth of over-burden sediment in the study area which has a very important significance as regards to the hydrocarbon generation potential of the area. Previous study showed the geology of the area to be associated with the marine Albian Asu River Group which commenced sedimentation in the middle Benue Trough.

A researcher [29] has pointed out in his work that if all other conditions for hydrocarbon accumulation are favourable, and the average temperature gradient of 1°C for 30 m obtainable in oil rich Niger Delta of the country is applicable, then the minimum thickness of the sediment to achieve the threshold frequency temperature of 115°C for the commencement of oil formation from marine organic remains would be 2.3 km depth. Thus on the basis of the average estimated magnetic basement depth of the study area from Table 3, and the probable marine source bed of Asu River group in the lower/ middle sector of the Nigerian Benue trough and the likely traps found in the basement surface of the study area, the area could be looked upon as favourable and very promising for hydrocarbon accumulation especially the Gboko axis. The shalow magnetic source depth from Table 3 is probably magmatic intrussion into the sediment and this could be responsible for the Lead-Zinc mineralization found in the study area. Active magmatism have been reported in the Benue trough and the work [30] have confirmed the close association between magmatism, mineralization and fractures in the Benue Trough.

These three independent techniques gave depth values that are similar to each other and comparable to those obtained by previous researchers who employed the old data.

The spectral analysis and Landsat Imagery of the adjacent lower Benue trough by [28] revealed an average depth of 3.574 km for the deeper magnetic source bodies, and average shallower magnetic source depth of 1.041 km. [11] estimated the magnetic basement depth of the middle Benue Trough to be between 2.00 km - 4.00 km, considering it to be the most prospective area within the trough. Depth to the magnetic basement from the spectral work of [13] vary from 1.513 km and 4.936 km. The gravity work of [10] over the upper Benue Trough estimated the thickness of the sediments between 0.9 km and 4.6 km.

Discusions and comparisons of the analysis techniques

Although the slope method is an old pre-computer method based on empirical constant developed out of enormous experience in magnetic interpretation acquired over the years, but it has given result that is comparable to the more modern automatic source depth method of Spectral and SPI. It's a graphical techniques which make use of the sloping flanks of profile to estimate depth. The slope techniques generally yields reasonable result for horizontal basement models with steeply dipping contact. It is much simpler and faster and provides more depth estimate than analysis by model curve fitting. This method is mostly used as a quick check on the validity of depth estimates.

A serious advantage of SPI over spectral analysis as automatic source depth techniques is that SPI is a profile or grid-based method for estimating magnetic source depths and does not require a moving window.In Spetral analysis, for small windows of data the limited number of grid nodes often leads to power spectra becoming jagged at the start or end. This is the reason for omitting the first point in the automated determination of the deepest straight-line segment of the power spectra. SPI method makes the task of interpreting magnetic data significantly easier as depth can be displayed as an image as earlier shown (Figure 7). The depth estimate can be summarized on one map independent of an assumed model which can be interpreted even by non-expert in the field, and the computation time is relatively short. However, the interpretation of the gravity and magnetic data is preferred in frequency domain because of simple relations between various source models and field [31]. The power spectrum of the surface field can be used to identify average depths of source ensembles [21]. The technique of spectral analysis provide rapid depth estimates from regularly- spaced digital field data, no geomagnetic or diurnal correction are necessary here as these remove only low wave number component and do not affect depth estimate which are controlled by high wave number component of the observed field [32]. The output of this operation is the radial spectral energy and frequency. These can be plotted in straight line graphs and further analysis like gradients calculation done before depth could be calculated from these gradients as already shown.

CONCLUSIONS

The preliminary qualitative interpretation of the anomalies in the 3-D and 2-D residual map revealed areas of high frequency anomaly sources related to deep seated bodies and areas of low frequency anomaly related to shallow seated bodies. Several undulation and likely traps is noticeable on the basement surface. Predominant NESW with surbodinate E-W contour trendings, an attribute of the Pan–African Orogeny trends was equally found. SPI, spectral and slope techiques was now used to quantitatively determine the depth of the deep and shallow seated bodies. The magnetic basement depth values from these techniques compares favourably with themselves and that of other researchers who used the old 1970's map. The study area has been confirmed to be favourable for oil accumulation if other conditions are met especially the Gboko area with good geological features and likely traps. However, these magnetic basement depth gotten using the the high resolution 2009 data is more solid than that gotten using the 1970's data.This is because of the improved flight parameters, imporovement in technology and various error correcting softwares used during the acquisition of the digitized 2009 data.

REFERENCES

1. Nwosu OB, Umego MN, Onuba LN (2013) Spectral Re-evaluation of the Magnetic basement depth over parts of Middle Benue Trough Nigeria using HRAM. International journal of scientific & technology research 2: 111-119.
2. Nwosu OB, Onuba LN (2013) Evaluation of the magnetic basement depth over parts of middle Benue trough Nigeria by empirical depth rule based on slope techniques using the HRAM. International journal of scientific & technology research 2:19-29.
3. Nwosu OB (2014) Determination of magnetic basement depth over parts of middle Benue trough Nigeria by source parameter imaging (SPI) technique using HRAM. International journal of scientific & technology research 3:262-271.
4. Obaje NG (2009) Geology and Mineral resources of Nigeria. Springer Dordrecht Heidelbug London, New York.
5. Modified from NGSA (2003) Nigerian geological survey agency.
6. Olade MA (1975) Evolution of Nigerian Benue trough (Aulocogen): A tectonic model. Geol Mag 112: 575 -583.
7. Cratchley CR, Jones GP (1965)An interpretation of Geology and Gravity anomalies of the Benue Valley Nigeia.Geophy Pap Overseas Geol Surv London.
8. Burke K, Dessauvagie TFJ, Whiteman A(l970) Geological history of the Benue Valley and Adjacent areas. In:Dessauvagie TFJ, Whiteman AJ African Geology. University of Ibadan Press, Nigeria.
9. Offodile ME (1976)The geology of the middle Benue Nigeria. Disertation, Univ of Uppsala Sweden.
10. Osazuwa IB, Ajakaiye DE,Verheijen PJT (1981) Analysis of the structure of parts of Upper Benue rift valley on the basis of new geophysical data. Earth evolution Sci 2: 126-135.

11. Nwachukwu JI (1985) Petroleum Prospects of Benue Trough, Nigeria. AmAssoc Pet GeolBull 69: 601-609.
12. Ofoegbu CO (1984) Aeromagnetic anomalies over the lower and middle Benue Trough Nigeria; Nigeria. Journal of mining and geology 21: 103-108.
13. Ahmed NM (1991) Spectral Analysis of Aeromagnetic Data over the Middle Benue Trough, Nigeria; research publication; M.Sc. thesis University of Nigeria Nsukka.
14. Grant FS, Dodds J (1972) MAGMAPFFT processing system development notes. Paterson Grant andWatson Limited.
15. Blakely RJ (1995) Potential Theory in Gravity and magnetic application. Cambridge: University Press70: 285-3003.
16. Peters LJ (1949)The direct approach to magnetic interpretation and its practical application. Geophysics 14: 290-320.
17. Nettleton LL (1971) Elementary Gravity and magnetic for Geologist and Seismologists.Society of Exploration Geophysicists Geophysical Monograph Series 1: 113-115.
18. Spector A, Parker W (1979) Computer compilation and interpretation of geophysical data, in Geophysics and geochemistry in the search for metallic ores:Geol rep 31: 527-544.
19. Telford WM, Geldart LP, Sheriff RE(1990) Applied geophysics(2ndeds). Cambridge University Press 113-114.
20. Bhattacharrya BK (1966) Continuous Spectrum of the total magnetic field anomaly due to rectangular prismatic body. Geophysics 31: 97-121.
21. Spector A, GrantFS (1970) Statistical models for interpreting aeromagnetic Data. Geophysics 35: 293-302.
22. Mishra DC, Naidu PS (1974) Two-dimentional Power spectral analysis of aeromagnetic fields. Geophysics Prop 22: 345-534.
23. Ofoegbu CO, Onuoha KM (1991) Analysis of magnetic data over the Abakaliki Anticlinorium of the Lower Benue Trough, Nigeria. Marine and PetrGeolo 8: 174-183.
24. Bath M (1974) Spectral analysis in Geophysics Elsevier Amsterdam. Bulletin of the seismological society of America 2: 633-634.
25. Thurston JB, Smith RS(1997) Automatic conversion of magnetic data to depth, dip, and susceptibility contrast using the SPITM method. Geophysics 62: 807-813.
26. Nabighian MN (1972)The analytical signal of two- dimensional magnetic bodies with polygonal cross-section: its properties and use for automated anomaly interpretation. Geophysics 37: 507-517.
27. Ajakaiye DE (1981) Geophysical investigation of Benue Trough; A review. Earth Earth Sci 1: 126-136.
28. Onyewuchi RA, Opara AI, AhiarakwemCA, Oko FU (2012) Geological Interpretations Inferred From Airborne Magnetic and Landsat Data: Case Study of Nkalagu Area, Southeastern Nigeria. Int J SciTechnol 2:178-191.
29. Wright JB, Hastings D, JonesWB,Williams HR (1985) Geology and Mineral resources of West Africa. George Allen and Urwin, London.
30. Ofoegbu CO,Odigi M (1989) Basement structures and ore mineralization in the Benue Trough, In the Benue Trough: Structure and Evolution. Earth EvolSci 239-248.
31. Keary P, Brooks M (1984)An introduction to Explorational Geophysics. Blackwell Scientific publications.
32. Naidu PS, Mathew MP (1998) Analysis of geophysical potential fields, A digital signal processing approach. Elsevier Science publishers, Amset erdam.

CHAPTER 8

Spatial Monitoring of Groundwater Drawdown and Rebound Associated with Quarry Dewatering Using Automated Time-Lapse Electrical Resistivity Tomography and Distribution Guided Clustering

J.E. Chambers [a], P.I. Meldrum [a], P.B. Wilkinson [a], W. Ward [b,c]. Jackson [a], B. Matthews [c], P. Joel [c], O. Kuras [a], L. Bai [b] , S. Uhlemann [a,d] , D. Gunn [a]

[a] *Geophysical Tomography Team, British Geological Survey, Nottingham NG12 5GG, UK*

[b] *University of Nottingham, School of Computer Science & Information Technology, Nottingham NG8 1BB, UK*

[c] *Lafarge Tarmac, Stancombe Lane, Flax Bourton, Bristol BS48 3QD, UK*

[d] *ETH-Swiss Federal Institute of Technology, Institute of Geophysics, Sonneggstr. 5, 8092 Zurich, Switzerland*

ABSTRACT

Dewatering systems used for mining and quarrying operations often result in highly artificial and complex groundwater conditions, which can be difficult to characterise and monitor using borehole point sampling approaches. Here automated time-lapse electrical resistivity tomography (ALERT) is considered as a means of monitoring subsurface groundwater dynamics associated with changes in the dewatering regime in an operational sand and gravel quarry. We considered two scenarios: the first was unplanned interruption to dewatering due to a pump failure for a period of several days, which involved comparing ALERT monitoring results before and after groundwater rebound; the second involved a planned interruption to pumping over a period of 6 h, for which near-continuous ALERT monitoring of groundwater rebound and drawdown was undertaken. The results of the second test were analysed using distribution guided clustering (DGC) to provide a more quantitative and objective assessment of changes in the subsurface over time.

ALERT successfully identified groundwater level changes during both monitoring scenarios. It provided a more useful indication of the rate of water level rise and maximum water levels than piezometer monitoring results. This was due to the piezometers rapidly responding to pressure changes at depth, whilst ALERT/DGC provided information of slower changes associated with the storage and delayed drainage of water within the sediment. By applying DGC we were able to automatically and quantitatively define changes in the resistivity sections, which correlated well with the direct observations of groundwater at site. For ERT monitoring applications that generate numerous time series, the use of DGC could significantly enhance the efficiency of data interpretation, and provide a means of automating groundwater monitoring through assigning alarm thresholds associated with rapid changes in groundwater conditions.

Keywords: Electrical resistivity tomography (ERT) Mining Dewatering Groundwater monitoring Clustering Image analysis

1. INTRODUCTION

Quarry dewatering requires careful groundwater management, using extraction boreholes, ponds, trenches and well-point systems (Wardrop et al., 2001). This can result in highly artificial and dynamic groundwater conditions that can present challenges both for quarry management and the protection of groundwater resources outside of the mineral extraction operations (Brunetti et al., 2013; French, 2009; Mayes et al., 2005). Monitoring approaches typically rely on manual or automated measurements of groundwater levels in boreholes or surface waters. However, borehole monitoring networks provide sparsely distributed point information, which in complex settings can be insufficient to understand the groundwater system.

Geophysical monitoring approaches have the potential to complement conventional point monitoring by providing spatial information relating to changing groundwater conditions (e.g., Binley et al., 2002; Kuras et al., 2009). Electrical resistivity tomography (ERT) is of particular relevance due to the close relationship between resistivity and pore saturation and pore fluid quality (e.g., Binley et al., 2002; Brunet et al., 2010). This coupled with recent advances in monitoring instrumentation (e.g., Ogilvy et al., 2009; Wilkinson et al., 2010; Chambers et al., 2014a; Supper et al., 2014) and time-lapse inversion techniques (e.g., Hayley et al., 2011; Karaoulis et al., 2011; Kim et al., 2009; Loke et al., 2014) have facilitated automated remote monitoring activities relating to a range of hydrogeophysical applications (Loke et al., 2013). However, to the best of our knowledge these automated approaches have not yet been developed for application to quarry dewatering activities. An additional challenge for the application of automated ERT monitoring is the very large or numerous time series data sets generated by these systems, which are difficult to process manually. Automated image analysis using computer vision techniques such as gradient based edge detectors (e.g., Chambers et al., 2012, 2013; Elwaseif and Slater, 2012; Hsu et al., 2010) and clustering techniques (e.g., Audebert et al., 2014; Chambers et al., 2014b; Ward et al., 2014) is beginning to be applied to ERT images. For time-series data these techniques have the potential to provide an automated means of interrogating monitoring results.

In this study, we aim to assess a combination of automated time-lapse electrical resistivity tomography (ALERT) and distribution guided clustering (DGC) as a means of monitoring groundwater dynamics in an active quarry with well-point dewatering. Two monitoring scenarios are considered; the first concerns an extended unplanned interruption to quarry dewatering monitored using ALERT, and the second is a planned interruption to pumping, which was monitored at a higher temporal resolution using a combination of ALERT and DGC to provide detailed information on groundwater rebound and subsequent drawdown over a much shorter time period. Resistivity monitoring results are compared to direct observations of groundwater levels within the monitoring boreholes, and the potential benefits of image analysis for the automated interpretation of resistivity images are considered.

2. SITE DESCRIPTION

Rock Common Quarry (Fig. 1) is located approximately 11 km to the north of Worthing, West Sussex (British National Grid reference 512500, 113500). Quarrying activity has been concentrated in the Folkestone Formation of the Lower Greensand Group (Cretaceous), which dips to the south beneath the clays of the Gault Formation (Cretaceous). The Folkestone Formation comprises fine to medium, and occasionally coarse, cross-bedded sand. This formation is underlain by the Sandgate and Hythe Beds, also of the Lower Greensand Group, which in the area of Rock Common Quarry include a fine grained upper unit called the Marehill Clay.

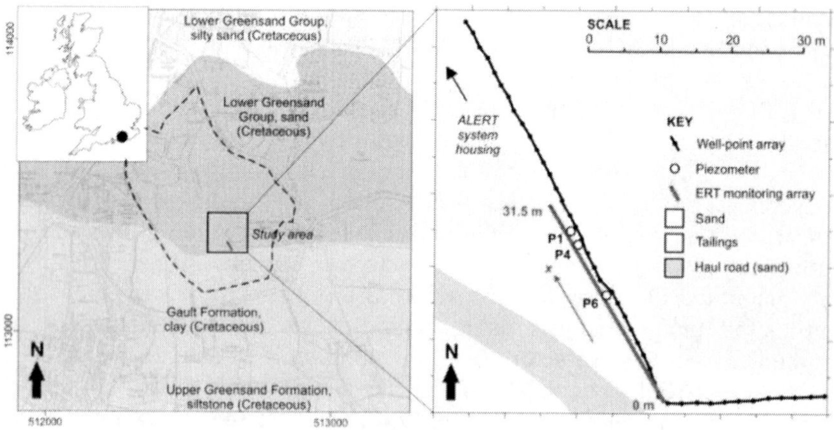

Fig. 1. Setting — geological map with Ordnance Survey base layer (© Crown copyright and database rights 2015 Ordnance Survey 100021290), showing quarry boundary (dashed black line) and the study site location, with site location at national scale inset (left). Schematic field layout showing monitoring array, piezometers and well-point system (right).

The quarry has been worked since the 1920s, and at the time of this study extended to a depth of approximately 40 m at its deepest point. In recent years dewatering of the quarry has been carried out using well-point arrays (e.g., Shaqour and Hasan, 2008; Chiocchini and Castaldi, 2011) connected to a network of pumps installed across the base of the quarry. This has been supplemented with extraction from ponds or sumps at times of higher rainfall. The well-point system comprises linked arrays of shallow boreholes from which groundwater is abstracted using a network of pumps. For this quarry the well-point extraction boreholes extended to 5 m below ground level and were positioned at approximately 2 m intervals (Figs. 1

and 2). Continuous pumping maintained the water level between approximately 0.5 to 2 m below the lowest point of the quarry floor during normal conditions, and approximately 20 m below the naturally occurring water table observed in the land surrounding the quarry. The hydrological regime was extremely dynamic, with flash flooding across the quarry floor due to heavy rainfall events particularly during winter months. Likewise, groundwater levels have been observed to be very sensitive to changes in the pumping regime; e.g., when pumping is reduced groundwater rebounds of tens of centimetres occur within a few hours.

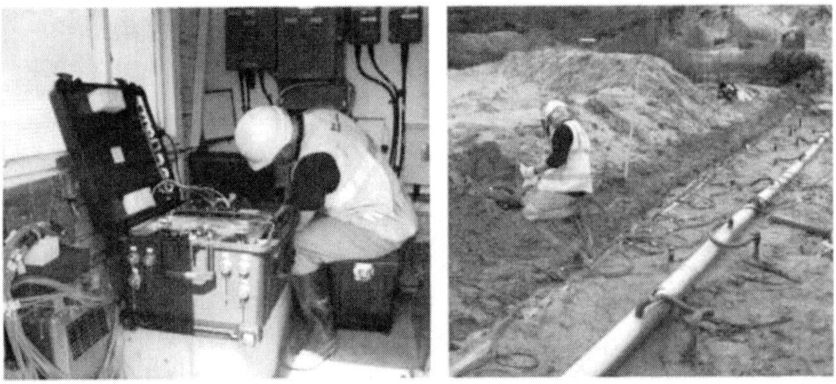

Fig. 2. ALERT instrument installation (left). Electrode monitoring array installation (right) — the well-point groundwater extraction system can be seen running parallel to the electrode array.

The monitoring array was installed within the deepest part of the quarry (Fig. 1). This location was chosen for a number of reasons: (1) it was in close proximity to the pump control cabin that was used to house the ALERT system; (2) it was within an area of the quarry that was not going to be significantly impacted or disturbed by quarrying operations for the duration of the study; (3) it was directly adjacent to one of the active well-point groundwater extraction array, thereby providing the opportunity to observe groundwater drawdown and rebound directly associated with small changes to the pumping regime. Routine monitoring at the site was undertaken at the site over a period of 12 months, during which ERT measurement sets were typically collected at a frequency of one per day, though periods of higher frequency monitoring were also undertaken to monitor the active interruption to pumping.

3. METHODOLOGY

3.1. Monitoring scenarios

Two contrasting monitoring scenarios were considered in this study, the first of which involved changes between equilibrium conditions at high and low water levels monitored at a low temporal resolution, and the second concerned dynamic, continually changing conditions monitored at higher temporal resolution.

Scenario 1 involved an unplanned interruption to groundwater extraction resulting from the failure of one of the quarry pumps for a period of 5 days during April 2010. Water extraction continued at a reduced rate for this period, during which groundwater levels rapidly reached a new equilibrium state at a higher elevation within the subsurface. ALERT results are compared before and after pump failure.

Scenario 2 concerned a planned temporary pump switch off for a period of ~ 6 h during May 2010. The purpose of this test was to monitor the rebound and subsequent drawdown of groundwater levels using the ALERT system operating at a much greater temporal resolution, with near-continuous monitoring being undertaken immediately before, during and after pump switch-off.

3.2. Automated time-lapse electrical resistivity tomography (ALERT)

3.2.1. Instrumentation

The resistivity distribution in the subsurface was monitored using an ALERT system (Fig. 2), which is described in detail by Ogilvy et al. (2009) and Wilkinson et al. (2010). In brief, the system uses wireless telemetry (a wireless cellular router in this case) to communicate with a database management and control system located on a remote server (Fig. 3). This system controls the scheduling, storage, inversion and delivery of the data and resulting tomographic images. Once installed, no manual intervention is required; data is transmitted automatically according to a pre-programmed schedule and specific survey parameters, both of which can be modified remotely as conditions change. The system is powered by 12 V batteries, which were charged at this site using mains power. It is a 200 W direct current (DC) resistivity system and supports 10-channel simultaneous potential difference measurements

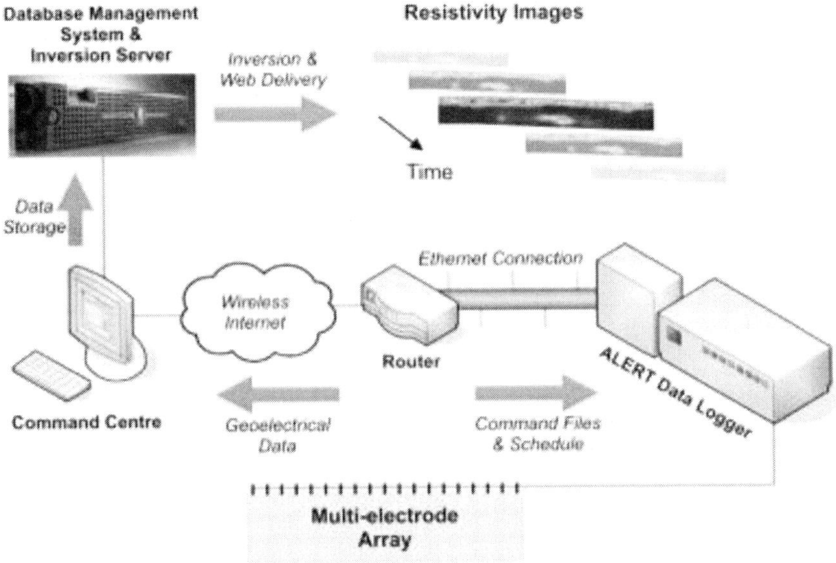

Fig. 3. ALERT system concept showing both site (multi-electrode array, data logger, router) and office based components (command centre and server). Wireless telemetry is used to schedule measurements at the remote field site, which are automatically undertaken and transferred back to the command centre. Data retrieved from the field site are stored on the server and processed to generate resistivity image time-series.

The resistivity monitoring array comprised 64 electrodes positioned at 0.5 m intervals. The array was constructed using stainless steel rod electrodes, which were connected to the ALERT instrument by multi-stranded cable arrays. The array was positioned within a shallow trench, no more than 15 cm deep, which was backfilled after installation (Fig. 2). Electrodes were bedded with a small quantity of clay (kaolin), which was intended to reduce contact resistances and maintain a good galvanic contact between the ground and the electrodes.

3.2.2. Data collection, processing and inversion

A 64-electrode dipole–dipole array (normal and reciprocal) with dipole sizes (a) of 0.5, 1, 1.5, 2, 2.5, 3, 3.5, and 4 m, and dipole separations (na) with $n = 1$ to 8 was used for monitoring the unplanned interruption to pumping (Scenario 1).

For the planned interruption (Scenario 2) a subset of the array was used comprising 48-electrodes ($x = 8$ to 31.5 m) and a dipole–dipole array (normal and reciprocal) with dipole sizes (*a*) of 0.5, 1, 1.5, 2, and 2.5 m, and dipole separations (*na*), n = 1 to 10 was used. By reducing the number of electrodes and dipole sizes measurement time was reduced (i.e., from 160 min to 50 min per set) thereby improving the temporal resolution and reducing motion blur (Rucker, 2014) during a period of rapidly changing groundwater levels associated with the pumping experiment.

Time-lapse inversion of the ERT time-series data was undertaken using an L_1-norm time constraint (Kim et al., 2009; Karaoulis et al., 2011). A robust constraint (L_1-norm) was imposed on the resistivity model, and on the data discrepancy to minimise the effects of outlying data. Prior to inversion data with a reciprocal error of > 5% were removed (Wilkinson et al., 2012). The agreement between the modelled and observed data was generally very good, with mean absolute differences converging to < 1.3% for Scenario 1 and < 0.9% for Scenario 2.

3.3. Automated image analysis

For this study DGC was applied to define resistivity changes associated with water level rise. It has previously been successfully applied to resistivity image analysis (e.g., Chambers et al., 2014b), and is adapted here for use with time-lapse data to highlight subtle change over time. In this case clustering is applied to the log resistivity ratio ($\log(\rho_i/\rho_0)$) rather than the log resistivity ($\log(\rho)$). The method for DGC is extended from that described in Ward et al. (2014) for clustering volumes of differing resistivity in temporally stationary ERT models. This approach is based on the probability density of the data on a log scale.

To define regions, and assign data, some measure of similarity must first be declared. Typically, for so-called centroid-based clustering methods, data is assigned to a cluster based on some distance measured between each datum and the geometric mid-point (i.e., centroid) of the cluster. Each datum is assigned to a cluster based on minimising the distance to its centroid so that each group is made up of the close neighbourhoods in the dataset.

It is common to first define the number of clusters based on some user observation and input, then centroids for each cluster are generated in some manner, e.g., random assignment. For the *k*-means method, some *k* defined centroids are used to initialise the data into groups. From these groups, new geometric centres are found by means of a weighted average. The new set of centroids is used to re-sort the data and this process of updating and reassignment is repeated until some convergence criteria are met.

To avoid this approach of random initialisation, and remove the necessity of an iterative scheme for solving the clustering problem, it is possible to incorporate known information about the data. One such approach is to approximate the probability distribution function of the data. This can be used to identify distinct modal regions representing regions with similar values within the data. Modelling these regions by fitting normal distributions gives statistical information about likely clusters, which can be used directly in the assignment (Fig. 4). The number of modelled distributions also corresponds to the number of clusters.

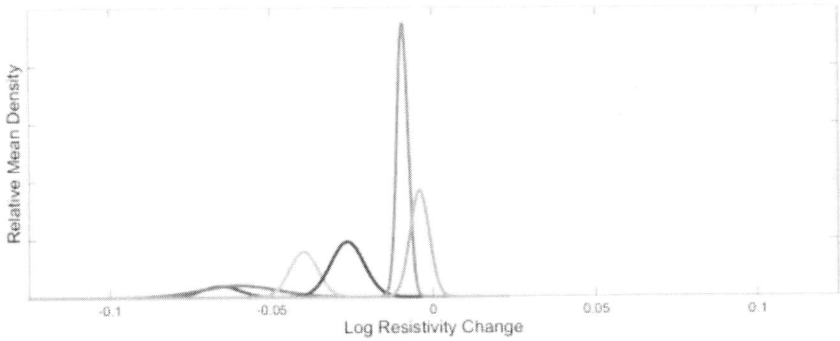

Fig. 4. Normal distributions fitted to the mean density function of Rock Common quarry log resistivity ratio model, created by summing the density function at each time slice and dividing by the total number. Each independent distribution forms the distance and centroid information for a cluster.

For each cluster, a function can be defined to assign some measure of "membership" of each datum in the model:

$$u_i(r) = \left[\sum_{j=1}^{k} \left(\frac{d_i(r)}{d_j(r)}\right)^2\right]^{-1}$$

$$d_i(r) = |r - \bar{r}_i|(1 - G_i(r)). \tag{1}$$

Subscripts i and j correspond to the index of a cluster and its distribution information $1 \leq i, j \leq k$. G_i refers to the fitted normal for that ith cluster/modal region centred on mode $\bar{r}_i$. σ^2 is the variance calculated for the mode.

$$G_i(r) = \exp\left(-\frac{(r-\bar{r}_i)^2}{2\sigma_i^2}\right) \tag{2}$$

To cluster the time-lapse model, which is made up of a collection of ERT models (in this case 2D slices), the clustering was applied to the collection as a whole. The assignment was applied to each point, independent of spatial and temporal information, and resulting cluster indices mapped back to their corresponding positions and times.

3.4. Direct groundwater monitoring

An array of piezometers was installed to provide calibration and validation for the ALERT monitoring data. Three piezometers were used during this study, which are referred to as P1, P4, and P6, extending to 1.61 m, 1.42 m, and 1.86 m below ground level (bgl) respectively (Fig. 1). The holes were fully screened, with the lower 1 m comprising slotted well screen. Each piezometer was instrumented with a groundwater level logger (Solinst Level Logger Gold M5), which was set to record a measurement at 5 minute intervals. The sampling interval was reduced to 60 s during the planned interruption to pumping.

4. RESULTS

4.1. Scenario 1: unplanned interruption

The failure of the pump during the unplanned interruption resulted in a significant rise in groundwater levels. The levels recorded in piezometers P1 and P6 (Fig. 5) rose from 0.9 m and 1.5 m bgl to 0.3 m and 0.2 m above ground level respectively during the period of reduced pumping. Although water levels in the piezometers rose above ground level the quarry floor was not flooded during the event. This is because these levels are indicative of the change in the pressure head over the screened section of the piezometers, and not of the movement of the phreatic surface. This response has also been observed in a number of nearby open well-points.

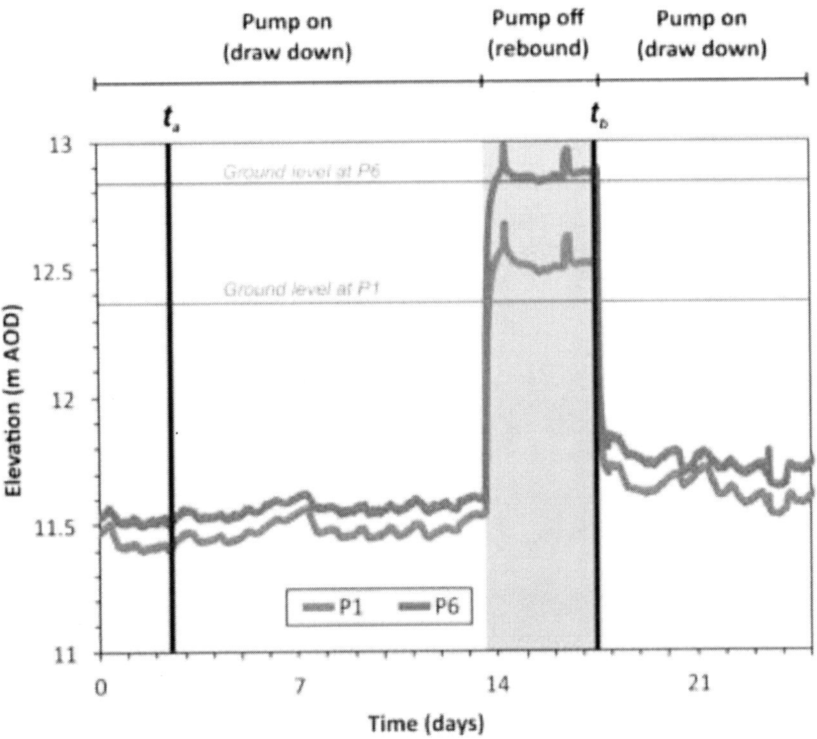

Fig. 5. Groundwater level records for piezometers 1 and 6 for the period of the unplanned interruption to pumping (Scenario 1). Shaded region indicates period of pump failure.

The changing moisture content associated with the rise in groundwater level was clearly seen in the ALERT data. The monitoring 'snapshot' in Fig. 6, shows the changes in resistivity between the initial (t_a) and reduced (t_b) pumping regime. Although changes are clearly seen in the resistivity sections, their extent and magnitude are most clearly defined in the log resistivity ratio plot, which removes the geological heterogeneity that is included in the resistivity sections. The dark blue linear feature in the ratio plot, representing a decrease in resistivity, shows the area of the section that went from unsaturated to saturated. The bottom of this feature therefore defines the pre-interruption water level, whereas the top defines the raised water level. The lower level in particular shows significant variability that may be related to differences in the performance of the respective well-points (i.e., due to silting up or local variation in the permeability of the formation). The lower level defined in the resistivity images corresponds closely to the level indicated by the water level loggers in P1 and P6, i.e., ~ 0.9 and 1.2 m bgl respectively. There is, however, a slight discrepancy between the upper groundwater level defined by the logger and the ERT section. The water level indicated by ALERT in this case closely matches the true water level, which was visually observed at or just below the ground surface.

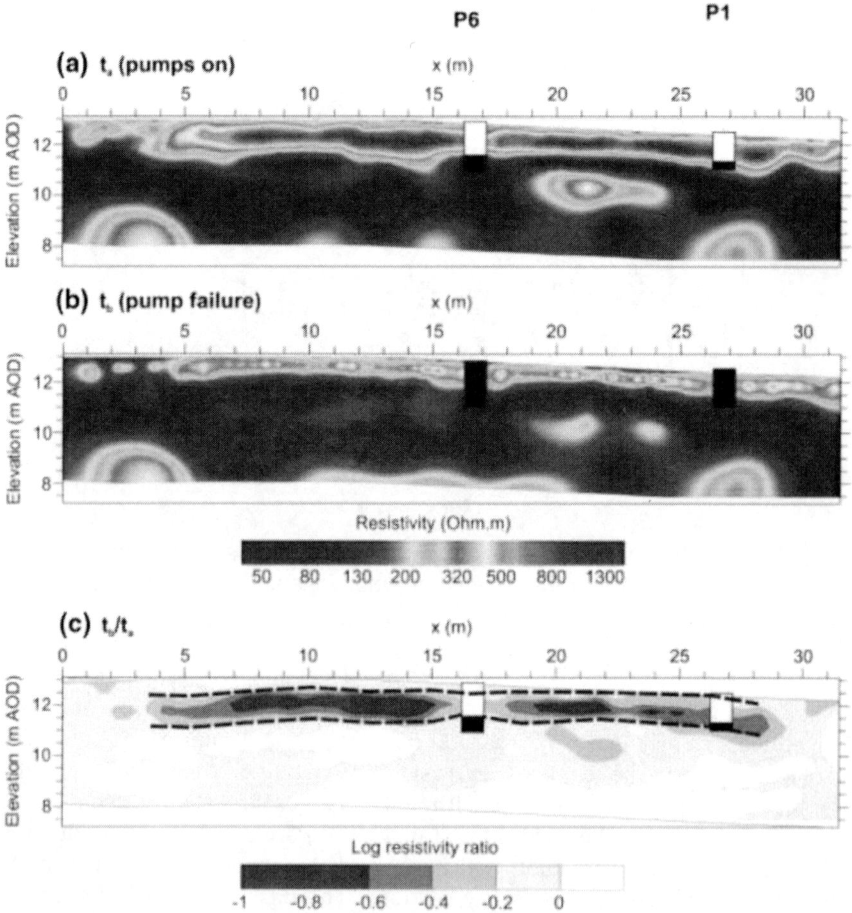

Fig. 6. ALERT monitoring results for the unplanned interruption to pumping (Scenario 1) at (a) $t_a$and (b) t_b (see Fig. 5), and (c) log resistivity ratio plot (t_b/t_a) showing subsurface change. Water levels shown for peizometers P1 and P6. Dashed lines show the minimum and maximum water levels estimated from the log resistivity ratio section.

The feature defining the rise in water level is concentrated between $x = 6$ and 28 m. At each end of the section this feature is either poorly defined or absent. There are a number of potential reasons for this observation. The first is that diminishing image resolution towards the end of the lines has prevented water table change being detected in these areas. Secondly, it may be due to limited resistivity change towards the ends of the section related to either an elevated water table or compositional differences associated with the presence of clay minerals. Small scale variations in water level are indicated throughout the imaging sections (e.g., at $x = 17$ m), and could be related to the variable performance of individual well points or due to proximity to the southern limit of the well point array (i.e., $x = 0$ m) where groundwater head gradients are likely to be greater — however, this does not explain the lack of change towards $x = 31.5$ m. Compositional differences towards the end of the section are probably more relevant. A greater concentration of clay minerals (e.g., due to mixing or local peaks in the elevation of the top surface of the Marehill Clay) would reduce the magnitude of resistivity changes due to greater water retention in the fine grained material and electrical conduction on the clay mineral surfaces. The resistivity sections indicate that the Marehill Clay is very close to the surface in this area of the quarry floor, as evidenced by low resistivity material (i.e., clay) underlying a thin (~ 0–2 m) higher resistivity surface layer (sand) observed in fully saturated conditions (Fig. 6, t_b), and is particular close to the surface towards the ends of the sections.

4.2. Scenario 2: planned interruption to pumping

During the planned pump switch-off ground water levels rebounded by up to ~ 0.9 m as recorded by the groundwater level loggers (Fig. 7). Water level change was also indicated by ALERT monitoring, shown in Fig. 8 as time-lapse resistivity and log resistivity ratio sections, and in Fig. 9 as plots of the log resistivity ratio principal cluster distributions.

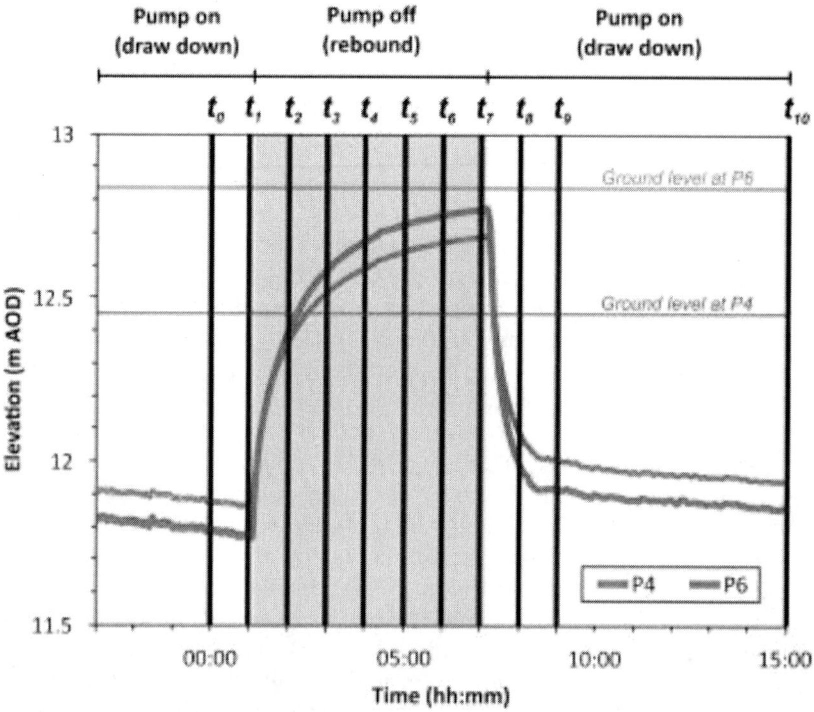

Fig. 7. Groundwater level records for piezometers 4 and 6 for the period of the planned interruption to pumping (Scenario 2). Shaded region indicates period during which pumping was reduced.

The mean density function calculated for the ratio image was modelled with distributions for 6 distinct populations, as shown in Fig. 4. Note that major regions in the relative density were located about 0 log ratio change, which correspond to the regions of little change within the image. The regions with largest negative changes represent the areas where there was a large drop in resistivity from the baseline — the corresponding cluster is then most likely to identify the area through which the water table rises.

Using the fitted distributions for the distance function described in (2), data within the entire ratio image can be given membership values for each cluster. Each datum is assigned the cluster index for which u_i is maximised. From the resulting clustering, by taking the boundary of some major cluster that is non-background and represents the greatest magnitude change, a contour can be created that highlights significant resistivity change from the baseline. The results of this contouring, taken around the cluster representing the maximum absolute change is shown Fig. 9. The threshold range captured by the major cluster is log resistivity ratio [− 0.546,

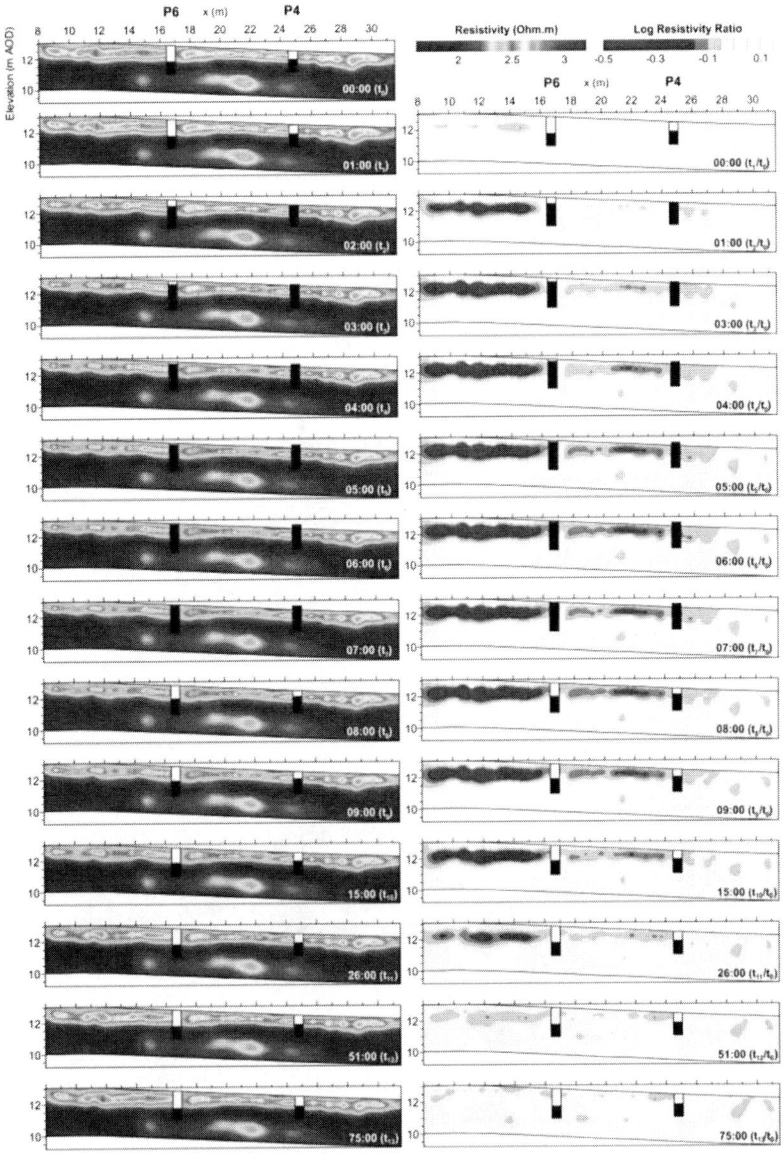

Fig. 8. Left column — resistivity monitoring results for the planned interruption to pumping (Scenario 2) between t_0 and t_{13} (see Fig. 7). Right column — log resistivity ratio plots showing change from t_1 and t_{13} relative to base line conditions, t_0. Time is given as hours and minutes (hh:mm). Water levels shown for peizometers P4 and P6.

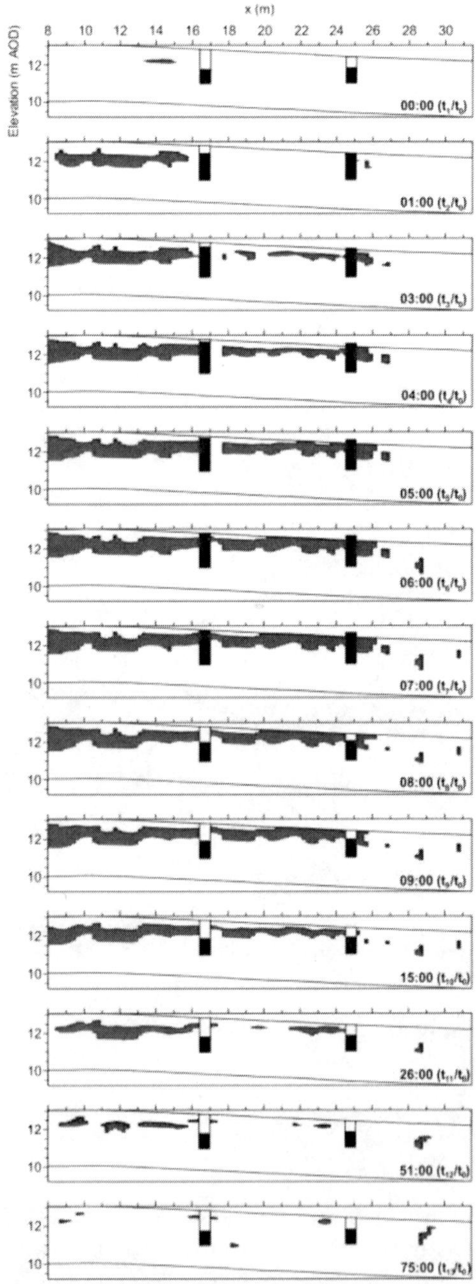

Fig. 9. Contour plots of major change cluster from t_0 to t_{13}, with water level records for piezometers 4 and 6.

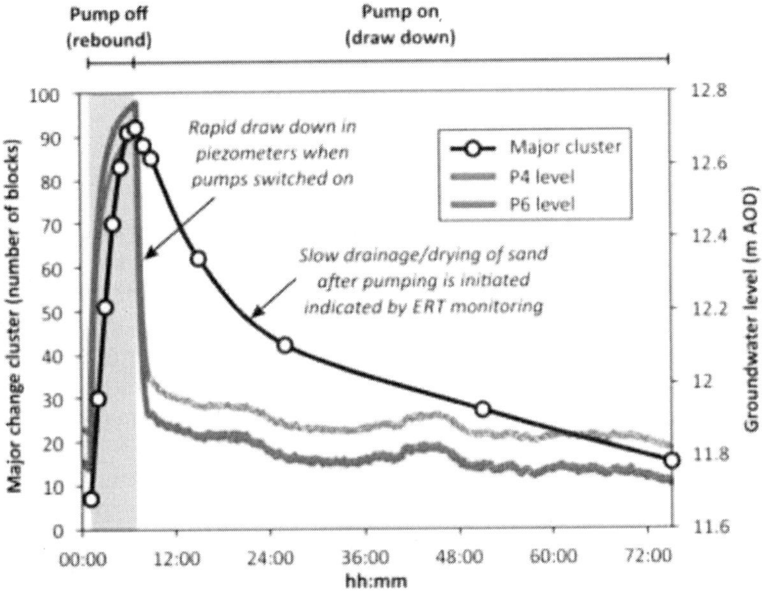

Fig. 10. Major change cluster area verses water level records from piezometers 4 and 6 during the planned interruption to pumping (Scenario 2). Shaded region indicates period during which pumping was reduced.

– 0.064]. The volume of the principal cluster at each time slice is plotted in Fig. 10 along with the piezometer monitoring records of P4 and P6, showing the increase in the volume of the saturated zone that gradually decreases back to baseline level after the pump was switched back on.

Although ground moisture changes observed with ERT are in broad agreement with the piezometer data, level changes indicated by the resistivity monitoring appear to lag those recorded in the piezometers — e.g., see Figs. 8 and 9, t_2 (rise lag) and t_8(fall lag). The lag between ERT derived change and piezometer results is also shown in Fig. 10, where the major cluster area change shows a significant lag particularly during the drawdown phase. Moreover, the artesian levels observed in the boreholes were not observed across the ground surface — confirming that water level rise in the piezometers proceeded more rapidly than the saturation of the surrounding material. Minor water seepages at the ground surface were observed in the vicinity of P1 and P4 from t_4, providing direct confirmation of water level, which agreed very well with ERT and DGC derived groundwater levels. The greatest rebound in groundwater level is seen towards $x = 8$ m, which marks the southern limit of the well-point system and is the closest part of the array to the quarry edge and, hence, the steepest hydraulic gradients.

4.3. Discussion

4.3.1. Comparison of scenarios 1 and 2

Comparison of the log resistivity ratio sections shows that the magnitude of the resistivity changes associated with groundwater level change was greater for Scenario 1 (Fig. 6) than for Scenario 2 (Fig. 8). Establishment of equilibrium conditions over a period of several days during pump failure may have provided more time for the sediment in Scenario 1 to become saturated as air entrained in the pore space was lost. In addition, near equilibrium conditions established in Scenario 1 eliminated motion blur effects (e.g., Rucker, 2014), whereas in Scenario 2 significant level changes occurred during the time it took to collect a measurement set, which would inevitably result in temporal and spatial blurring of the image. A consequence of this would be smoothing and potential underestimation of the peak resistivity changes during water level rise.

Temperature changes are not thought to have significantly affected the resistivity results in either scenario. Temperature records from the water level data loggers recorded changes of less than 1 °C for each test period, which would have a negligible impact on resistivity (Hayley et al., 2007). Moreover, given the brevity of the tests (i.e., 16 days for Scenario 1 and 4 days for Scenario 2) seasonal temperature variations will have had little influence. Similarly, diurnal temperature changes will have only effected the top few centimetres of the subsurface (e.g., Chambers et al., 2014a) with little influence on ALERT results.

4.3.2. Direct versus non-invasive monitoring

Three principal differences can be observed between the direct (water level logger) and non-invasive (ALERT) monitoring approaches: (1) spatial and temporal resolution; (2) water level response times to changes in pumping; (3) maximum water level elevation.

The temporal resolution (or sampling frequency) that can be achieved using water level loggers is very much greater than can be achieved for resistivity monitoring (i.e., seconds verses tens of minutes). Likewise, the vertical resolution of the in-hole loggers is considerably greater (i.e., sub-centimetric verses decimetric). The ALERT results provide only an approximate indication of the rise and fall of water levels. In particular, water level rise at t_2 in Scenario 2 is observed towards $x = 8$ m (Fig. 8), with little indication of rise towards $x = 32$ m despite borehole water levels having increased at both ends. It is likely that the rise towards $x = 32$ m was too small to be resolved. However, at the following time step (t_3) the water level rise has extended across the right hand side of the section towards $x = 32$ m as the newly saturated layer became thick enough to be resolved. A key advantage of ERT monitoring is that greater horizontal resolution can be achieved, with the results in this case indicating significant variability in degree of saturation and magnitude of rebound across the section. Further benefits are likely to be realised for larger scale deployments with sparser borehole distributions.

The piezometers responded more rapidly to changes in the pumping regime than the ALERT monitoring results (Fig. 8). This was observed during the rebound phase (pumping off), but was particularly marked during the drawdown phase (pumping on) (Fig. 10). The piezometers react quickly to the onset and cessation of extraction because they respond predominantly to the change in the *pressure head* in the system, which propagates rapidly through the saturated aquifer; the speed of this response is controlled by the hydraulic conductivity of the sediments, and by the compressibility of the matrix and of water, termed *specific storage*. By contrast, the changes in the resistivity images are caused by the changes in the degree of saturation of the pore space of the unsaturated zone above the phreatic surface. This is a slower process because the storage associated with wetting up or drainage of pores, the *specific yield*, is typically a number of orders of magnitude greater than that of specific storage. In this case the ALERT and DGC results have provided a more representative record of true subsurface moisture changes compared to piezometer records. The piezometers respond very rapidly to changes in pressure head, but over the periods of the extraction borehole pumping/recovery tests, they are not indicative of wetting up/drying of the unsaturated zone and the associated movement of the water table.

The maximum water levels observed in the boreholes exceeded the maximum levels observed by ALERT monitoring in both Scenarios 1 and 2. Direct visual observations of the ground surface confirmed that maximum water levels were at or just below ground level, whereas the piezometers displayed artesian levels (Figs. 5 and 7). The ALERT/DGC results were consistent with these observations (Figs. 6, 8 and 9), and therefore provided a more useful indication of subsurface moisture conditions than the piezometers. The discrepancy between piezometer and ALERT records for Scenario 2 can be explained by the lag between the pressure response observed in the boreholes compared to the slower changes in sediment saturation detected in the resistivity images. However, for Scenario 1, where new steady state conditions were established after pump failure, the piezometric head observed in the boreholes did not equilibrate with the water levels in the sediment (this is supported from visual observations on the quarry floor and the ALERT/DGC monitoring results) — even after a period of several days (Figs. 5 and 6). This may be due to semi-confined behaviour of the shallow aquifer due to lower permeability materials near the surface. In the vicinity of the arrays cemented sands were observed at the ground surface along with clayey materials washed in from previous flood events, which would have had a lower permeability than the underlying unconsolidated sands. It should also be noted that groundwater conditions in the quarry were highly artificial, with significant topographic variability, and continued pumping in other areas of the quarry during Scenarios 1 and 2. This included groundwater extraction from a surface pond ~ 40 m to the northwest of the ALERT electrode array, which could have also significantly influenced subsurface flow conditions.

4.3.3. Automated interpretation of ERT monitoring

Distribution guided clustering provides a means of automatically detecting and quantifying spatial changes within ERT images. In this case DGC successfully identified the zone in which water levels rose (validated with reference to initial levels observed in the boreholes and observations of maximum water levels at the ground surface). Manual intervention was required at the outset to select the principal cluster that corresponded best to the observed level changes — however, after selection of the cluster, image analysis can be automatically applied across the whole time series. For ongoing monitoring applications it is anticipated that alarm thresholds linked to change areas automatically detected by DGC could be used to generate alarms indicating significant changes in groundwater conditions.

5. CONCLUSIONS

Groundwater monitoring using ALERT has revealed significant spatial changes in subsurface sediments related to changes in the dewatering regime in an active quarry. In particular, resistivity change images and DGC results have been useful for identifying pre-rebound water levels, which correlated well with levels in observation wells, and peak groundwater levels, which correlated well with observations at the ground surface (i.e., seepages and very shallow excavations). A significant lag was observed between piezometer and ALERT derived groundwater responses during rebound and drawdown, which we have attributed to the slower wetting up or drying out of unsaturated zone, and associated movement of the water table, compared to the more rapid pressure head change response observed in the piezometers. Crucially, for peak water levels the ALERT monitoring results provided a more reliable means of estimating water level than the piezometers — this was probably due to semi confined aquifer conditions affecting water levels in the boreholes. DGC provided a means of quantitatively assessing changes in the log resistivity ratio sections; this approach, which requires no manual intervention once the principal change cluster has been identified, will allow the automated interrogation of resistivity monitoring data for long term monitoring operations.

ACKNOWLEDGEMENTS

The research was funded by Defra through the MIST Programme (grant MA/7/G/1/007), and in-kind contributions from project partners. The authors wish to thank the project partners for their contributions to the project. This paper is published with the permission of the Executive Director of the British Geological Survey (NERC).

REFERENCES

1. Audebert et al., 2014 M. Audebert, R. Clément, N. Touze-Foltz, T. Günther, S. Moreaua, C. Duquennoi**Time-lapse ERT interpretation methodology for leachate injection monitoring based on multiple inversions and a clustering strategy (MICS)**J. Appl. Geophys., 111 (2014), pp. 320-333
2. Binley et al., 2002 A. Binley, G. Cassiani, R. Middleton, P. Winship**Vadose zone flow model parameterisation using cross-borehole radar and resistivity imaging**J. Hydrol., 267 (3-4) (2002), pp. 147-159
3. Brunet et al., 2010 P. Brunet, R. Clement, C. Bouvier**Monitoring soil water content and deficit using Electrical Resistivity Tomography (ERT) — a case study in the Cevennes area, France**J. Hydrol., 380 (2010), pp. 146-153
4. Brunetti et al., 2013 E. Brunetti, J.P. Jones, M. Petitta, D.L. Rudolph**Assessing the impact of large-scale dewatering on fault-controlled aquifer systems: a case study in the Acque Albule basin (Tivoli, central Italy)**Hydrogeol. J., 21 (2013), pp. 401-423
5. Chambers et al., 2012 J.E. Chambers, P.B. Wilkinson, D. Wardrop, A. Hameed, I. Hill, C. Jeffrey, M.H. Loke, P.I. Meldrum, O. Kuras, M. Cave, D.A. Gunn**Bedrock detection beneath river terrace deposits using three-dimensional electrical resistivity tomography**Geomorphology, 177 (2012), pp. 17-25
6. Chambers et al., 2013 J.E. Chambers, P.B. Wilkinson, S. Penn, P.I. Meldrum, O. Kuras, M.H. Loke, D.A. Gunn**River terrace sand and gravel deposit reserve estimation using three-dimensional electrical resistivity tomography for bedrock surface detection**J. Appl. Geophys., 93 (2013), pp. 25-32
7. Chambers et al., 2014a J.E. Chambers, D.A. Gunn, P.B. Wilkinson, P.I. Meldrum, E. Haslam, S. Holyoake, M. Kirkham, O. Kuras, A. Merritt, J. Wragg**4D electrical resistivity tomography monitoring of soil moisture dynamics in an operational railway embankment**Near Surf. Geophys., 12 (2014), pp. 61-72
8. Chambers et al., 2014b J.E. Chambers, P.B. Wilkinson, S. Uhlemann, J.P.R. Sorensen, C. Roberts, A.J. Newell, W.O.C. Ward, A. Binley, P.J. Williams, D.C. Gooddy, G. Old, L. Bai**Derivation of lowland riparian wetland deposit architecture using**

geophysical image analysis and interface detection Water Resour. Res., 50 (2014), pp. 5886-5905

9. Chiocchini and Castaldi, 2011 U. Chiocchini, F. Castaldi**The impact of groundwater on the excavation of tunnels in two different hydrogeological settings in central Italy**Hydrogeol. J., 19 (2011), pp. 651-669
10. Elwaseif and Slater, 2012 M.E. Elwaseif, L. Slater**Improved resistivity imaging of targets with sharp boundaries using an iterative disconnect procedure**J. Environ. Eng. Geophys., 17 (2012), pp. 89-101
11. French, 2009 C. French**Hydrological monitoring of an alluviated landscape in the lower Great Ouse valley at Over, Cambridgeshire: the quarry restoration phase**Environ. Archaeol., 14 (2009), pp. 62-75
12. Hayley et al., 2007 K. Hayley, L.R. Bentley, M. Gharibi, M. Nightingale**Low temperature dependence of electrical resistivity: Implications for near surface geophysical monitoring**Geophys. Res. Lett., 34 (2007)
13. Hayley et al., 2011 K. Hayley, A. Pidlisecky, L.R. Bentley**Simultaneous time-lapse electrical resistivity inversion**J. Appl. Geophys., 75 (2011), pp. 401-411
14. Hsu et al., 2010 H.L. Hsu, B.J. Yanites, C.C. Chen, Y.G. Chen**Bedrock detection using 2D electrical resistivity imaging along the Peikang River, central Taiwan**Geomorphology, 114 (3) (2010), pp. 406-414
15. Karaoulis et al., 2011 M.C. Karaoulis, J.-H. Kim, P.I. Tsourlos**4D active time constrained resistivity inversion**J. Appl. Geophys., 73 (1) (2011), pp. 25-34
16. Kim et al., 2009 J.-H. Kim, M.-J. Yi, S.-G. Park, J.G. Kim**4-D inversion of DC resistivity monitoring data acquired over a dynamically changing earth model**J. Appl. Geophys., 68 (2009), pp. 522-532
17. Kuras et al., 2009 O. Kuras, J.D. Pritchard, P.I. Meldrum, J.E. Chambers, P.B. Wilkinson, R.D. Ogilvy, G.P. Wealthall**Monitoring hydraulic processes with automated time-lapse electrical resistivity tomography (ALERT)**Compt. Rendus Geosci., 341 (2009), pp. 868-885
18. Loke et al., 2013 M.H. Loke, J.E. Chambers, D.F. Rucker, O. Kuras, P.B. Wilkinson**Recent developments in the direct-current geoelectrical imaging method**J. Appl. Geophys., 95 (2013), pp. 135-156
19. Loke et al., 2014 M.H. Loke, T. Dahlin, D.F. Rucker**Smoothness-constrained time-lapse inversion of data from 3D resistivity surveys**Near Surf. Geophys., 12 (2014), pp. 5-24
20. Mayes et al., 2005 W.M. Mayes, A.R.G. Large, P.L. Younger**The impact of pumped water from a de-watered Magnesian limestone quarry on an adjacent wetland: Thrislington, County Durham, UK**Environ. Pollut., 138 (2005), pp. 443-454
21. Ogilvy et al., 2009 R.D. Ogilvy, P.I. Meldrum, O. Kuras, P.B. Wilkinson, J.E. Chambers, M. Sen, A. Pulido-Bosch, J. Gisbert, S. Jorreto, I. Frances, P. Tsourlos**Automated monitoring of coastal aquifers with electrical resistivity tomography**Near Surf. Geophys., 7 (2009), pp. 367-375
22. Rucker, 2014 D.F. Rucker**Investigating motion blur and temporal aliasing from time-lapse electrical resistivity**J. Appl. Geophys., 111 (2014), pp. 1-13
23. Shaqour and Hasan, 2008 F.M. Shaqour, S.E. Hasan**Groundwater control for construction purposes: a case study from Kuwait**Environ. Geol., 53 (2008), pp. 1603-1612
24. Supper et al., 2014R. Supper, D. Ottowitz, B. Jochum, J.-H. Kim, A. Roemer, I. Baron, S. Pfeiler, M. Lovisolo, S. Gruber, F. Vecchiotti**Geoelectrical monitoring: an innovative method to supplement landslide surveillance and early warning**Near Surf. Geophys., 12 (2014), pp. 133-150
25. Ward et al., 2014W.O.C. Ward, P.B. Wilkinson, J.E. Chambers, L.S. Oxby, L. Bai**Distribution-based fuzzy clustering of electrical resistivity tomography images for interface detection**Geophys. J. Int., 197 (2014), pp. 310-321
26. Wardrop et al., 2001D.R. Wardrop, C.C. Leake, J. Abra**Practical techniques that minimize the impact of quarries on the water environment**Trans. Inst. Min. Metall. Sect. B Appl. Earth Sci., 110 (2001), pp. B5-B14

27. Wilkinson et al., 2010P.B. Wilkinson, P.I. Meldrum, O. Kuras, J.E. Chambers, S.J. Holyoake, R.D. Ogilvy**High-resolution Electrical Resistivity Tomography monitoring of a tracer test in a confined aquifer**J. Appl. Geophys., 70 (2010), pp. 268-276
28. Wilkinson et al., 2012 P. Wilkinson, M.H. Loke, P.I. Meldrum, J.E. Chambers, O. Kuras, D.A. Gunn, R.D. Ogilvy**Practical aspects of applied optimised survey design for Electrical Resistivity Tomography**Geophys. J. Int., 189 (2012), pp. 428-440

CHAPTER 9

Volumetric Assessment through 3D Geostatic Model for Abu Roash Reservoir in Amana Field-East Abu Gharadig Basin-Western Desert-Egypt

Abu-Hashish MF[1]* and Ahmed Said[2]

[1]*Geology Department, Faculty of Science, Menoufiya University, Egypt*

[2]*Qarun Petroleum Company, Cairo*

ABSTRACT

The main objective of constructing a geostatic model is to determine the 3-D geometry of the reservoir rock and assess its hydrocarbon volumetrics. To achieve this goal, the Amana oil field is taken as a real example. An integrated methodological approach was applied starting with data collection and quality control and then followed by interpretation of the available geological, geophysical and petrophysical data. The field area is located in the most eastern trough of the Abu Gharadig basin in the East Bahariya Concession in the Western Desert of Egypt. The source rock in this area is the Upper Jurassic Khatatba Formation that was deposited in a continental to inner-middle shelf environment. The reservoir rock is Abu

Roash "G" sand, one of the seven lithologic members of the Abu Roash Formation. Interpretation of seismic data together with well logs revealed the presence of a horst block, acting as a good structural trap, meanwhile the Abu Roash "F" carbonate and Abu Roash "G" shale are the seal rocks. Well data have shown that the reservoir rock in the Amana field is the Abu Roash "G" Member, which comprises three sand lithologic zones; namely the upper, meddle and lower zones. Of these zones, the middle one is the most attractive and has the best reservoir quality. The shale content in this sand is 8% compared to 13% and 26% in the upper and lower zones. In addition, the net to gross thickness ratio is less than 18%, more than 35%, and about 10% in the upper, middle, and lower zones, respectively. Data analysis also indicates that the upper and lower zones are appreciably water wet. However, the middle sand zone appears to be prospective. The net pay thickness in this zone varies between 10 and 32 ft, porosity 19-22%, water saturation 18-40% and average permeability 40 md. Based on the geostatic model of the Amana field reservoir, it is concluded that this area is a positive prospect in the most eastern part of the Abu Gharadig basin. The volume of oil (STOLIIP) in Amana reservoir is calculated as 10 million barrels, with an initial recoverable oil of 1.4 million barrels.

Keywords: Geostatic model; Petrophysics; Seismic interpretation; Western desert abu gharadig basin; Egypt

INTRODUCTION

Abu Gharadig basin is considered the most petroliferous basin in Western Desert as far as hydrocarbon production and potential. It is a deep E-W trending asymmetric graben, has dimensions of 300 km E-W and 60 km N-S, with an area of about 17500 km2, and basement at depths over 10 km. Its structure has been recognized as a major rift basin in which there are numerous localized highs that in NE-SW oriented plunging anticlines that are believed to be fault-controlled folding [1]. The northern margin of this basin is marked by a major border fault zone which up-throws the basement to about 10,000 feet forming Sharib-Sheiba ridge, and the southern boundary is called Sitra platform [2]. Amana oil field is located in the most eastern trough of the Abu Gharadig basin (Figure 1), exactly in East Bahariya concession that exists in Mubarak sub-basin.

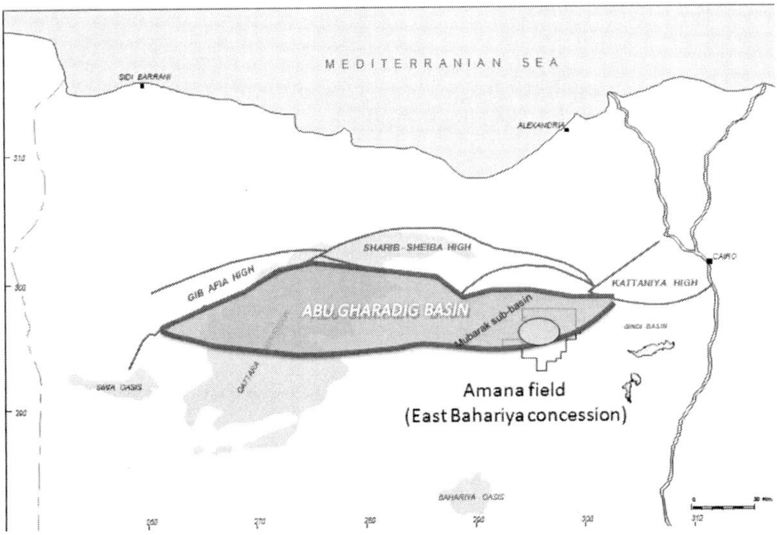

Figure 1: Location map of Amana field relative to Abu Gharadig basin.

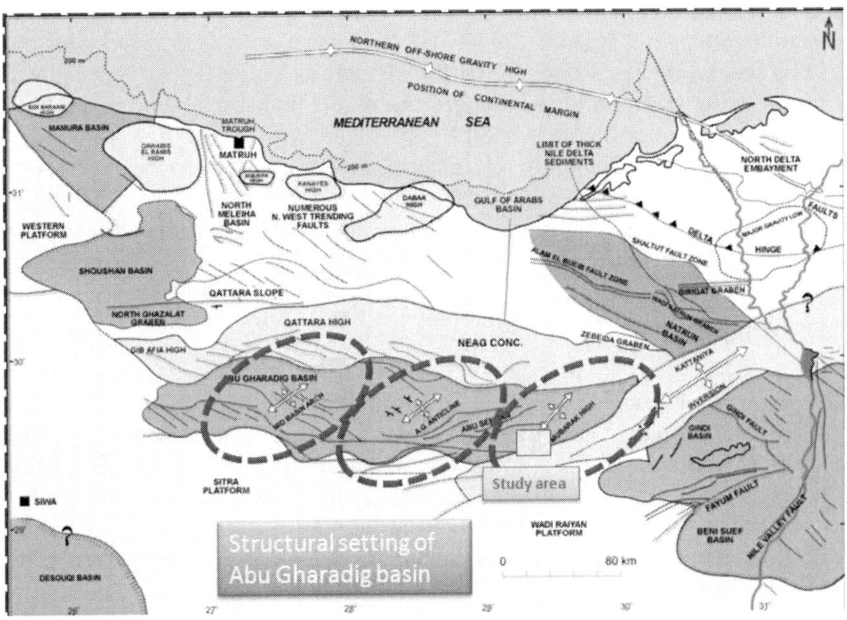

Figure 2: The main structural divisions of Abu Gharadig basin, modified after (Bayoumi, 1996).

Tectonic framework and structural settings

Abu Gharadig Basin may have begun life as a pull-apart basin between two right-lateral wrench faults, its development began during the Jurassic and Cretaceous and the tectonic activity reached a peak during the Upper Cretaceous to Eocene interval [3]. It is subdivided into three structural units from east to west; Mubarak sub-basin, Abu Gharadig basin and Qattara depression including Mubarak High, Abu Gharadig Anticline and Mid basin Arch (Figure 2).

Litho-stratigraphy

The study area is related to the unstable shelf which covers the northern belt of Egypt, which is characterized by a thick stratigraphic succession that ranges, in age, from Pre-Cambrian Basement to Holocene time, varying in lithology (Figure 3). Abu Roash Formation is characterized by a cyclic alteration of deltaic flood plain sandstones, coastal sandstones and shales, and shallow marine shales and limestones [4]. The formation has been divided into seven units (members) “A” to “G”. Units “B”, “D” and “F” are mainly carbonates, units “A”, “C”, “E” and “G” contains variable amounts of detrital materials [5]. Abu Roash “G” shale acts as a very good top and lateral seal while Abu Roash “F” carbonate acts as a lateral seal, especially in the central part Amana field , where the major fault throw exceeds 300 ft. The pay sand zone in Amana field is involved within Abu Roash “G” Mbr. which is composed mainly of marine shale intercalated by carbonate streaks, with three cycles of sand facies considered as the main reservoirs in Abu Gharadig basin, and named (from top to base) as Upper sand zone, Middle sand zone and Lower sand zone) (Figure 4).

Methodology and Workflow

In order to perform a reliable 3D Geostatic Model, it is necessary to quality control and manages all types of the data, starting from seismic data, well logs and well test results. Conducting the Model is starting through seismic data to be interpreted explaining the areal extent of the interpreted surfaces and the regime of the faults running in the area, and then suppose the horizontal gridding in which the model spreads the values, with determination of the trends of the modeled area. After that, a high light should be focused on the reservoir bed by Zoning stage, frequently, the tying with well data and well correlation should be processed to help in facies modeling layering the reservoir bed for more accuracy. The facies and petrophysical properties are distributed (i.e. interpolated and extrapolated) to assign a value for each cell in volume calculation stage as shown in Figure 5.

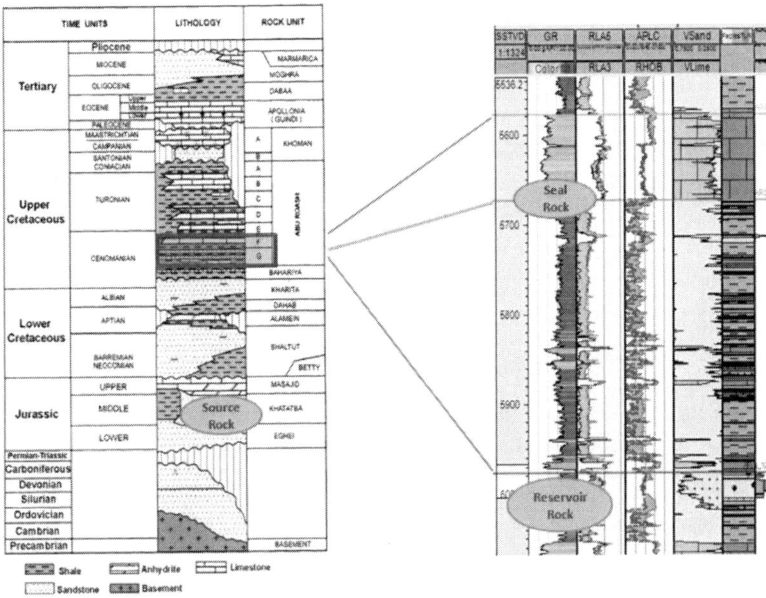

Figure 3: Generalized lithostartigraphic column in Abu Gharadig basin with high light on reservoir and seal beds in Amana field [14].

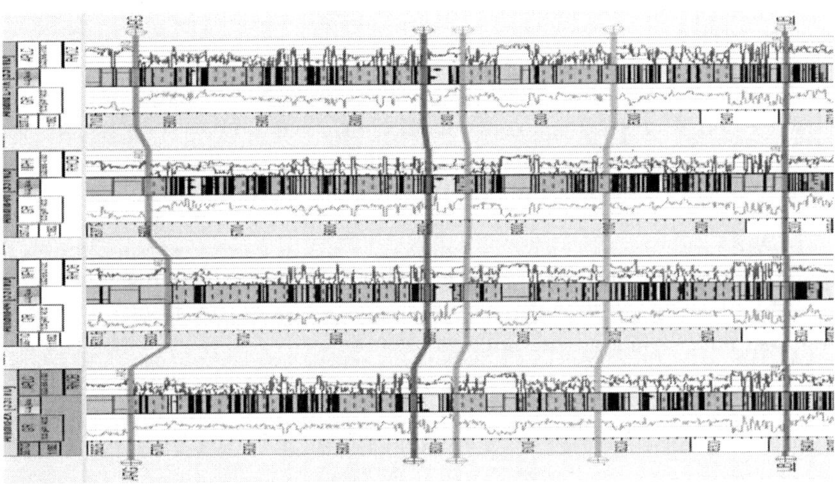

Figure 4: Stratigraphic correlation for Abu Roash "G" Member through Amana wells with key map.

DATA COLLECTION AND QUALITY CONTROL

In order to acquire a reliable 3D geostatic model, it is necessary to use all available data related to the study area. The data should be quality checked, and environmental corrected, in addition to removing the spikes or any unreliable readings. Almost all seismic data are enhanced and qualified for interpretation after passing through the stage of processing. Some complicated corrections should have been conducted on the well logs to be suitable for log analysis. These corrections are varied in certainty levels and have various types of mechanism. The most common log corrections that have been run are:

Temperature and pressure effect removal.

Mud weight, salinity and mud resistivity calibration.

Gamma Ray log corrections for mineral and organic matter effect, borehole washout, and mud weight.

Resistivity logs' corrections for mud salinity, borehole size, and mud resistivity.

Neutron, Density and sonic logs' corrections for borehole size, drilling fluids and formation fluids, mud cake thickness.

Well log analysis

In order to conduct the 3D geostatic model, it is necessary to analyze the well logs to obtain the most interesting properties needed for spatial distribution. After the logging acquisition has been done, petrophysicist will start the correction and the quality control on the data, and then will interpret and evaluate the logs resulting in a traditional petrophysical evaluation. This petrophysical evaluation is the most reliable stage in the static reservoir study, since it generates a series of vertical profiles, for each well in the reservoir, describing the main properties of the reservoir pore system [6], such as Vshale, porosity, fluid saturation, estimated permeability from logs, and net to gross ratio.

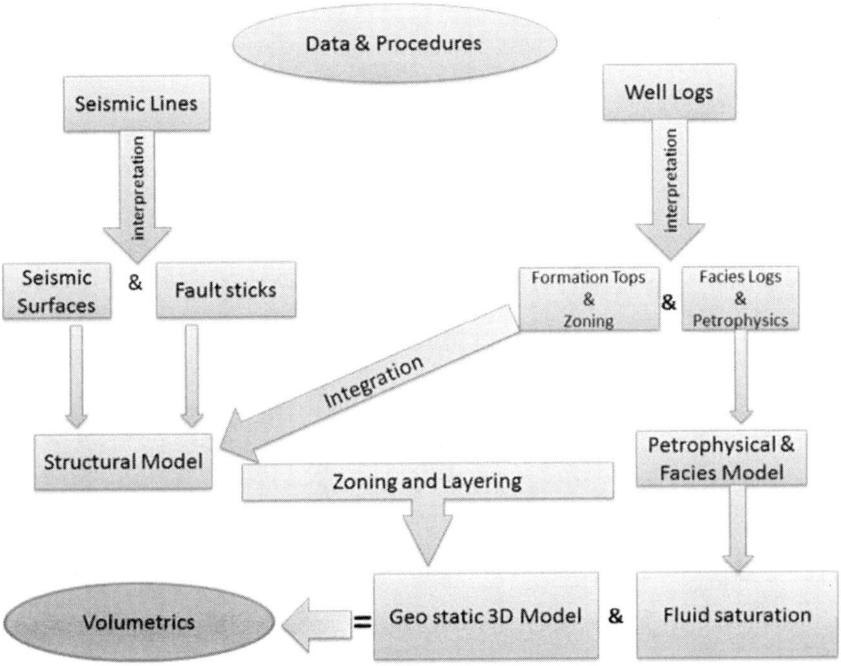

Figure 5: A sketch illustrating Model construction work flow.

Shale volume

The estimation of the volume of shale and clay minerals within the reservoir interval has an essential impact on the effective porosity [7]. In this study gamma Ray logs (wireline or LWD) are the main indicator for determining the volume of shale for Abu Roash "G" reservoir. And it was found that the middle zone is considered a clean sand zone with average volume of shale less than 8%, and it is clearly recommended to use Archie's equation in determining water saturation for this zone while the upper zone has an average volume of shale more than 13% and the lower zone shows 23% average volume of shale (Figure 6).

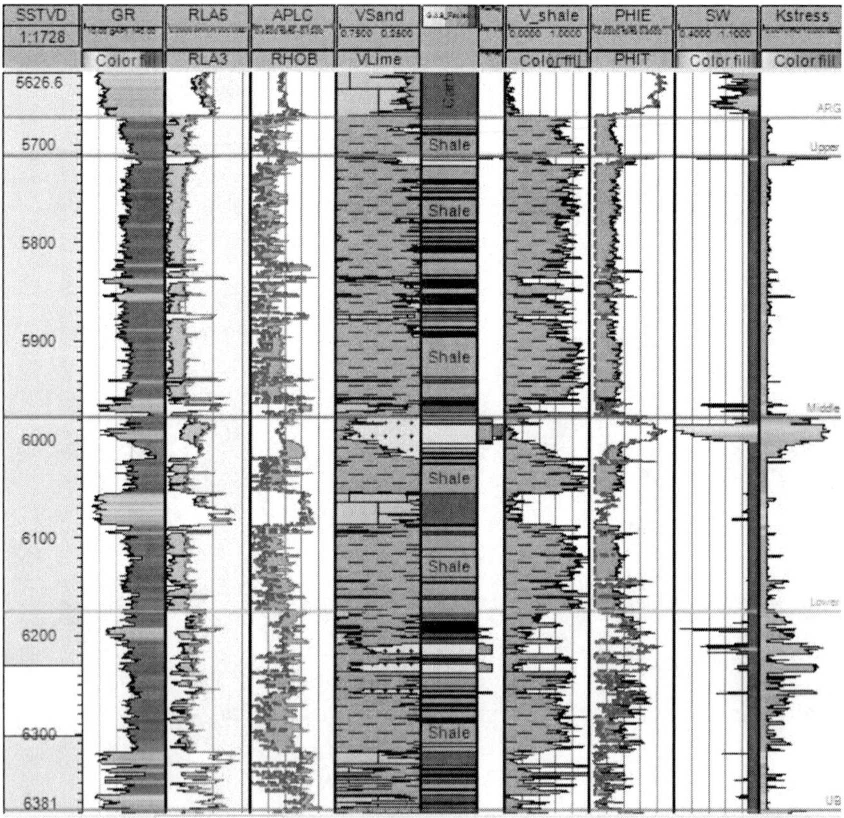

Figure 6: Log interpretation in Amana field.

Total porosity

Determination of porosity can generally be considered the least complex stage in the petrophysical interpretation but the most important one because it defines the quantity of the hydrocarbons present in the reservoir [8]. The most frequently used methods in determining porosity are those based on the interpretation of well logs. Actually, none of the logging tools measure porosity directly, but the interpretation is carried out by indirect measurements [6], even neutron technique, which detects the hydrogen resides in the pore spaces (as a function of water and/or hydrocarbon), and the logging unit used this hydrogen index in order to simply deduce the formation porosity, thus the total porosity can be directly read from the neutron display.

Effective porosity

Not all the fluid presented in the rock is movable, because of the existence of isolated pores (not connected together) and wettability, thus during determining porosity, it should be the effective one regardless the immobile fluid. The effective porosity is determined clearly by using total porosity and volume of shale from this formula:

$$\Phi_{eff} = \Phi_{total} * (1\text{-}Vsh)$$

Abu Roash "G" Mbr. has a wide variation in effective porosity, especially among its three zones, where the Upper zone has an average effective porosity of 11%, Middle zone has 21% and the Lower zone displays about 14% Φeff. (Figure 6).

Water saturation

The porous system of a reservoir rock is filled fluids, typically water and hydrocarbons with percent depending upon the chemical and physical properties of the rock and the fluids as well as the interaction between rock and fluid (rock wettability). The water saturation influences not only the volume of hydrocarbons, but also the productivity of the wells. For the purposes of reservoir studies, water saturation is mainly predicted on the basis of log interpretation for open holes using the famous Archie equation for clean formation [9].

$$Sw = \sqrt{\frac{F*Rw}{Rt}}$$

Where: Sw ..... is water Saturation

Rw ..... is Formation water resistivity

Rt ...... is true resistivity

F...... is the formation resistivity factor

Calculating water saturation for Abu Roash "G" Mbr. has indicated that the whole member is a water bearing zone Sw is more than 90%, even the Upper and Lower Zones, except the Middle zone which is subsequently considered as a hydrocarbon bearing zone with average water saturation of 30% (Figure 6).

Permeability

The productivity of the wells and reservoir ability to feed drainage areas are the function of the permeability which is the most difficult parameter to be determined because of the extreme spatial variability among wells. The measurement of permeability derives from Darcy's law in case of available core data, but if the core data is absent, it will be estimated from the well logs by Timur equation [10].

Timur permeability: Timur permeability is given by equation

$$K = \left(\frac{100 * \Phi e^{2.25}}{Swirr} \right)^2$$

Estimation of permeability for Abu Roash "G" Mbr. showed that the permeability average of the Middle zone is about 100 mD and more, while all the rest of the member is ranging from 1 to 40 mD (Table 1).

Table 1: Average petrophysical properties for Abu Roash "G" Reservoir in Amana Field

Well Name	Total Porosity Φt (%)	V _ Shale (%)	Eff. Porosity Φeff (%)	Water Saturation Sw (%)	Net Pay (Ft)	Estimated Permeability K (mD)
Amana-1X						
Middle zone	24	8-Feb	22	18	32'	100 - 250
Lower zone	15	20 - 33	14	52	wet	0.5 - 4
Amana-2						
Middle zone	22	8-Jun	22	28	28'	110 - 150
Lower zone	20	23-Oct	16	50	wet	20 - 30
Amana E-1X						
Middle zone	23	7-Apr	21	30	20'	100 - 300
Lower zone	21	20 - 30	17	55	wet	Feb-40
Amana-3						
Middle zone	22	16-Jan	19	41	10'	50 - 150
Lower zone	19	20-Dec	15	90	wet	30 - 50

Net to gross ratio (N/G)

For more detailed description of reservoir, the gross and net thickness should be calculated. Some geologists depended on the porosity cut-off to distinguish between the gross and net thickness, and others are depending on permeability cut-off, but a more rigorous approach to net thickness determination is on the basis of a detailed analysis of the rock properties [11,12]. The net to gross is a relation compiling the gross thickness and net thickness in one function obtained from dividing the net thickness by the gross thickness as a decimal fraction or percentage. The N/G ratio was calculated for the three zones in Abu Roash "G" Mbr. as shown in the figure below, illustrating that the N/G ratio increasing in the Middle zone in the central part of the study area (Figure 7).

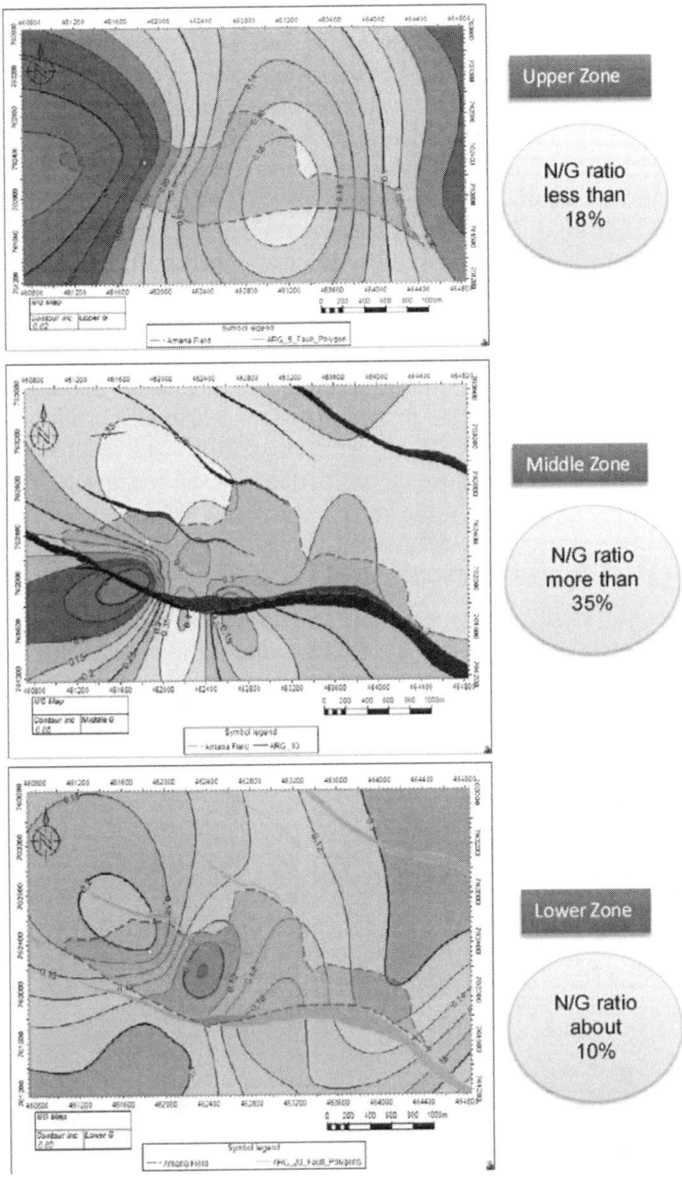

Figure 7: N/G map for the three zones of Abu Roash "G" Reservoir.

Structural modeling

Structural model is the initial point of the reservoir model construction. It is considered as the backbone of the reservoir model. The structural model for the reservoir is basically conducted through two major procedures; mapping the structural geometry of the top of reservoir, and defining the set of faults running through the reservoir. These two procedures are ideally performed through the stage of seismic interpretation, where this stage involves two modules of interpretation; Horizon interpretation which is acquired on the horizons of interest, and Fault interpretation for the faults affect the reservoir [13,14]. In Horizon interpretation, all seismic lines and section used for covering all the area of interest, while in Fault interpretation, the cross lines (the lines that are perpendicular to the major fault trend) should be selected at the first in order to clearly pick the faults with their real throw and trend. Interpretation of the seismic lines in Amana field explains the structural setting affected the reservoir and formed the hydrocarbon trap in the area. As shown in the figure below, Amana oil field displays a horst block that is considered a good structural trap for accumulating the hydrocarbon in it.

The direct results for seismic interpretation are summarized in the lateral explanation of the fault system affecting the reservoir as shown in Figure 8, in which the tectonic setting shows a flower structure that indicates a wrench fault system resulting in a horst structure, and mapping the structural geometry of the top of reservoir that displays NW-SE trending normal faults as shown in Figure 9.

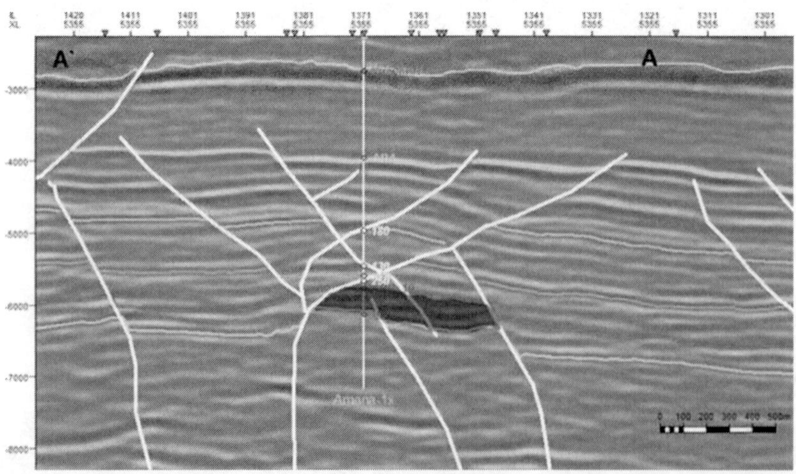

Figure 8: A Seismic X-line showing the structural setting in Amana block.

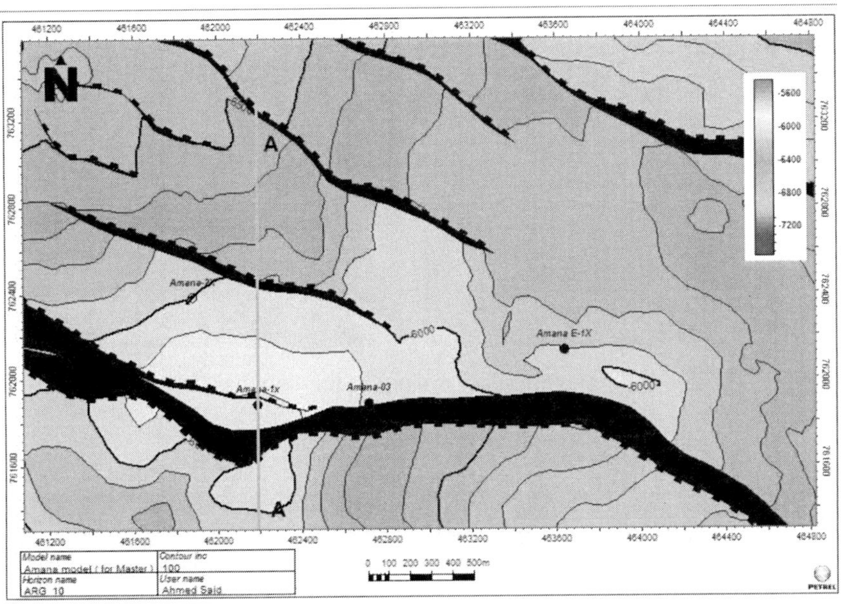

Figure 9: Structural map on top Abu Roash "G" Middle reservoir in Amana field.

The structural model of the reservoir is the result of compiling the output of Horizon interpretation and Fault interpretation, thus the indirect results for seismic interpretation are the basic inputs for constructing structural model, such as fault sticks (the fault interpretation in lateral view), also defined as sets of line data that represent the fault plane, and seismic surfaces (the horizon interpretation), so that an interpretation has been carried out on Abu Roash "F" "G" and Upper Bahariya surfaces (Figure 10).

The structural model is carried out through four steps as follow.

Fault model: This step converts the fault sticks (the output from seismic interpretation) into key Pillars which are easily editable in 3D grid. This step is considered as a fine tuning for the faults to be more fitting in the structural model as the key Pillars connecting two faults, are used for extending a short one, making cross faults, or dividing one to two faults (Figure 11).

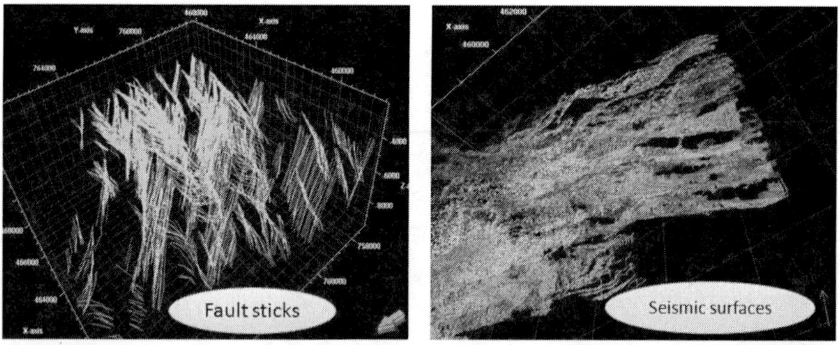

Figure 10: Showing seismic data as input for constructing the structural model.

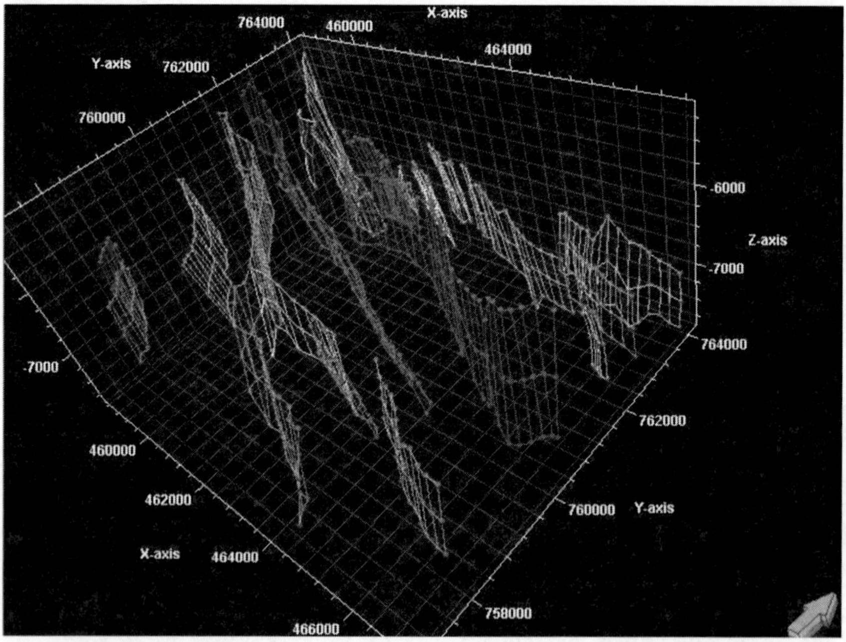

Figure 11: 3D window showing the Key Pillars in fault model step

Pillar gridding: The construction of the Model is proceeded after that with Pillar Gridding step in which the trending is determined as I, J, and K directions (Figure 12). In this step we build the horizontal gridding of cells in the Model and determine their size which will aid, after that, in volume calculation.

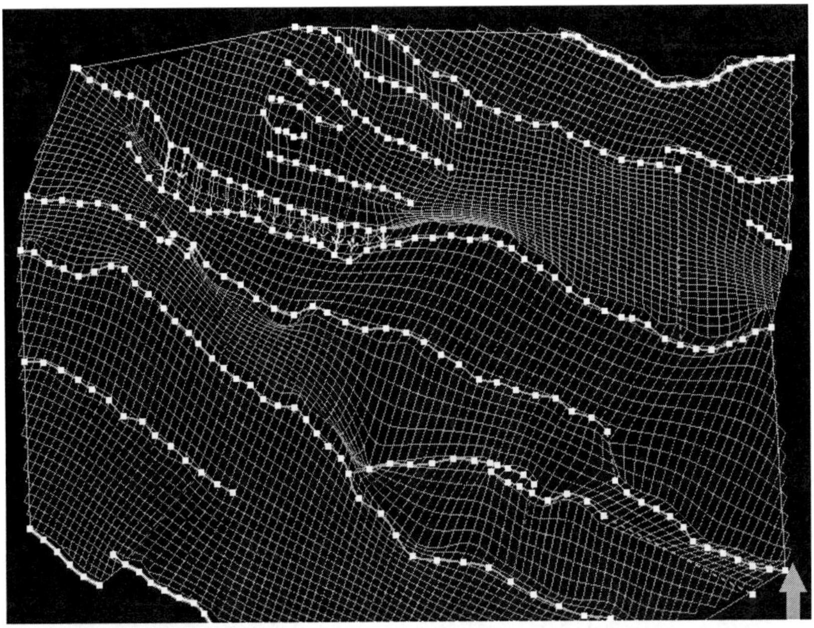

Figure 12: The I and J trends with horizontal Pillar gridding in Amana field.

Another important function of Pillar gridding step, is converting the faults in key Pillars into fault surfaces that will be displayed as cell walls in the final static model (Figure 13).

Making horizons: Regarding to the faults, all of them were treated, edited and smoothed in the previous stage, but in this stage we are dealing with the seismic surfaces and 2D grids. This step deals with the seismic surfaces resulted in seismic interpretation and uses them in order to make horizons fitting for constructing the structural model (Figure 14). The resulted horizons from this stage are celled horizons in which each cell takes a value of coordinates (X and Y) and a value of each property. The resulted horizons should be matched with the formation tops from well data at wells' locations.

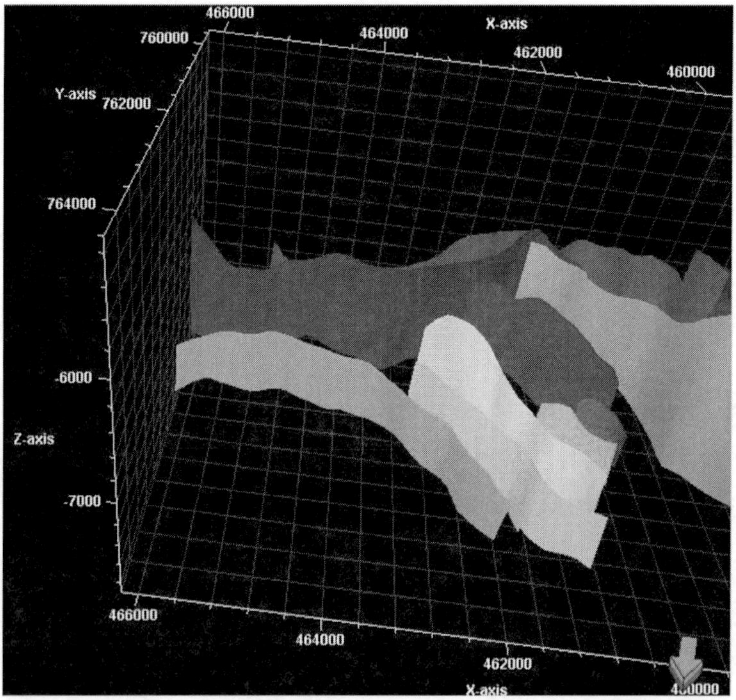

Figure 13: Showing the fault surfaces converted from Key Pillars.

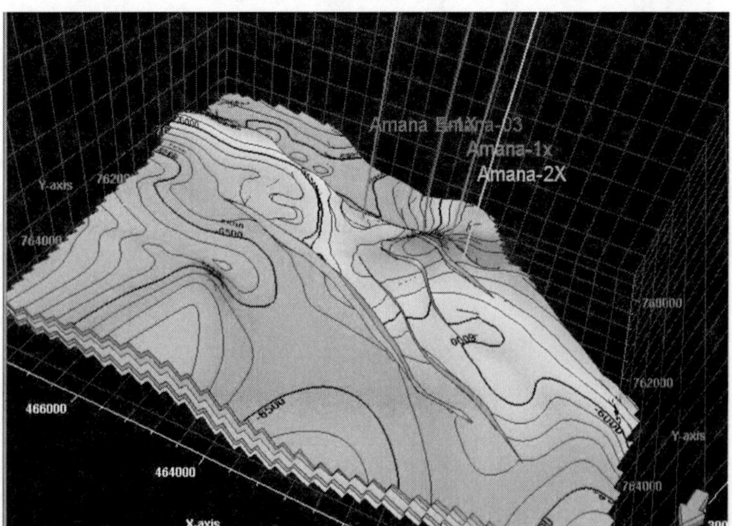

Figure 14: 3D modeled surfaces after making horizons stage.

The making horizons stage may be considered the final step for structural model, but in most cases, the resulted horizons are for major surfaces only, thus it is recommended to be zoned and detailed.

Zonation and layering: Because only major seismic surfaces can be interpreted, and frequently, just major horizons will be involved in the model, so it is necessary to consider the minor surfaces and sub-zones. By using well tops of formations and sub-zones, it is possible to make sub-zones between the major modeled surfaces for high lightening the pay zone involved within the reservoir bed as shown in Figure 15.

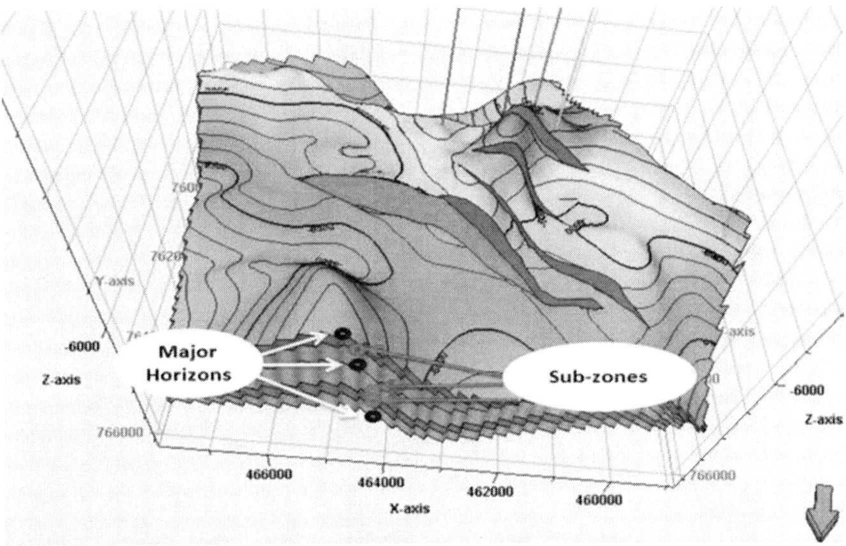

Figure 15: 3D zonation between the major horizons.

After the zoning process for the reservoir in order to focus on pay zone, the pay zone itself has vertically varied values of the properties, so this pay zone should be subdivided into a number of small layers in to obtain reliable calculations, this process is called layering and had been performed for the middle zone reservoir resulting in ten layers where each one having a reliable value for each property as shown in Figure 16.

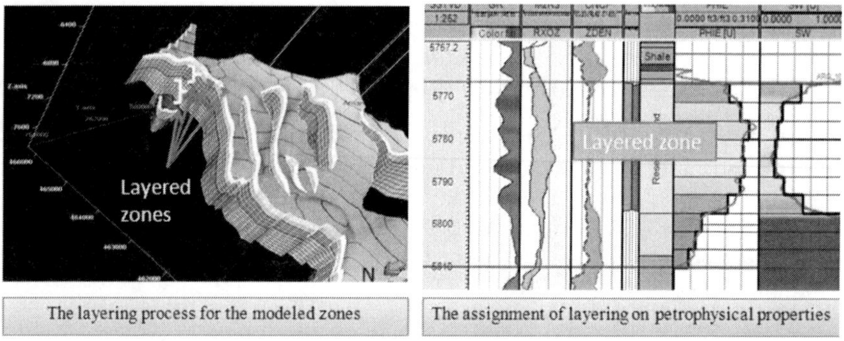

Figure 16: The layering process impact on vertical distribution of the data.

Facies and petrophysical modeling: After Make zones stage and Layering, we use the facies log for each well and distribute the facies over all the area of interest among the wells, to predict the lithology in the non-penetrated area, this stage is named facies Modeling and is followed by Petrophysical Modeling in which the Petrophysical parameters are distributed through the pay facies (Figure 17).

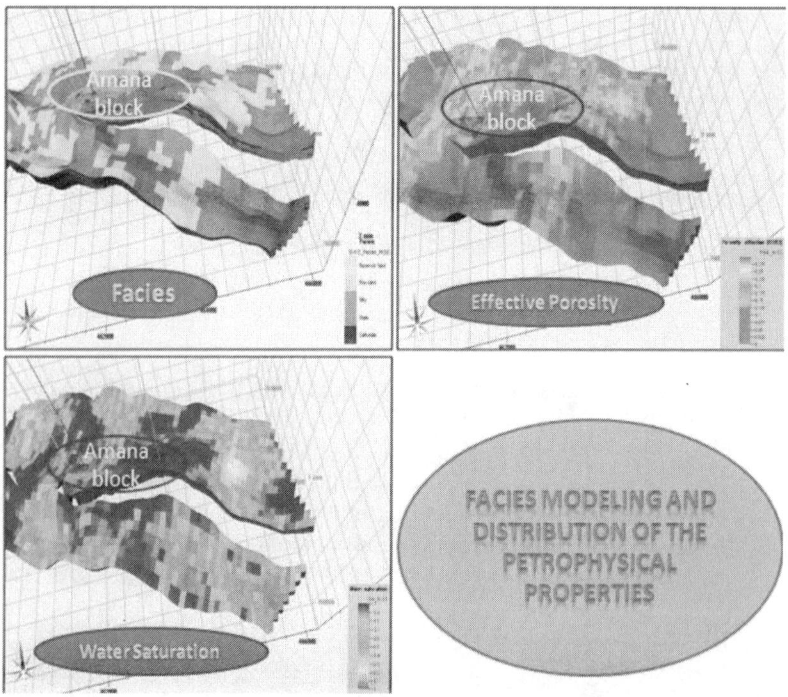

Figure 17: Spatial distribution through Amana block.

Estimating properties for undrilled well locations: The main target behind this work is to estimate facies and Petrophysical properties for the undrilled locations in order to drill new wells for increasing productivity and enhancing the field recovery. Figure 18 shows the area of interest (green area) above the oil water contact where the probability of oil presence is of a value, so it is recommended to drill more wells within the green area lying (far from the blue one), as the Geographic system, between:-

Latitudes: 29°33’20”N and 29°33’48”N

Longitudes: 29°25’17”E and 29°26`01”E

and as the Metric system, it lies between:

X: 462000 m E and 463200 m E

Y: 761800 m N and 762600 m N

Reserves and volumetric calculations: The final step (for completing this work) is volume calculation, where the well logs and seismic surfaces are used indirectly for calculating area, pay thickness, porosity, and fluid saturation in order to determine the OOIP and Reserves. In the 3D Modeling, as a result of facies distribution and calculating Effective Porosity and Water Saturation within the area of interest in the Model, the volume of oil in place could be calculated easily, and by adding the recovery factor, the recoverable reserve is determined.

The area of interest (above the oil water contact) is defined in a function of trap geometry delineated from the top by top seal and from the base by fluid contact as shown in Figure 18.

The net pay thickness is calculated from the well logs, petrophysical properties such as porosity and water saturation are averaged, thus, the application of recoverable reserve equation is achieved as follows [15]:

$OOIP = 7758 * A * H * \Phi * So$

Where:

OOIP: Original Oil Initially in Place (reservoir barrel).

7758: conversion from (acre.ft) volume to barrel (bbl) volume.

A: the areal extent of reservoir (acre).

H: the net pay thickness (feet).

Φ: Porosity (%).

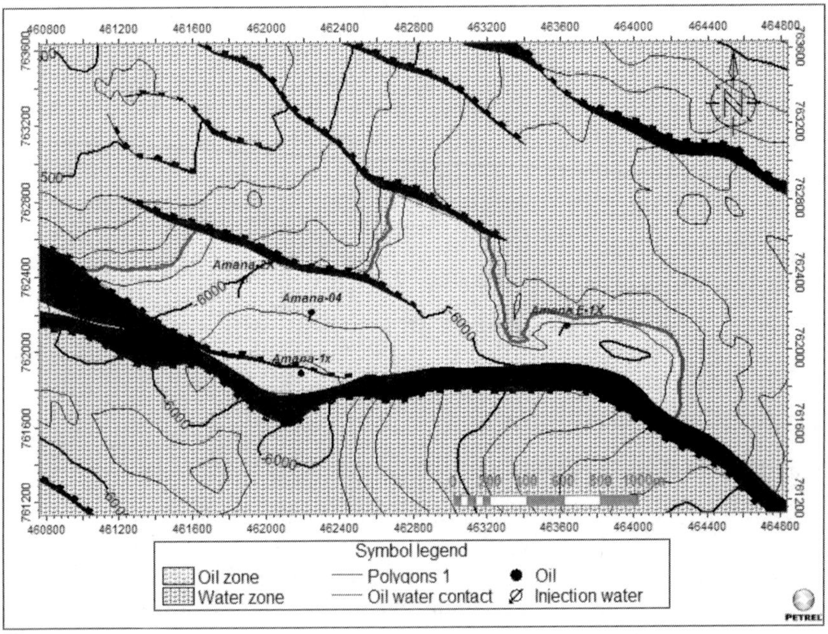

Figure 18: Promise area in Amana block.

Table 2: The volume calculation results.

Case	Net volume[acre.ft]	Pore volume[RB]	HCPV oil[RB]	STOIIP [STB]	Recoverable oil[STB]
Middle Zone	12331	14,513,846	10,159,692	9,152,875	1,372,931

So: Oil saturation (1 - Sw) (%).

$$RR = OOIIP * FVF * RF$$

where:

RR: Recoverable Oil (stock tank barrel).

FVF: Formation Volume Factor (1/Bo). Bo is reservoir barrel multiplied by stock tank barrel (VR/VS).

RF: Recovery Factor (%).

The calculated volume of hydrocarbons in Abu Roash "G" member in Amana Field is displayed in Table 2.

CONCLUSION

The study area (Amana field) represents a positive prospect included in the most eastern part of Abu Gharadig basin where the proven oil that produced from Amana wells proved the presence of petroleum system in the area.

The seismic interpretation for the reservoir horizon and fault system in the area indicated that there is a structural petroleum trap (Horst style) that could be able to accumulate the oil in it.

Well logging interpretation was very useful in order to determine the reservoir and its lithology by constructing the facies log that resulting in the existence of reservoir sandstone in middle zone in Abu Roash "G" Member which shows an average porosity more than 22% and permeability more than 100 mD.

Seismic interpretation integrated with well log analysis, indicats that Abu Roash "G" shales are typical top seal and lateral with Abu Roash "F" carbonate for the present reservoir and explained that the main structural features in Amana field appear to have had their maximum of development during the Late Cretaceous (base Khoman deposition), thus the timing of oil expulsion and primary migration (that had commenced during the Campanian age) relative to the structure trap formation should be favorable.

The reservoir areal extent and vertical thickness, in addition to, petrophysical properties are recorded in the 3D Model, so the volume of the original oil in place and the recoverable reserves have been easily calculated showing about 10 million barrels STOIIP and 1.3 million barrels initial recoverable oil without former of any stage of EOR(enhancing oil recovery).

3D petrophysical modelling yielded an area of promise in order to increase Amana field productivity, this area lies in the central northern part in the field, so it is recommended to drill more producer wells in this part and drill water injector wells in the most northern edge.

REFERENCES

1. Schlumberger (1995)Well Evaluation Conference Egypt.
2. Enayet O (2002) Geology and petroleum potentiality of western desert- Egypt.
3. Meshref WM, Beleity A, Hammouda H, Kamel M (1988) Tectonic evaluation of the Abu Gharadig basin, AAPG Mediterranean Basins Conference.
4. Shora MM (2012) Sandstone as a reservoir rock.

5. Qarun Petroleum Company, Amana NW-1X well recommendation.
6. Krygowski D (2003) Guide to petrophysical interpretation, Austin, Texas, USA.
7. Alberty M (1992) Standard interpretation. In: Thompson DM, Woods AM (eds.) Development geology reference manual: AAPG Methods in Exploration Series 10: 180-185.
8. Cosentino L (2005) Oil field characteristics and relevant studies. In Exploration, Production and transport. Encyclopedia of hydrocarbons.
9. Archie GE (1942) The electrical resistivity log as an aid in determining some reservoir characteristics. American Institute of Mining, Metallurgical, and Petroleum Engineers. Transactions 146: 54-62.
10. TimurA (1968) An investigation of permeability, porosity, and residual water saturation relationships for sandstone reservoirs 9.
11. Gaynor GC, Sneider RM (1992) Effective pay determination. In: Thompson DM, Woods AM (eds.) Development geology reference manual, AAPG Methods in Exploration Series 10: 286-288.
12. Worthington P, Cosentino L (2005) The role of cut-off in integrated reservoir studies, In: Proceedings of the Society of Petroleum Engineers annual technical conference and exhibition, Denver (CO), 5-8 October, SPE 84387.
13. Bayoumi T (1996) The influence of interaction of depositional environment and synsedimentary tectonics on the development of some Late Cretaceous source rocks, Abu El-Gharadig basin, Western Desert, Egypt.
14. Mahsoub M, Abul-Nasr R, Boukhary M, Abdel-Aal H, Faris M (2012) Bio- and Sequence Stratigraphy of Upper Cretaceous - Paleogene rocks, East Bahariya Concession, Western Desert, Egypt.GeologiaCroatica 65.
15. MurthaJA, Peterson SK (2001) Another Look at Layered Prospects. Presented at the SPE Annual Technical Conference and Exhibition, New Orleans, Louisiana, 30 September-3 October 2001. SPE-71416-MS.

CHAPTER 10

Engineering Geology Maps for Planning and Management of Natural Parks: “Las Batuecas-Sierra de Francia” and “Quilamas” (Central Spanish System, Salamanca, Spain)

Antonio Martínez-Graña [1,*], José Luis Goy [1], Caridad Zazo [2] and Mariano Yenes [1]

[1]*Geology Department, Science Faculty, University of Salamanca, Plaza Merced s/n, Salamanca 37008, Spain*

[2]*Section Geology, National Museum of Natural Sciences, Street José Gutiérrez Abascal 2, Madrid 28006, Spain*

ABSTRACT

Presented herein is a cartographic procedure that is easy to utilise and at low-cost, which facilitates the first stages of planning and management of a naturally protected space and considers the geotechnical parameters that influence human activity. This procedure uses geographical information systems technology by combining the cartographies for the most influential parameters on the stability of the area (lithology, hydrogeology, geomorphology, slopes, lineament/fractures and seismicity) with geomechanical mapping generated from geotechnical parameters obtained through field and laboratory tests. This geotechnical mapping facilitates the division of a territory into zones according to each type of problem and generates a cartography for natural hazards. Using this information, it is possible to produce a cartography of constructive conditions or geotechnical hazards. This methodology has been validated by application to two natural protected spaces, "Las Batuecas-Sierra de Francia" and "Quilamas". The validation confirmed that the cartography procedure described herein is a preventive, and not a structural measure. It is a tool that delimits areas with different constructive use recommendations and limitations, and therefore, is useful for natural space managers.

Keywords: geotechnical cartography; natural hazard; territorial planning; GIS techniques; protected nature reserve

1. INTRODUCTION

From a geotechnical perspective, engineering soils consist of the natural accumulation of mechanically separable mineral particles [1], occurring with varying thicknesses. From a geomorphological perspective, they are produced by the weathering and sedimentation of unaltered bedrock, and occur at shallow depths. The geotechnical approach determines the engineering soils behaviour from its mechanical properties, when loads are applied by human structures, or by dynamics from its intrinsic properties in the natural environment [2].

Different human activities are distributed geographically and depend on social, economic and environmental criteria, and geotechnical parameters are rarely considered during the first phases of territorial planning for soil use. This phenomenon is particularly important for natural parks, which comprise important resources (landscape, geology, fauna and vegetation as well as archaeological deposits). These spaces must be promoted as environmentally important through the construction of different facilities, such as classrooms in natural parks, bridges, didactic pathways; additional constructions for traditional soil uses; and the promotion of sustainable activities (agriculture and livestock). Urban development in these sectors, which are naturally fragile, must necessarily consider the geotechnical character of the different sectors in the territory.

The aim of this paper, is to prepare a set of engineering geology maps primarily to guide decision making on planning and management of two natural protected areas but also so that necessary urban and infrastructure development can be carried out with fewer impacts on fragile environments as well as reducing hazards and damage to constructions with consequent savings to society and the economy.

Engineering geology mapping includes basic previously derived information for urban development decision making in a territory, and it can facilitate important socioeconomic savings if *a priori* decisions consider the area's natural hazards and spatial distribution. It is necessary to understand the resistance and deformability of each geological material as well as the geomechanical behaviours on a short-to-medium timescale to establish *a priori* the potential natural risks from physical maps (geomorphology, hydrology-hydrogeology, lithology and geotechnic).

The precedent in Spain for geotechnical mapping began in the 1970s [3,4,5,6,7] and comprised geotechnical zoning by describing the following factors: bedrock lithology, superficial formations, geomorphology and hydrology/hydrogeology. In addition, geotechnical mapping includes the geotechnical conditions for each territorial sector, including the physical and mechanical characteristics of the soils and rock massifs. These maps, which were scaled at 1:200,000, were used to analyse the aptitude for numerous civil engineering projects because they included aspects such as topography, morphology, seismicity, lithological competence, mechanical properties, groundwater levels, vegetation, soil uses and climatology. The methodology followed in this mapping was promoted by UNESCO in Europe [8]. Later, in the 1990s, geotechnical mapping was developed and included natural hazards for certain cities wherein urban development growth was notable; these maps were at a 1:25,000 scale and, in some sectors, a 1:5,000 scale [9]. Worldwide, geotechnical cartographies have been applied to natural risks in large cities and regions [10,11,12,13,14] or for certain infrastructures or movements in the area [15,16,17].

The geographic information system (GIS) that includes geotechnical mapping considers many natural variables that are applied to planning urban areas and most wide territorial sectors [18,19,20].

Herein, a methodological cartographic procedure is established that facilitates geotechnical characterisation of a region, which different administrations (national, regional government and/or municipal) use in the first phases of planning region. This procedure includes zoning that spatially differentiates and distributes the geological and geomorphologic factors with constructive incident, that can be attributed especially to the lithology, geomorphologic domains, hydrogeological units and geotechnical properties. In addition, this geotechnical, basic and preliminary characterisation facilitates estimation of the foundation conditions and typology as well as analysis and geotechnical classification of the soils in each territorial sector. This characterisation complements analysis of the physical environment on having integrated the different susceptibilities (lithological and hydrogeological, among others) that affect the geotechnical aspects. Without this characterisation, studies and geotechnical reports may not be fully used for each constructive typology considered for development. Further, the geotechnical mapping will aid in establishing the natural associated dangers or potential risks for each zone, which will preserve constructive activities in the above sectors or establish the appropriate structural measurements for different infrastructures (building houses, road construction and recreation areas, among others).

To verify and to validate this cartographic procedure, it was applied to two natural protected parks, "Las Batuecas-Sierra de Francia" and "Quilamas", which are located in the spurs of the Central Spanish System south of the Autonomous Community of Castile and León and the province of Salamanca (Figure 1). Nevertheless, though validation of this methodology was successful using these two natural parks, the methodology is global and applicable to any natural space. This methodology is different from other existing methodologies because it details certain factors with greater precision. Thus, for example morphological mapping in the geotechnical maps published in Spain is performed at 1:200,000, whereas here the geomorphological analysis and the synthetic mapping of geomorphological domains are performed at a working scale of 1:10,000 and 1:18,000, and are presented at a scale of 1:50,000 with the other maps. The structural and slope factors were obtained with greater precision since we were working with a digital terrain model created from a grid of individual geodesic reference points and level curves at a scale of 1:10,000. This enhanced precision is a result of the implementation of GIS technologies (mapping of slopes from the digital terrain model, and the mapping of alignments and fractures) and the elaboration of a natural hazards map based on the interactions among the factors analyzed, and it offers a non-structural measure to protect these sectors from potential uses that promote instability in the area, and which are so discontinuous and random that they pass unnoticed. The non-structural measures are hazard mitigation measures, which do not involve engineering work, and are good prevention tools that may include legislation, planning strategies, or thematic maps, such as that reported here. These measures allow certain hazards to be defined and prevented in the early steps of the spatial planning of different projects, affording alternatives for correct emplacement without the need for structural measures involving civil engineering. The "Las Batuecas-Sierra de Francia" natural park comprises a 32,300 ha surface in a zone with a socioeconomic influence affecting 44,800 ha. The number of visitors in 2009 was 38,518. The "Quilamas" natural park comprises an 11,100 ha surface. These natural protected parks form part of the Central Spanish System in the Iberian Massif. Lithologically, they comprise metasedimentary materials from the schist-greywacke complex from the Upper Vendiense to Lower Cambrian, materials from the Ordovician and Devonian (Armorican quartzites, grey and black slates), and the granitic outcrops and Neogene sediments that have formed the foothills of the saws.

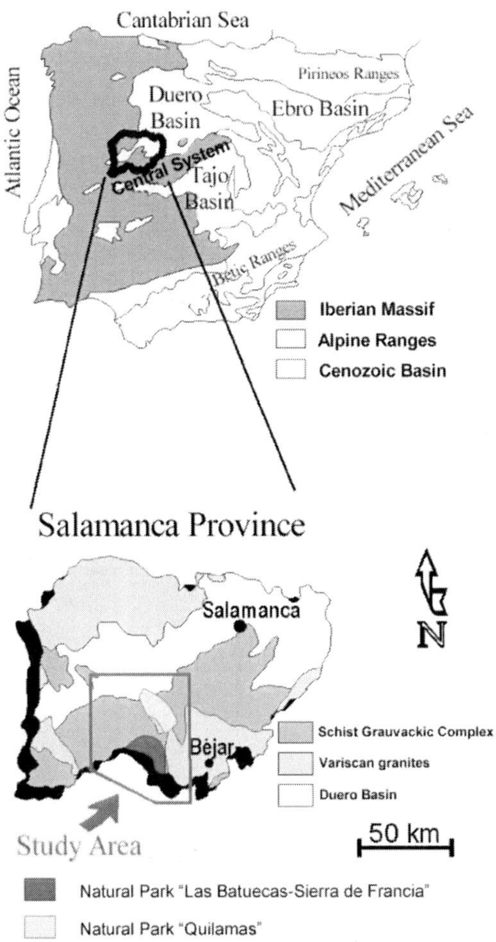

Figure 1. Study area: Protected natural park "Las Batuecas-Sierra de Francia" and "Quilamas".

2. METHODS

In the first phase (Figure 2), the basic cartographies are known, including the lithological (bedrock and superficial formations), hydrogeological and geomorphologic (geomorphologic units) cartographies from aerial photographs and field campaigns. Next, certain cartographies at a 1:50,000 scale are synthesised to overlap the different maps and facilitate a manageable number of categories for zoning the area of study; in this form, the geomorphologic unit mapping has been simplified to geomorphologic domains cartography.

The spatial information that corresponds to the digital topographic database (vector format) on a 1:10,000 scale aids in the generation of a digital model terrain using GIS technologies, ArcGIS version 10. This digital model terrain comprises marked points as well as level curves and more accurately delimits certain geomorphologic domains (controls) using slope mapping (format grid). Using the extensions for the GIS ArcGIS version 10: Spatial Analyst and 3D Analyst, we generated a reliable, shaded model that facilitates analysis of the limits for adding existing fractures, which were produced using digitalization and obtained from other cartographies published and observed in the field. The territorial sectors near these zones comprising weakness and/or breaks were increasingly susceptible to risk, and danger, and they determine the instability for the area. Finally, the mapping considered seismoresistance because seismic risk is a factor in activation of instabilities.

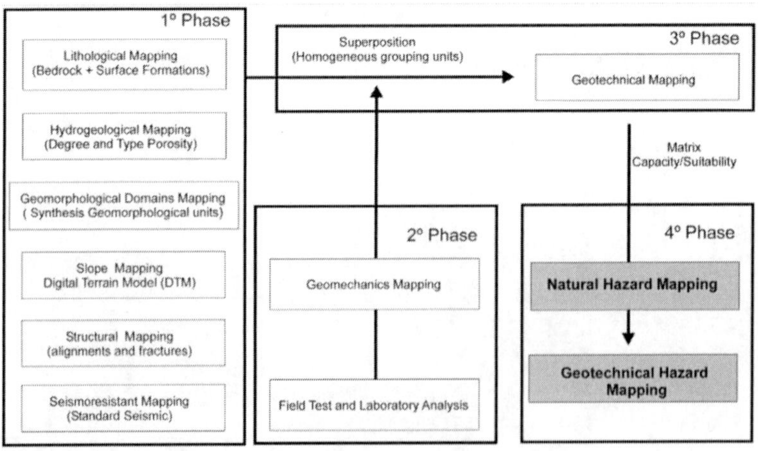

Figure 2. Methodological scheme for the production of the natural hazard cartography and geotechnical hazard cartography.

To generate the lithological mapping, we grouped the materials by related characteristics such as composition and geomechanical behaviour. Further, we differentiated two large lithological groups, the bedrock and superficial formations. To represent the bedrock, the geological national map (1:50,000) is used as the base, and the materials are grouped in broader categories such as composition and geomechanical properties, which creates homogeneous areas. The surficial deposits (Quaternary deposits) were mapped in their entirety, given the small amount of existing information in the official geological cartographies.

This method generates the first area division on the basis of different lithological groups. These areas were assigned with roman numbers: area I, granites; area II, metamorphic rocks; area III, carbonate rocks; and area IV, detritic rocks. Next, the lithology for each area or group differs, and considering these characteristics and compositional variations, the differences for the lithological II area that correspond to metamorphic rocks include the following three types of materials: II_1 slates, schists, graywackes and sandstones; II_2 black and micro-striped slates; and II_3 armorican quartzites. In the lithological area from detritic rocks IV, the following three materials differ: IV_1 arkosic conglomerates, sands and muds; IV_2 reddish-colour conglomerates, sands and mires; and IV_3 alluvial floor gravels and sands (Figure 3a).

The hydrogeological characteristics were then analysed (hydrogeological units) using hydraulic parameters, which reclassified the previous areas (I, II, III and IV) by differentiating the more concrete lithological aspects that include the degree (high place: $>1 \times 10^{-1}$ cm/s, middle: 1×10^{-1}–1×10^{-5} cm/s and low: $<1 \times 10^{-5}$ cm/s) [21] and type of permeability (for porosity, fracturation, dissolution and alteration) for the different hydrogeological units. Permeability due to dissolution is limited to the limestone unit. Permeability due to fracturing is observed in all the lithologies, but the W sector has a lower density of fractures. Permeability due to porosity is mainly distributed in Quaternary detritic materials associated with water courses. Finally, permeability due to alteration is governed by the differential alteration of the various materials, their thickness and alteration intensity differing according to the environmental conditions of each sector. Thus, for example, the granite in the locality of La Alberca shows an alteration profile of dozens of meters, generating an arenization with low-medium permeability due to weathering and the porosity of the sandy detritic sediments. In contrast, the presence of slates has generated a lower degree of alteration, leading to clayey materials, which are less permeable. This methodology generated a synthetic mapping, with eight zones differentiated by their hydrogeological characteristics (Figure 3b).

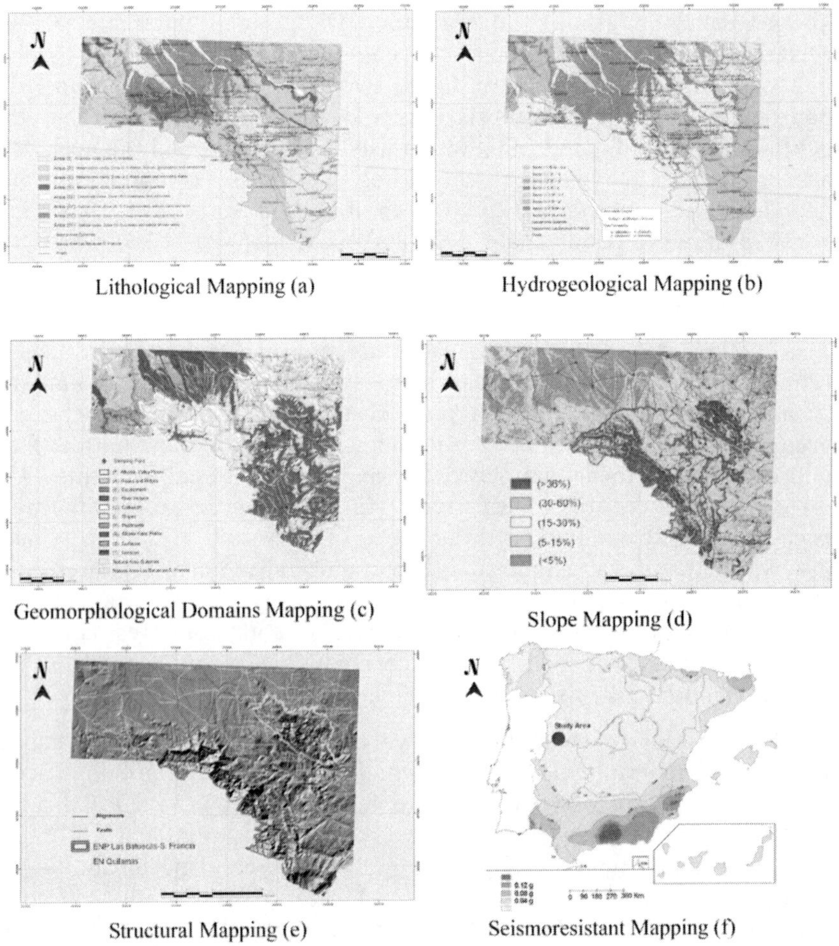

Figure 3. Parametric cartography for the sector of study: (**a**) Lithological mapping; (**b**) hydrogeological mapping; (**c**) geomorphological domains mapping; (**d**) slope mapping; (**e**) structural mapping and (**f**) seismoresistant mapping.

Further, the synthetic map of geomorphologic domains was generated and includes fundamental relief aspects to identify the most relevant geomorphologic aspects [22,23] that influence the geotechnical characterisation; "sectors" were created for morphologically similar and lithologically different areas (Figure 3c). Likewise, the slopes that concern each geomorphologic domain come from the mapping of slopes generated from the digital model terrain (Figure 3d) and the structural mapping (lineaments and fractures) (Figure 3e) that illustrates the weak zones in the area. The lineaments reflect a structural arrangement: alignments of the topography, straight lines on the drainage network, and changes in or a noteworthy absence of rocky outcrops. Its causes may be due to lithostructural modelling or to deformation of the materials or hidden faults. These lineaments were obtained from the interpolation of topographic data of the terrain from the digital terrain model.

The R.D 997/2002 seismo-resistance norm for construction—antiseismic building code [24]—establishes the level of seismicity by means of seismo-resistance mapping (Figure 3f), on the basis of the basic seismic acceleration (the value of the horizontal acceleration of the surface of the terrain) and bearing in mind the amplification of seismic waves and the elastic response of the geotechnical characteristics of the terrain. As regards construction, sectors with values lower than 0.04 g, where "g" is gravity; do not have a seismic risk.

In the second phase, the geotechnical characterisation of the different lithological materials includes characterisation of different rocky massif outcrops according to the RMR geomechanical classification (Rock Mass Rating) [25] (analysis of discontinuities: orientation, spread, continuity, ruggedness, resistance for sclerometer or Schmidt's impact testers, openings, landfills and filtrations), characterisation of soils by withdrawing samples for laboratory analysis (granulometric analyses, textures of soils, mineralogical characterisation, Atterberg's limits and simple compressions) and in-field tests (boreholes and dynamic penetrations). This characterisation consists of the pressures acting on the area (the degree of alteration of the rock unit), of the most common bedrock types in the study area (such as isolated or continuous footings), of the ease of excavation (altered non-rocky formations or young Quaternary sediments), and of the formations of the unaltered bedrock. Non-rippable Neocene arkoses and metasediments are excavated using machines, while other lithologies, such as granites, quartzites, slates, schists and limestones, are un-rippable and are therefore excavated using explosives. This characterisation aids in the compartmentalisation of the geomechanical characteristics in our studied zone and the generation of the geomechanical mapping (Figure 4g).

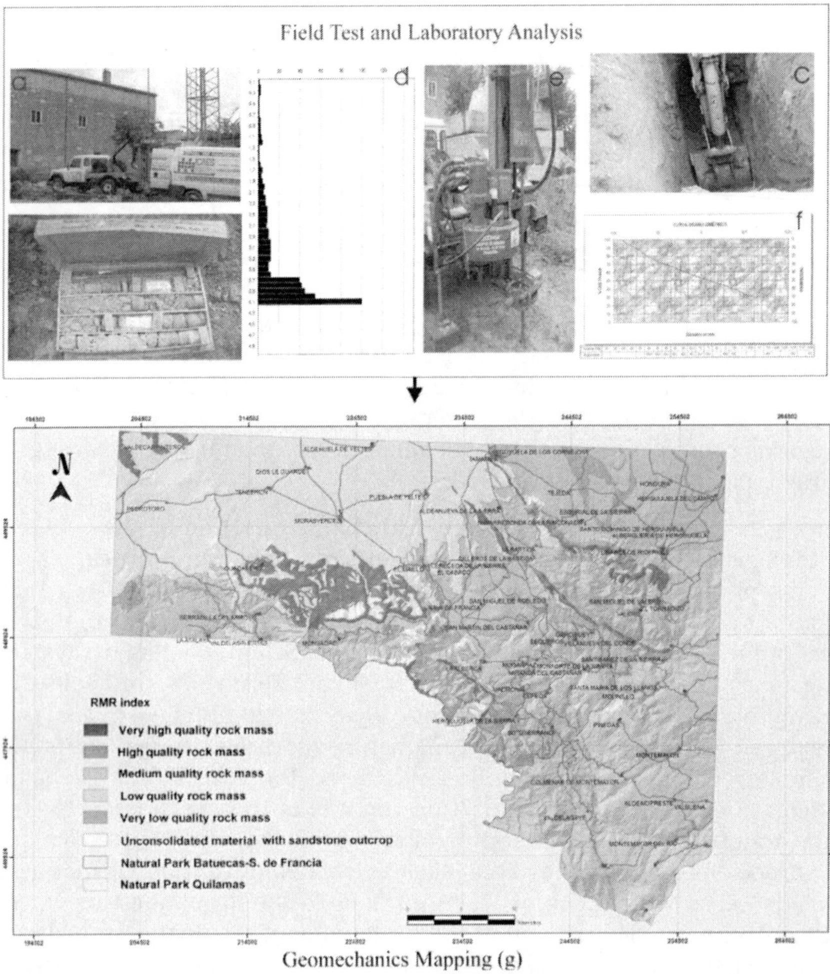

Figure 4. Field Test: (**a**) borehole; (**b**) borehole sample box; (**c**) pits; (**d**) dynamic penetration graph result; (**e**) dynamic penetration machine; (**f**) granulometric analysis and (**g**) Geomechanics mapping.

In a third phase, the overlapped maps and materials were grouped by homogeneous characteristics that allowed us to establish zones in which the response of the area was similar, according to the geotechnical behaviour, which generated the geotechnical mapping (Figure 5). This geotechnical mapping analyzes the lithological (bedrock and surficial deposits), hydrogeological (fractures-diaclases and permeability), geomorphological (domains and slopes) and geotechnical (geomechanical and seismicity) characteristics, which allowed the stability conditions of the terrain to be established (Table 1). Analysis of the W sector of the study zone, e.g., the locality of Pedrotoro, covers area IV (formed by Neogene arkoses and Quaternary detrital materials made up of alluvial fans and terrace deposits associated with river banks, hillside deposits colluvial-, scree-, and alluvial sediments on valley floors), and zone IV1 comprises gravels, and arkosic sandstones and slates. The superposition of the hydrogeological layer indicates a low permeability (B) due to porosity, an absence of fractures in the structural mapping being observed. The superposition of the geomorphological domain mapping indicates a domain of colluvial hillsides. Finally, the geomechanical mapping shows that the zone has unconsolidated materials and sandstone outcrops, the geomechanical characteristics being obtained from field work (a borehole, Figure 4a,b) with a depth of 10 m that provided a single continuous core; two cuts of 4 m depth performed with a back hoe (Figure 4c), and continuous dynamic penetration, constrained to the greatest depth mechanically possible (Figure 4d,e), together with laboratory assays (granulometric analysis Figure 4f and Atterberg limits). The assays revealed that the degree of alteration of the arkose, reached a depth of 4 m, as confirmed by the recoil of the dynamic penetration device, pointing to a consolidated arkose as from that depth. Above a depth of 4 m, substrate is seen in the first two meters, corresponding to gravels of the colluvial hillside materials, with a sandy matrix from the weathering of sandstone. Between 2 and 4 m, the substrate is mainly made up of sands with a fine silt fraction, as seen from the granulometric curves.

In this geotechnical cartography, the 10 points for the sampled soil are indicated. The parameters obtained for these points were the depth of the unaltered bedrock; the typology of soil according to the unified system of classification of soils; and the type of dominant clay (illite, montmorillonite, kaolinite, smectite, chlorite or goethite).

Finally, in the fourth phase, using the previous geotechnical characterisation, it was possible to establish the stability conditions of the terrain (Table 1) and a mapping of natural hazards was generated (Figure 6a). The mapping of natural hazards illustrates the different processes that can produce lithological, geomorphologic, hydrological and/or geotechnical problems. These are described for the different zones in Results. This cartographic analysis considered the recommendations of the international commission of geotechnical

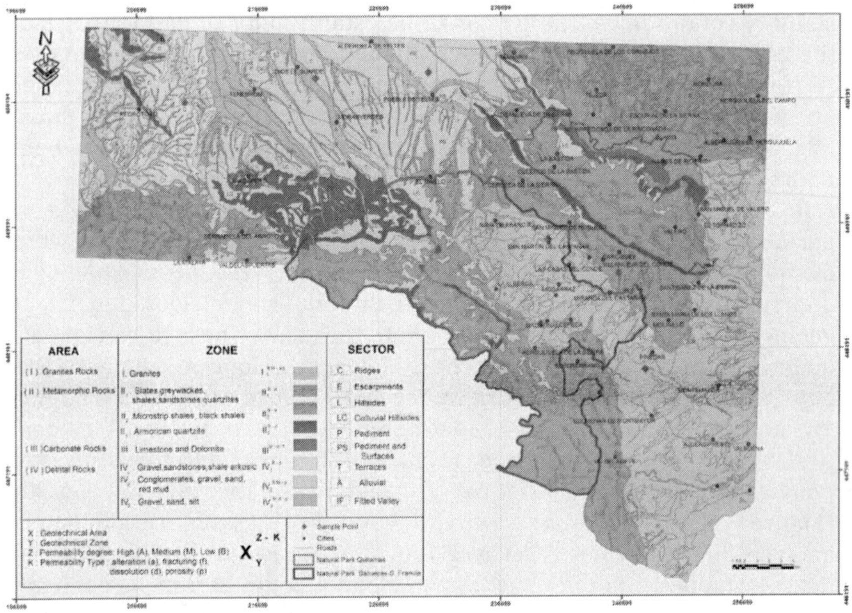

Figure 5. Geotechnical mapping.

mapping [26,27,28,29], which is a basic mapping used for the anticipation risks [30]. Additionally, the matrix of the conditions permitting construction was considered (Table 2), in which the different zones are contrasted with the sectors (geomorphological domains: escarpments, foothill and alluvial fans (raña), hillsides, colluvial hillsides, terraces, fitted valleys, alluvial river beds and ridges: favourable (F), acceptable (A) and unfavourable (U) conditions are shown, and this allowed us to establish the mapping of geotechnical hazards (Figure 6b). The classification criteria used to define the conditions viable for construction came from the analysis of the stability obtained for each geomorphological domain on the basis of all the parameters used in the natural hazard mapping. This mapping allowed us to determine which sectors were favourable for the emplacement and installation of different construction works (buildings, schools, and infrastructures (roads), installation of power lines systems *etc.*), affording, in preliminary phases, knowledge about the behaviour of the terrain to be integrated in order to take advantage of the capacity and stability of certain sectors of the natural park.

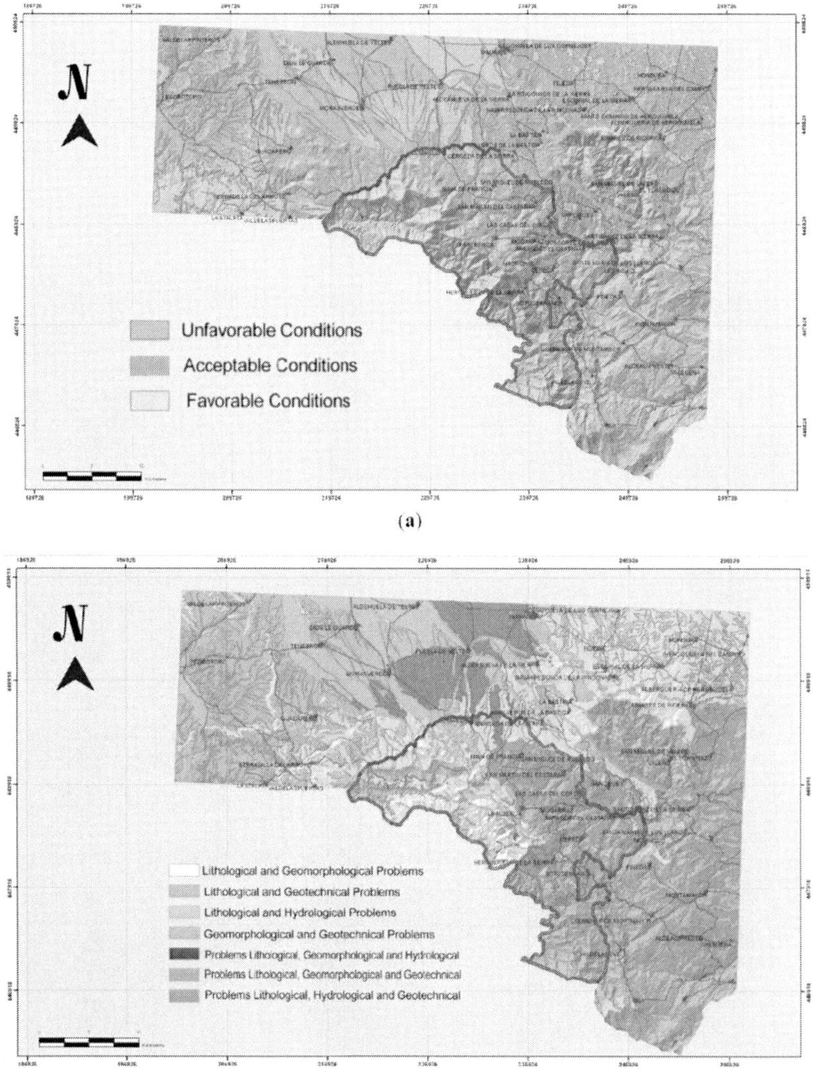

(a)

(b)

Figure 6. (**a**) Natural hazard mapping and (**b**) geotechnical hazard mapping.

Table 1. Area and subarea with the lithological, hydrogeological, geotechnical and stability characteristics.

Area/ Subarea		Characteristics			
		Lithological	Hydrogeological	Geotechnical	Stability
I		Granites	Permeability Low-Medium due to alteration or porosity	High load capacity and acceptable settlement	Stable at summits Unstable on colluvial hillsides
II	1	Slates, schists, greywackes, sandstones and quartzites	Low permeability for alteration	Low load capacity in altered zones and different degrees of settlement	Unstable on hillsides on gentle slopes
	2	Microstriped slates and black pelitic with phosphates and pyrites	Very Low-Low Permeability due to alteration		Stable on pediments on gentle slopes
	3	Armorican quartzite	Very Low Permeability due to porosity	High load capacity and acceptable settlement	Stable at summits Unstable on hillsides and escarpments
III		Limestones, dolomites, and calcareous gaps	Medium Permeability due to dissolution or fracture	High load capacity in unaltered zones and low-dissolution zones and different degrees of settlement	Stable in unaltered zones Unstable due to dissolution: subsidence and collapses
IV	1	Gravels, sandstones and arkosic shales	Low permeability due to porosity	Medium-High load capacity and without settlement	Stable on pediments Unstable on hillsides and escarpments
	2	Conglomerates, gravels, sands and red muds	Low, medium permeability due to porosity	Low load capacity due to ground water and a medium degree of settlement	Unstable due to gravity and the effects of groundwater
	3	Gravels, sands and free silt	Medium-high permeability due to porosity	Very low load capacity and a low of degree of settlement	Stable on terraces Unstable on hillsides and flood plains

Table 2. Matrix capability suitability: (U) unfavourable, (F) favourable and (A) acceptable.

Geomorphologic domains/areas	I	II			III	IV		
		II_1	II_2	II_3		IV_1	IV_2	IV_3
Escarpment	U	U	U	U	U	U	U	U
Alluvian Fan (Raña)	U	A	A	A	F	F	F	F
Slopes	U	U	U	U	F	F	U	U
Colluvium	U	U	U	U	U	U	U	U
Pediment	F	F	F	F	F	A	A	A
Terraces	A	A	A	A	A	A	A	A
Fitted valley	A	U	U	U	A	A	A	A
Alluvial	U	U	U	U	U	U	U	U
Ridges	U	U	U	U	U	A	A	A

Notes: (I) granites. (II) metamorphic rocks: II_1 slates, schists, graywackes and sandstones; II_2 black and micro-striped slates; and II_3 armorican quartzites. (III) carbonate rocks and (IV) detritic rocks: IV_1 arkosic conglomerates, sands and muds; IV_2 reddish-colour conglomerates, sands and mires; and IV_3 and IV_3 alluvial floor gravels and sands.

3. RESULTS

The cartographic procedure followed generated a geotechnical mapping for the zone of study, which differentiates the following areas.

Area I. Formed by granitic materials. These plutonic rocks comprise aplite and pegmatite intrusions that can resist erosion and disintegration from weathering. From the hydrogeological perspective, this area comprises impermeable materials in which permeability limits the area to fractures and jointing, favouring their alteration. The alteration of this granitic mass produces an intense sand alteration that promotes intergranular permeability. Therefore, the permeability is considered low to moderate because of the alteration and/or fracturation. The geotechnical characteristics are suitable, with granular, altered soils with the granite at a low depth, generating zones with high load capacity and acceptable settlement. The domains of summits and motions have high load capacity and minimal settlement, with saprolitic compact of granitic alteration and nearness of the fresh rock. The domains comprising colluvial hillsides are unstable at the superficial level against side thrust forces that promote slopes.

Area II. Formed by materials metasediments composed of the complex schist-greywackes. The lithology is primarily formed by interleaving slates, graywackes and sandstones as well as by interleaving quartzite, black slates and conglomerates. From the hydrogeological perspective, they are materials with low permeability with infiltration processes that are favoured by alteration and fracturation. This impermeability is promoted by clay formation due to alteration of these lithologies and can entirely or partially "seal" the fissures therein. From the geotechnical perspective, certain zones are highly altered and comprise different settlement in superficial foundations with isolated footings. Construction conditions are acceptable, in unaltered, or slightly upset rock. Banks with middle and high slopes can induce slides that favour the foliation planes in the slates. The quartzite and schists may yield problems with detachment or falling blocks. In this area, three zones differ, II_1, II_2 and II_3, depending on the varying lithological and hydrogeological parameters (degree and typology of permeability).

The zone II_1 is formed by slates, schists, graywackes, sandstones and quartzites; it has low permeability and slow drainage that favours the alteration processes. Zone II_2 is composed of microstriped slates and black pelitic with phosphates and pyrites as well as permeability conditions little lower than the previous ones and drainage favoured by the alteration processes. Zone II_3 comprises the most resistant lithology in this area: the armorican quartzite with continuous crest projections.

This area has low permeability and is limited to fracturation zones. The domains comprising hillsides and scarp with high slopes generate take-offs that favour stratification planes and middle settlement in zones of intense alteration. This zone is stable in the domains that comprise summits and motions with few alterations and soft slopes.

Area III. This area is formed by carbonate rocks, limestones (primarily), dolomites, and calcareous gaps. This is an area with a minor extension in the studied zone and is limited in outcrops to the zones with the highest levels of Quilamas's Saw (limestones of Tamames). From the hydrogeological perspective, they are considered materials with moderate permeability and the most dominant processes that affect this area are dissolution and fracturation. The load capacity is high on the compact rock with slightly significant processes of karstification; however, collapses or structure subsidence may be a problem.

Area IV. This area is formed by Neogene and Quaternary detrital materials. The primary materials that compose this area are arkoses of the Miocene with former alluvial fans covered by extensive surfaces that comprise deposits superposed with alluvial fans from the Pliocene–Quaternary (locally, "rañas") and terrace deposits from the Pleistocene associated with fluvial riverbeds, hillside deposits (colluvial and screes), and alluvial sediments in valley funds. In the domains comprising bedrock arkoses in summits and motions with stretched slopes, the load stability and capacity is high for acceptable settlement.

Zone IV_1 comprises gravels, sandstones and shales that are arkosic; this zone has low permeability and drainage that promotes intergranular porosity. This high degree of consolidation generates compacted and stabilised natural conditions; the load capacities are moderate-discharge; and this area does not contain settlement. The arcosas, with hanging middle-discharges, concentrate the rain run-off, which generates instability for the gullies and gulchs.

Zone IV_2 comprises deposits of conglomerates, gravels, sands and red muds that compose Neogene alluvial fan deposits that compose an additional formation: the raña of Pliocene–Quaternary with clay light slopes that can generate gravel movement in favour of the overlain levels. The permeability of these materials is low-moderate with drainage for intergranular porosity. The groundwater level may interfere especially at the sludge levels, the load capacity is low, and the generated settlements are middle.

Zone IV_3 includes materials that shape the alluvial plain deposits and valley funds. The lithology is formed by gravels, sands and free silt with low and very low slopes. The permeability of these materials is moderate-high with drainage for porosity. In the flood plains and intermittent riverbeds, the groundwater level is at a low depth and there are potential flood zones with low load capacities and likely significant settlement. The terrace deposits comprise zones with acceptable load capacities and small settlement because the detritic reorganisation adapts rapidly to the applied stresses. The colluvial and screes materials destabilise wide hillsides.

From this geotechnical mapping, by overlapping and crossing with technologies (GIS), natural hazard mapping (Figure 6a) was generated, which shows zones according to the levels of natural danger. The high hazards are limited, from the geomorphologic perspective, to the domains comprising peaks, ridges, and summits. These can generate problems from gelifraction that produce fissures, which can induce detachment and cracking for different rock movements. The colluvial materials and falls, associated with slopes, constitute zones of high risk, given the instability of these materials and the high slopes. Given the riverbeds and river margins in these zones, the potential flood areas are Morasverdes, Tenebrilla and Yeltes, which generate hydrological problems. These zones have characteristic lithologicies, groundwater levels at low depths, and low load capacities. They can generate ponds in partially depressed zones, including level surfaces because the clay accumulation induces deficient drainage. The fluvial scarps can produce erosional problems associated with the occasional displacement of the principal riverbed, which generates instability and lateral erosion of the fluvial riverbed. The limestones can produce lithological and geotechnical problems, mainly karstification and hollowing from dissolution. Clay soils are located above the metasedimentary materials, in slate alterations, and in flood plains along the fluvial margins, where they demonstrate a certain plasticity that can generate geotechnical problems. This plasticity is especially unfavourable in geomorphic domains that promote mobility (fluvial incisions and hillsides in slates) and depends on the disposition of the cleavage and the degree of alteration.

Moderately hazardous sectors present an important level of competence and high load capacity. However, localized geomorphologic problems could generate detachments and cause clay slides from highly sloping altered areas as well as creeping in moderately sloping areas. The depressions in this area and in certain sectors favour localized hydrological risks.

The low-hazard areas include detritic formations without significant clay levels and constitute stable zones; the settlements are immediate and acceptable. Materials therein include alluvial fans with stretched slopes, where the hydrological and geomorphologic problems are compatible with the most frequent uses of these territories. The stability is high, and the load capacity high in the granitic zones with little fracturing and weathering, as well as in geomorphologic domains comprising motions over metamorphic rocks (e.g., quartzites and schists) with low levels of alteration.

4. CONCLUSIONS

The constant increase in different human activities as well as the commitment to protection and conservation of certain natural parks requires more sustainable development practices and integral planning for areas such that environmental risks are predictable and mitigated. To achieve this goal, the conditions and uses of the areas under scrutiny must be evaluated. This evaluation relies on the exhaustive knowledge of geodynamic external processes, which can generate different geological risks.

The geotechnical mapping in the "Las Batuecas-Sierra de Francia" and "Quilamas" Natural Parks comprises hazards that directly influence determination of geology (different resistance lithology as well as degrees and types of material permeabilities); geomorphology (areas capable of slide and creeping; lateral erosion in fluvial riverbed, detachments in screes with high slopes); hydrology in zones with hanging falls for avenues and floods; geotechnical risks from load capacity; settlement; and degree of plasticity. The generation of this natural hazard mapping demonstrates the importance of a detailed, hydrological analysis of the geomorphologic problems, lithology and geotechnology in the mapping of risks for the potential uses of the region in natural parks planning and management.

The cartographic procedure described here is a sufficiently effective, economical, and precise tool for initial decision making regarding the management of natural parks and their planning, and it is applicable to any natural space. These geotechnical mappings facilitate the demarcation of areas to which limitations and recommendations concerning construction activities should be applied. This method can be used for the prevention and mitigation of natural risks, and it is validated on the basis of the presence of active processes that create unfavourable conditions for construction, as illustrated by the mapping of natural hazards (Figure 6a) that corresponds to trigger factors for these risks and is reflected in the three categories of geotechnical risks (Figure 6b). Thus, in recent years different active processes have been recorded (landslides and rock falls, *etc.*) in the study zone. These movements have caused fatal injuries and damage to infrastructures (roads and bridges) and buildings. The field work, its publication in different journals, and the analysis of requests for help from the Regional Government by town councils have allowed us to make an inventory of these sectors and include them on the maps obtained here, validating the procedure since they coincide with zones that are unfavourable for construction.

The results of the present work allow us to confirm that the mapping of natural hazards from thematic maps at a scale of 1:50,000, characterised by a series of assays representative of each geotechnical area and zone, offers a useful tool for the governance of future human activity in natural parks. The results of the field work and laboratory assays of geotechnical reports, both public and private, offer a geotechnical database of geomechanical parameters that can be used in this procedure, thereby decreasing the number of assays that need to be performed and the economic costs involved.

In Spain, in recent years maps addressing hazards have been drawn up for certain areas of reduced extension, usually within large cities and their surroundings, but not at regional level, in which geotechnical maps at a scale of 1:200,000 are used. The present work improves the methodology for regional territorial sectors, and the results reported here are based on a combination of classic mapping techniques and GIS technologies. Detailed mapping of the geomorphogical domains, compiled from a precise photo-interpretation at scales of 1:10,000 and 1:18,000 is of great importance for the correct differentiation and demarcation of the different sectors for their evaluation in combination with other already available maps since it can be used to determine land sectors that are susceptible to instability.

The potential beneficiaries of these maps are the managers of natural parks and directors of protected natural spaces, since the maps will allow such agents to make decisions about the suitability of authorizing, or not, the emplacement of different types of construction (buildings and infrastructures). The municipalities affected, that are interested in doing so, will also be able to access maps for the prevention of possible risks, allowing a more rational planning of land use.

ACKNOWLEDGMENTS

The authors express their thanks to Projects CGL2012-33430/BTE and CGL2012-37581-CO2-01. Also, we appreciate three anonymous reviewers for their comments.

REFERENCES

1. ISRM, Suggested methods for rock characterization, testing and monitoring. In ISRM Suggested Methods; Brown, E.T., Ed.; Pergamon Press: Oxford, UK, 1981.
2. Waltham, A.C. Foundations of Engineering Geology; Blackie Academic and Professional; Chapman and Hall: London, UK, 1994.
3. Geological Society Engineering Group Working Party. The preparation of maps and plans in terms of engineering geology. Q. J. Eng. Geol. 1972, 5, 293–381. [CrossRef]
4. Dearman, W.D.; Fookes, P.G. Engineering geological mapping for civil engineering practice in the United Kingdom. Q. J. Eng. Geol. 1974, 7, 223–256.
5. Instituto Geológico y Minero de España (IGME), Mapa Geotécnico General de Salamanca 1/200.000 [in Spanish]; Ministerio de Industria: Madrid, Spain, 1976.
6. Peña Pinto, J.L. Los mapas geotecnicos. Tecniterrae 1977, 18, 46–63.
7. Abad Fernández, J.; Ayala Carcedo, F.J.; Pernía Llera, J.M. Cartographie géotechnique a grande échelle (1:1000–1:5000) de l'Institut Géologique et Minier d'Espagne pour la Planification Industrielle et Urbaine. IAEG Symposium "Engineering Geological Mapping For Planning, Design And Construction In Civil Engineering", Newcastle. Bull. Eng. Geol. Environ. 1979, 21, 83–91.
8. United Nations Educational Scientific and Cultural Organization (UNESCO), Engineering Geological Maps. A Guide to Their Preparation; UNESCO: Paris, France, 1976.
9. Instituto Tecnológico y Geominero de España (ITGE), Mapa Geotécnico de Ordenación Territorial y Urbana de la subregión de Madrid: Torrelaguna. E: 1/ 100.000 [in Spanish]; Ministerio de Industria: Madrid, Spain, 1991.
10. Toulemont, M. Évaluation et cartographie des risques geotéchniques de dégradation des ouvrages d`assainissement: Une aide á la gestion des réseaux [in Spanish]. Bull. Eng. Geol. Environ. 1978, 36, 89–100.
11. Rozos, D.; Apostolidis, E.A.; Xatzinakos, I. Engineering-geological map of the wider Thessaloniki area, Greece. Bull. Eng. Geol. Environ. 2004, 63, 103–108.
12. Que, J.S.; Wang, Q.; Chen, J.P.; Shi, B.F.; Meng, Q.H. Geotechnical properties of the soft soil in Guangzhou College City. Bull. Eng. Geol. Environ. 2008, 67, 479–483.
13. Okogbue, C.O.; Aghamelu, O.P. Comparison of the geotechnical properties of crushed shales from Southeastern Nigeria. Bull. Eng. Geol. Environ. 2010, 69, 587–597.
14. Solberg, I.L.; Hansen, L.; Steinar Rønning, J.; Haugen, E.D.; Dalsegg, E.; Fredrik Tønnesen, J. Combined geophysical and geotechnical approach to ground investigations and hazard zonation of a quick clay area, mid Norway. Bull. Eng. Geol. Environ. 2012, 71, 119–133.
15. Phipps, P.J. Geomorphological assessments for transport infrastructure projects. Proc. ICE Transp. 2003, 156, 131–143.
16. Parry, S.; Hart, J.R. Engineering geology and the reduction of geotechnical risk: Challenges facing the profession in Hong Kong. Q. J. Eng. Geol. Hydrogeol. 2009, 42, 499–510.
17. Ellis, L.A.; Harrison, E.; Bowden, A.J. Landslides on Gault: Geomorphological identification and qualitative risk assessment. Q. J. Eng. Geol. Hydrogeol. 2011, 44, 35–48.
18. Bottino, C.; Civita, M. A Computer Semi-Quantitative Model for Microzonation of Hazard from Interconnection of Engineering-Geological Features and Urban Sub-Service Network. In Proceedings of 5th International Association of Engineering Geology Congress, Buenos Aires, Argentina, 20-25 October 1986; 3, pp. 1731–1740.
19. Chacon, J.; Irigaray, C.; Fernández, T.; El Hamdouni, R. Engineering geology maps: Landslides and geographical Information Systems. Bull. Eng. Geol. Environ. 2006, 65, 341–411.
20. Sidi Mohammed, A.M.; Houmadi, Y.; Bellakhdar, K. Geotechnical Risks Map of Saïda City, Algeria. Electron. J. Geotech. Eng. 2010, 15, 403–413.

21. Terzagui, K.; Peck, R.B. Soil Mechanics in Engineering Practice; John Wiley and Sons: New York, NY, USA, 1967.
22. Cooke, R.U.; Doornkamp, J.C. Geomorphology in Environmental Management, 2nd ed; Oxford University Press: Oxford, UK, 1990.
23. Hutchinson, J.N. Reading the ground: Morphology and geology in site appraisal. Q. J. Eng. Geol. Hydrogeol. 2001, 34, 7–50.
24. Real Decreto 997/2002, de 27 de Septiembre, por el que se Aprueba la Norma de Construcción Sismorresistente: Parte General y Edificación [in Spanish]; BOE-A-2002-19687; Ministerio de Fomento: Madrid, Spain, 2002.
25. Bieniawsky, Z.T. Engineering Rock Mass Classifications; John Wiley and Sons: New York, NY, USA, 1989.
26. Abad Fernández, J.; PeñaPinto, J.L. Metodología de la Cartografía geotÉcnica a E 1: 200.000 en España. Mapa Geotécnico General. Tecniterrae 1975, 3, 44–54.
27. Echevarria Caballero, M.R.; Peña Pinto, J.L. La Cartografía Geotécnica en España. Tecniterrae 1978, 25, 19–24.
28. Matula, M.; Dearman, W.R.; Golodkovskaja, G.A.; Pahl, A.; Radbruch-Hall, D.; Sanejouand, R. Classification of rocks and soils for engineering geological mapping. Part I: Rock and soil materials. Bull. Int. Assoc. Engin. Geol. 1979, 19, 364–371.
29. International Association Enginering Geology and Environment (IAEG). Rock and soil description and classification for engineering geological mapping report by the IAEG commission on engineering geological mapping. Bull. Int. Assoc. Engin. Geol. 1981, 24, 235–274. [CrossRef]
30. Commission Staff Working Document: Implementation Plan of the EU Strategy for Supporting Disaster Risk Reduction in Developing Countries 2011–2014; SEC(2011) 215; European Commission: Brussels, Belgium, 2011.

CHAPTER 11

Hydro-Engineering Solution for a Sustainable Groundwater Management at a Cross Border Region: Case of Lake Nyasa/Malawi Basin, Tanzania

Yohannes Yihdego[1] , Andrew Paffard[1]

[1.] *Snowy Mountains Engineering Corporation (SMEC)SydneyAustralia*

ABSTRACT

A groundwater resources assessment has been carried out for the Lake Nyasa Basin east Africa, with reference to sub catchments further to the whole basin wise analysis, including quantification of potential yields from the aquifers. Numerical groundwater models, MODFLOW-SURFACT coupled with Visual MODFLOW, were used to assess sub-basin groundwater sustainability (Via "Zone Budget" software package). The model has been calibrated to observed hydraulic head and processed baseflow estimate. It is concluded that the aquifer system is sustained by episodic recharge and the long-term gaining storage which represents the maximum extractable volume. Future predictions of groundwater usage indicate that by 2035 the percentage of annual safe yield extracted will increase to between 11.3 and 103%. Model result suggests that there is a need to revise the current estimate of sustainable yield based on future climate change conditions and projected population growth, which would decrease spring flows substantially and decrease hydraulic head basin-wide. It also suggests that by 2035 some sub catchments will be nearly at or exceeding the annual safe yield leaving no room for socio-economic development, or the need to reduce existing socio-economic demands to meet domestic demands. Besides to improving the natural replenishment capacity through artificial recharge technique, the other option is to increase the percentage of total recharge allocated to the annual safe yield from 10% of the total annual recharge to 20% in these catchments. This provides a basis for management of individual groundwater scheme in sustaining livelihood activities or implications of policies and to develop a plan for potential groundwater extraction scenarios pertaining to water use and allocation.

Keywords: Water resources management Lake Malawi/Nyasa Tanzania East Africa Rift valley

BACKGROUND

Groundwater models are valuable tools for the catchment managers to take actions on the management of groundwater resources [1, 2, 3, 4]. Groundwater occurrence and movement are controlled primarily by the aquifer permeability and the lithology of the underlying strata [5, 6]. In groundwater-resource assessments, terms such as sustainability, sustainable yield, sustainable pumping, sustainable development, and safe yield are often used interchangeably or with little distinction between the concepts. Definitions and approaches to assess sustainability of potable drinking-water resources are summarized by Bredehoeft [7], Alley and Leake [8] and Devlin and Sophocleous [9] among others. Sophocleous [10], for instance defines sustainable development as development that meets the needs of the present without compromising the ability of future generations to meet their own needs. Specifically, optimal or sustainable groundwater-resource use requires setting upper limits on water withdrawal (or sustainable yield) to avoid compromising the source [11]. Previous researchers (e.g. [12]) defined safe yield as the maximum quantity of water which can be extracted from an underground reservoir, yet still maintain the supply unimpaired. Under natural conditions, recharge is balanced by discharge from the aquifers by evapotranspiration and/or exfiltration into streams, springs, and seeps [10]. It is widely recognized that there is a need to reserve a fraction of the recharge for the benefit of surface waters, related ecosystems and that fraction is difficult to quantify [13, 14, 15, 16].

In areas where potable water resources are limited, an accurate and reliable estimate of sustainable use or yield is critical. Yield estimates have to address not only decreasing water availability, as manifested by declining water levels, but also deterioration of water quality caused by, e.g., seawater intrusion or agrochemical use. In recent years, hydrological models with various degrees of sophistication, have been used in different settings to evaluate water budgets under several climate or land-use scenarios. Several approaches have been suggested for evaluation of sustainable yield [17]. Generally, sustainable yield of small-scale abstraction schemes is estimated on the basis of analytical solutions applied to constant-rate test data and forward modelling with the estimated transmissivity and storativity. The modelled abstraction rate is adjusted to limit the drawdown to a predefined level (usually the uppermost water strike) at the end of a designed production period. Difficulties arise in applying analytical solutions where the simplifying assumptions are not supported by field data [18]. Numerical models are used in hydrogeology to increase understanding of the complex systems that characterise aquifers [19, 20]. The results obtained from these models cannot be considered complete or totally accurate because of the inherent uncertainty of the models [21, 22]. However, the main trends of the hydrological behaviour may be identified, as long as the underlying conceptual model is a reasonable reflection of reality, and the parameters of the model have physically reasonable values [7, 23, 24]. Once these conditions have been met, the numerical models may be used as decision-making tools in water-management planning [25].

Lake Nyasa, located in the south western part of Tanzania is shared by Malawi and Mozambique. The study for the preparation of the Integrated Water Resources Management and Development Plan (IWRMDP) for the Lake Nyasa basin is only concerned with the lake catchment which falls within Tanzania (Fig. 1). The quantification of the groundwater resources of Tanzania has not yet been possible because of a lack of requisite data [26, 27]. Groundwater development has concentrated mainly on shallow wells for domestic purposes over a wide part of the country. They are also commonly used in the peri-urban fringes where there is no distribution network and places with unreliable supply [28]. The objectives of this study are to determine the availability of groundwater, demand on groundwater and its development potential on sub-catchment basis individually, using three dimensional numerical modelling. The data presented in this paper and interpretations have been sourced from SMEC [29] consultant's report. Further to the whole basin wise assessment, categorizing sustainable groundwater development analysis based on sub-basin/sub-catchment, offers a better understanding of the sub-basin among the stake holders and decision makers for an effective management.

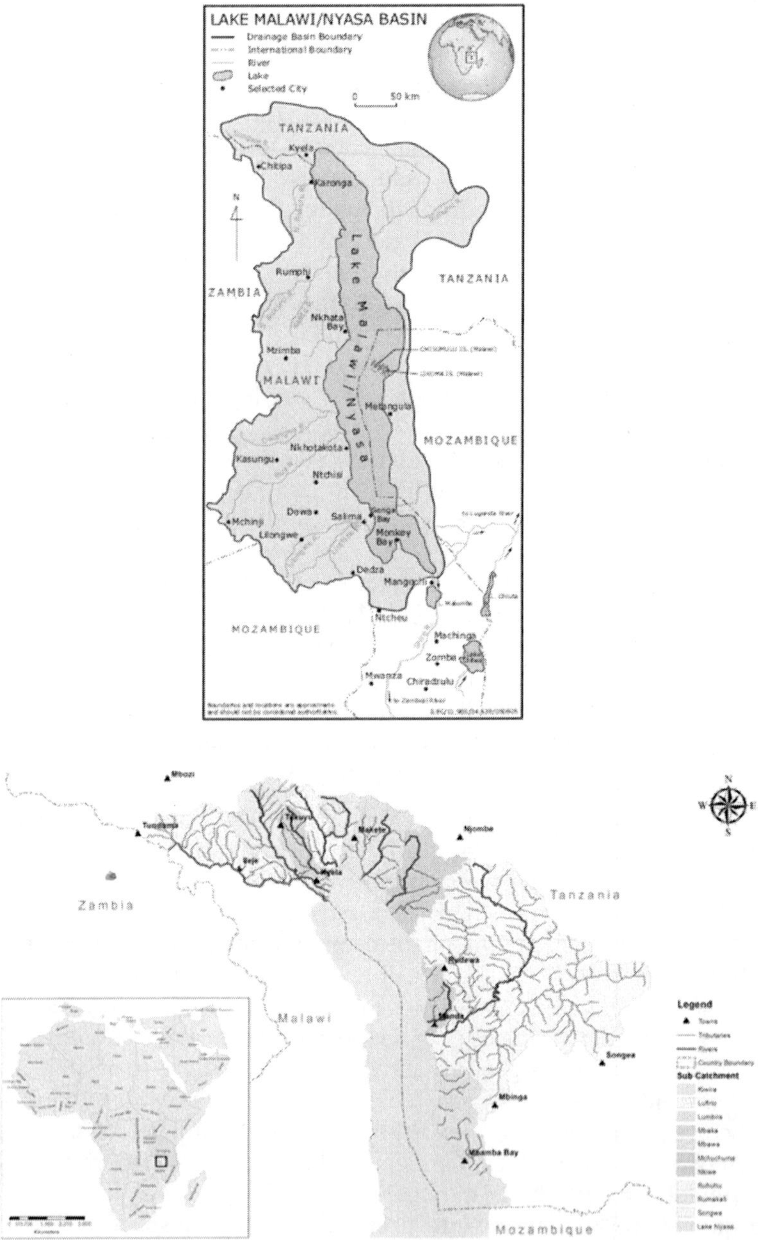

Fig. 1. Location map of the study area (Lake Malawi/Nyasa; after [30]), together with the Lake Nyasa sub-catchment names

Site description

Lake Nyasa Basin, situated in the south western part of Tanzania is part of the greater Zambezi Basin. Lake Nyasa is an African Great Lake and the southernmost lake in the East African Rift system. The lake is 560–580 km long, has a maximum width of 75 km and an average depth of 292 m and a maximum depth of 706 m. It drains via the Shire River in Mozambique which flow to the Zambezi River which discharges to the Indian Ocean. Lake Nyasa is the ninth largest and third deepest, fresh water lake on earth [30] with an estimated average volume of around 8000 km^3. The lake supports fisheries, livestock, agriculture, tourism and wildlife in all three countries [26, 31]. Lake Nyasa and the associated catchment basin is shared between Tanzania, Malawi and Mozambique and covers a total area of 165,109 km^2 including the lake. The lake catchment within Tanzania is calculated to be approximately 27,500 km^2. For this study, the portion of the basin which falls within Tanzania, will be referred to as the "basin". The basin within Tanzania has a large topographic range, being from approximately 470 masl at Lake Nyasa to 2960 masl at Mount Rungwe in the northern part of the basin. The elevation of Lake Nyasa varies annually by around 2 m depending on rainfall [32].

The climate of the basin is strongly controlled by the topography with orographic rainfall effects and a large mean temperature range [27]. The mean maximum and minimum temperature may range between 31.1 and 6.7 °C [29]. Rainfall ranges from 1000 mm to over 2600 mm (Fig. 2). The baseflow component was estimated to be as high as 85% using Baseflow Index for multiple years of data. The Baseflow Index has been found to be consistent and indicative of baseflow, and thus may be useful for analysis of long term base-flow trends [33, 34, 35].

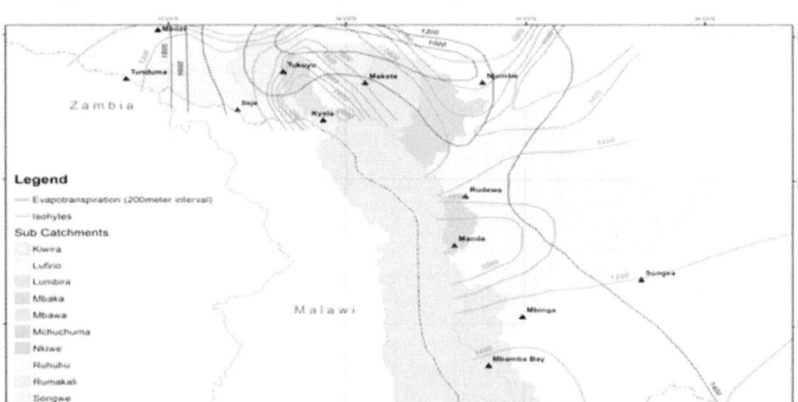

Fig. 2. Rainfall and evapotranspiration isohyet map

Geological setting

The geological framework of Tanzania [32, 36] reflects the geologic history of the African continent as a whole. Basement Complex rocks occupy most of the basin land area and the rest comprises Karroo Formation, volcanic and alluvial deposits (Fig. 3). Lake Nyasa area has been considered a potential oil and gas prospect.

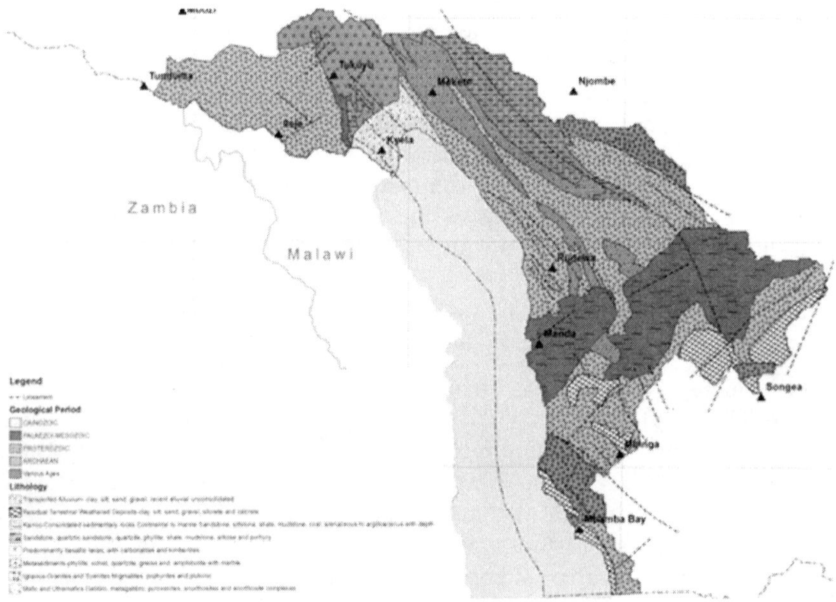

Fig. 3. Geology map of the study area

Population

Based on the National Bureau of Statistics (NBS) 2002 census the current (2012) population of the basin is estimated to be 1,294,500. This is projected to increase to 2,254,600 by 2035. The population distribution (Fig. 4) shows the greatest population concentrations are at the northern end of the basin around Kyela, Tukuyu and Tunduma and on the south eastern side around Mbinga and Mbamba Bay. This distribution closely correlates with the rainfall and evapotranspiration distribution (Fig. 2) with higher population in areas of high rainfall and associated better agricultural land.

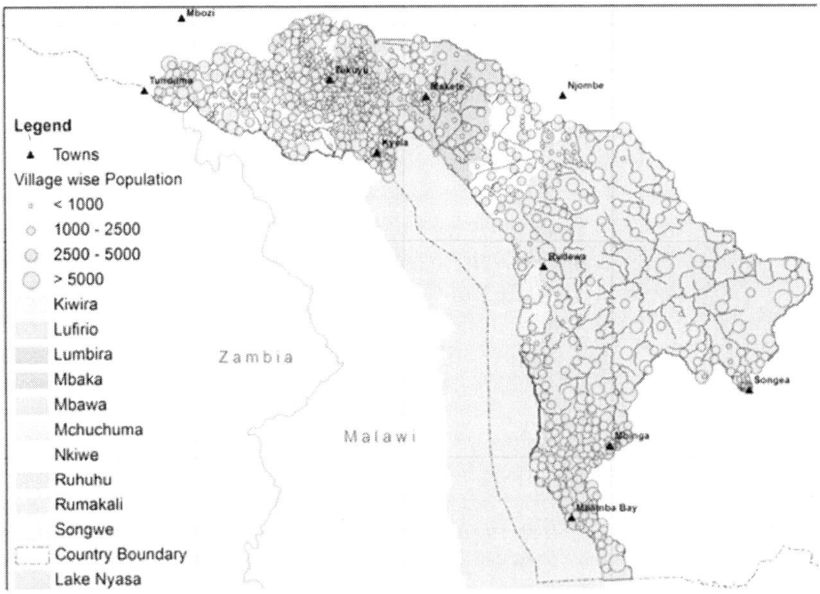

Fig. 4. Population density dot map

METHODS

Compilation of drilling completion, chemical analysis, climate data, soils, geology, topography, vegetation, landuse, soils and population information maps and existing reports were carried out for assessment of the geology, geomorphology, stratigraphy of the basin, establishment of groundwater resource inventory, groundwater potential evaluation, determine productivity for the basin and groundwater recharges mechanism using GIS (e.g. [5, 6]). A total of 316 borehole records were located. The deepest average Standing Water Levels (SWL's) are found in Ruhuhu, Mchuchuma and Kiwira catchments (Fig. 2), for weathered and fractured rock and range from 20 to 28 m below ground level (m bgl). The shallowest average SWL's are found in Mbaka, Kiwira alluvials, Lufirio, Songwe and Mbawa where the average SWL's are between 2.2 and 10 m bgl. It is also likely the depth to water is strongly controlled by topography and climate as opposed to geology.

The Kyela flood plain alluvium has the highest average yields at 21 m^3/h followed by Mchuchuma with 20.5 m^3/h. Average yields for all other areas, which are screened in weathered to fractured rock are low, being less than 8 m^3/h. The lowest yields are found in Mwaba with a range of 0.4–2.2 m^3/h and an average of 1.26 m^3/h.

AQUIFER HYDRAULIC PARAMETERS

Assessment of the pumping test data was made using Aqtesolv Pro® Version 4.5. The alluvials have a transmissivity and hydraulic conductivity range of 1.16–761 m^2/day and 0.01–7.6 m/day respectively. The transmissivity for the weathered and fractured rock have a range of 0.3–17.7 m^2/day. The permeability ranges from 0.01 to 1.0 m/day. Although these results indicate that the groundwater potential is very high for the alluvials, the permeability is generally low. On the other hand, while the groundwater potential for the basement rocks is low to moderate, the permeability is generally low to moderate.

Specific capacity was used to identify well or borehole potential, with high values indicating good potential. Typically the water level drawdown was recorded after 24 h of pumping to allow stabilisation of the water level and development of a cone of depression. Shorter time periods are not considered reliable. The specific capacity range of the alluvials and weathered/fractured rock are 129–4089 m^3/h/m and 186–11,640 m^3/h/m respectively. The specific capacity data indicates that alluvials can have nearly three times the potential of bores in the weathered and fractured rock aquifers [37]. This is in contrast to the yield information, but likely is a better reflection of the actual potential which is related to the thickness of the water bearing zone. This effectively links to aquifer types limiting the usefulness of the data. However assessment of the transmissivity from the data is considered reliable enough for providing a general basin wide estimate of aquifers potential.

Conceptual understanding

75% of Tanzania is underlain by largely Precambrian crystalline basement complex rocks with various composition and ages, which form the basement aquifers. Other aquifer types include sandstone, coastal sedimentary formation of limestone, and alluvial sedimentary sequences, and volcanic materials. The groundwater potential of each type of aquifer differs significantly at the local scale as well as at the basin scale. The occurrence of groundwater within the Basin is most influenced by the geology/geomorphology, followed by topography, rainfall, and all of which influence recharge and discharge from the aquifer systems. However development of groundwater within the basin is most influenced by topography and the associated drainage patterns, with the exception of the Kyela area (Fig. 1). The groundwater potential for the basin is generally considered low but locally moderate with no high yield systems, except the Kyela alluvials. However the lack of borehole data and poor distribution prohibits an accurate assessment of the potential.

Numerical simulation of hydraulic head and processed baseflow

The physical system described in the previous section provides a framework for setting up and calibrating a steady numerical groundwater flow model. The objective of groundwater modelling is to gain an understanding of the groundwater flow system in the Lake Nyasa basin and to help constrain and elucidate groundwater contribution to baseflow. The model is parameterized based on hydrostratigraphic zones and constrained using the aforementioned data (hydraulic head and baseflow). A three dimensional model is used for better reproduction of the complex topography and heterogeneity of the watershed. Head, river flow, precipitation data, and the geology and surface topography of the region are used to construct the model. Model output includes distribution of hydraulic head and water budget targets. Yield is calculated from the model output.

Model construction

The model was constructed based on background information provided and the assessment of available information in the previous sections. Modelling was undertaken using MODFLOW-SURFACT code [38], an advanced MODFLOW based code developed by HydroGeoLogic Inc. that handles complete desaturation and resaturation of grid cells, within the framework of Visual MODFLOW Version 4.6. The pseudo-soil function of MODFLOW-SURFACT was chosen to account for the saturated and unsaturated processes within the model, as it avoids the wet/dry non-linear iterations of standard MODFLOW. "Zone Budget" software package were used to assess sub-basin groundwater sustainability.

Spatial discretization and layering

The active model area covers approximately 27,500 km^2 (model area), and is the land catchment area defined by the surface water divide and Lake shore line. A constant head boundary has been used for the Lake Nyasa shore line (Fig. 5). An approximate 2500 by 2500 m grid has been applied across the Model Area (Fig. 5). The model has been set up with two layers:

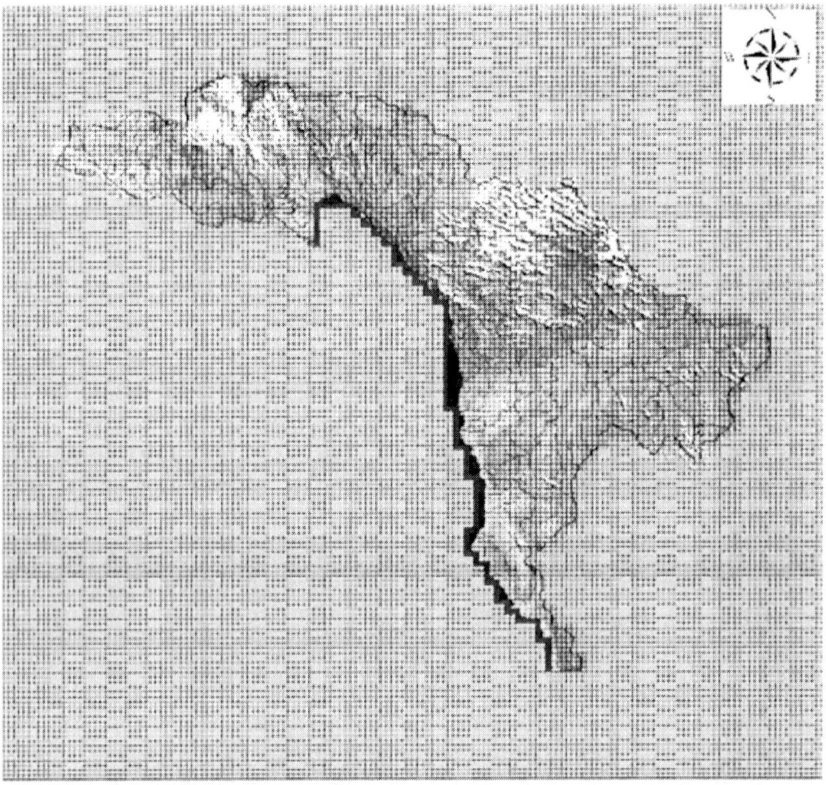

Fig. 5. Groundwater model grid. The *red cells* represent a constant head boundary along the lake shore line

- Layer 1: used to simulate the upper most aquifer including the alluvium, weathered basement and shallow fractured basement (Fig. 6) with a thickness of 5–250 m; and

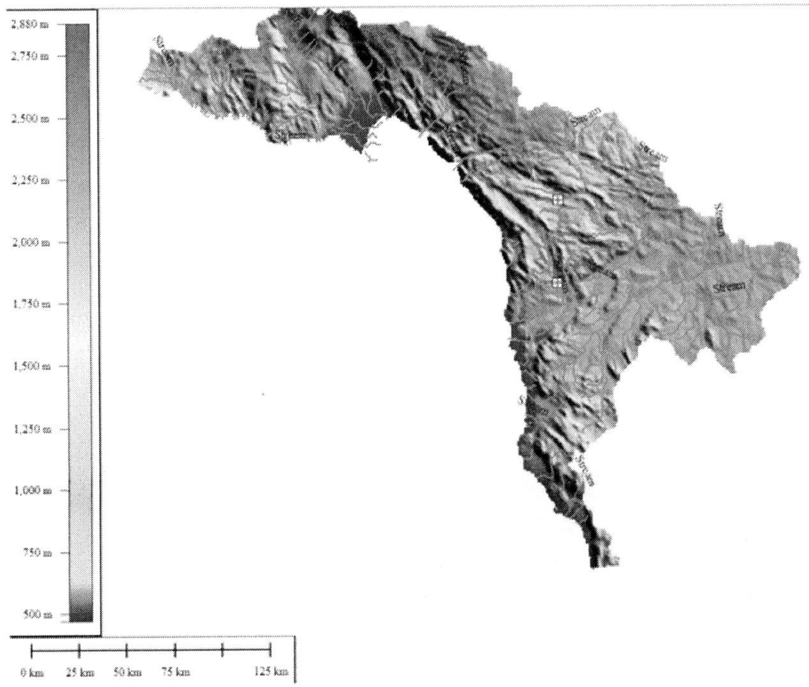

Fig. 6. Surface elevation (*top of layer 1*). The *cross line symbols* represent two stream gauging stations

- Layer 2: A non-uniform deep basement layer ranging in thickness from 400 m to 1500 m.

Boundary and initial conditions

For the purpose of modelling the basin, a simplified aquifer system has been used. The upper model layer is assumed to be one continuous interconnected aquifer with different permeability zones based on the geology. The initial potentiometric surface (Fig. 7) was estimated from borehole data where a water level and location were available. The groundwater elevation was estimated using the borehole location and digital elevation model.

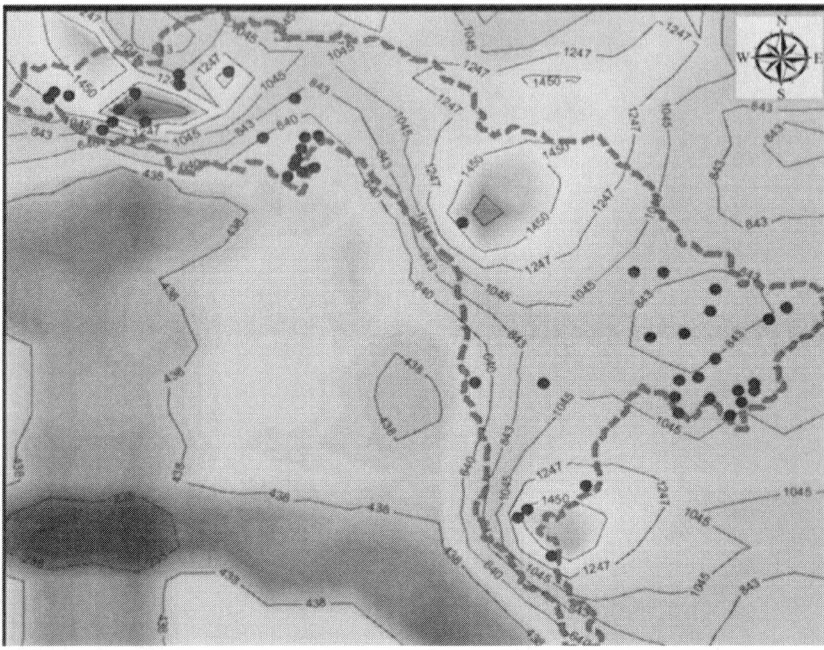

Fig. 7 Potentiometric surface of Lake Nyasa basin. The basin is shown as the *dashed red line* and *blue dots* represent boreholes with water levels

Recharge and discharge

Recharge is assumed to be predominantly from rainfall infiltration. This may occur as direct recharge through exposed bedrock fractures or via slow percolation through the weathered overburden. A small component of recharge is thought to occur via stream and river bed leakage especially in the lower reaches of the river systems. As there is no long term water level monitoring data recharge has been assumed to be a percentage of the total annual rainfall [39]. Rainfall infiltration rates in the order of 5–20% have been estimated during the model calibration. Recharge zoning was made based on the geology and rainfall isohyet map.

Groundwater discharge is via evapotranspiration, baseflow of streams and rivers (as river leakage which discharges to the Lake), pumping, and deep percolation discharge to Lake Nyasa. The component of deep percolation is generally assumed to be about 10% of recharge. In the elevated plateaus of the catchments, baseflow is assumed to be the major discharge mechanism. In the lower reaches, evapotranspiration is thought to play a larger role especially in the lower reaches of the Ruhuhu River and Kyela alluvials. Given the depth of groundwater is generally shallow in the model area, the evapotranspiration (ET) is considered significant. The ET rate was set at 10% of the estimated values shown on Fig. 2, with an extinction depth of 2 m. The ET rate was refined during the calibration process.

Hydraulic parameters

Aquifer properties were set based on the limited available information and accepted literature values (e.g. [40]). The properties were further refined (within acceptable limits) during model calibration and the adopted values are provided in Table 1. Initial attempts to calibrate the model using a single zone for the upper layer were unsuccessful. Zoning the upper layer (Fig. 8) was required for calibration adding confidence to the approach.

The initial head was interpolated from limited water levels recorded is used as the starting point for model calibration. The interpolation was used to establish precalibration steady-state conditions.

Model calibration

An attempt has been made to reproduce the observed water levels in the entire study area through the steady model. The heads may have been reproduced, but the solution is nonunique. The numerical model includes assumptions based on literature values and the experience of the

Table 1. Adopted aquifer properties layer 1 (for the zone color refer to Fig. 8)

Description	Color of unit	Horizontal hydraulic conductivity (K) (m/day)	Vertical hydraulic conductivity (K) (m/day)
Alluvium	Purple	5	0.5
Weathered deposit	Dark grey	2	0.2
Consolidated sedimentary rocks (Karroo)	Red	0.01	0.001
Granite/basalt	Light blue	0.002	0.0002
Sandstone	Green	0.005	0.0005
Basalt and mafic rocks	Blue	0.001	0.0001
Metasediments, undifferentiated, 1 and 2	Light grey and yellow	0.003	0.0003

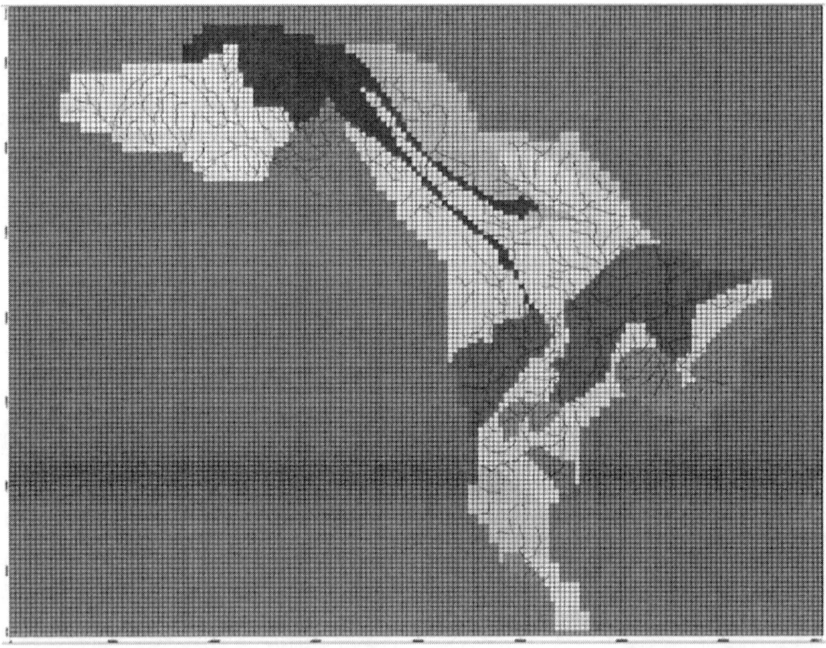

Fig. 8. Hydraulic conductivity zones used for the model calibration

modeller due to the limited amount of hydrogeological data. Limitations exist with respect to the amount of water level data available to constrain the calibration and aquifer recharges rates and spatial distribution of recharge areas. Calibration was accomplished by applying a set of hydraulic parameters; boundary conditions and stresses that produce computer generated simulated hydraulic heads that match actual field measurement within an acceptable range of error. Model calibration was performed using manual (trial and error) methods. The model was calibrated using observed groundwater heads and processed baseflow estimate. To improve calibration, the hydraulic conductivity, recharge and river conductance were adjusted until the modelled head elevations were able to match observed head elevations to an acceptable level of accuracy with a correlation coefficient of 0.97. The root mean squared error and residual mean error are 1.6 and −0.66 m respectively. This indicates a good correlation between observed and calculated groundwater heads [41, 42]. Analysis of the calibrated steady state model output indicates that the model simulates the groundwater elevation, hydraulic gradient and flow direction (Fig. 9) across the active model area to an acceptable level. The model calibration is considered acceptable, considering the correlation coefficient, mass balance 0.0% discrepancy) (Table 2) and spatially random residual errors [41, 43]. Sensitivity analysis carried out prior to steady state model calibration showed that the recharge flux parameter and hydraulic parameter of top layer are most sensitive.

Processed baseflow

Baseflow measurements in streams and rivers may be used as additional calibration targets as they allow calibrating to a flow rate as a head distribution, which may greatly increase the uniqueness or confidence of the model calibration. For this study, stream flow measurements from two locations on the Kitiwaka River recorded during an extended dry period on the 16th December 1983 were selected for the analysis (indicated by a cross line symbol on Fig. 6). The difference between these flow measurements is the net baseflow to the stream, assuming evaporative losses are negligible.

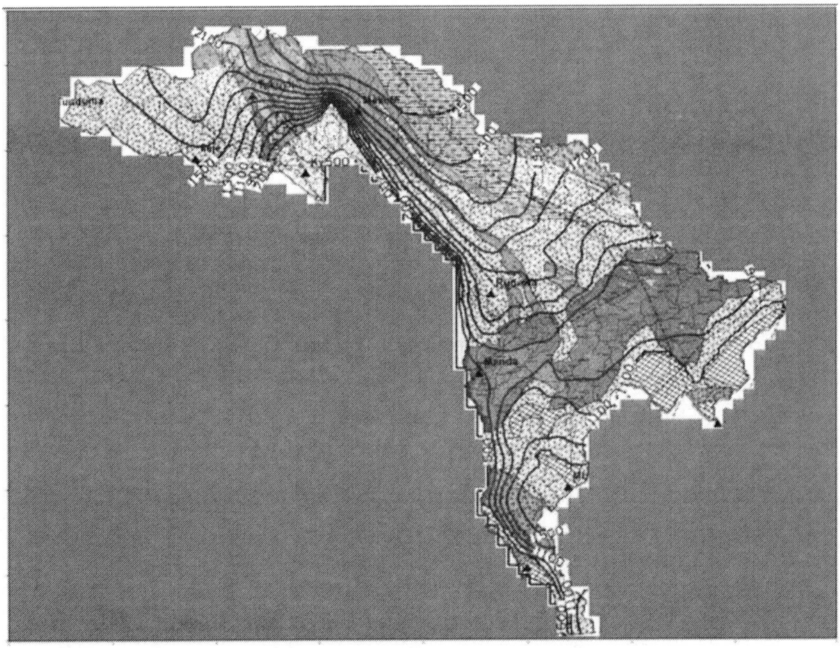

Fig. 9. Steady state calibrated potentiometric surface for layer 1

Table 2. Volumetric budget for entire model at end of the time step 1 in stress period

Cumulative volumes	L^3	Rates for this time step	L^3/T
In		In	
Storage	0.0000	Storage	0.0000
Constant head	1165.4406	Constant head	1165.4406
Wells	0.0000	Wells	0.0000
Recharge	1,070,256.5000	Recharge	1,070,256.5000
ET	0.0000	ET	0.0000
River leakage	0.5978	River leakage	0.5978
Total in	1,071,422.5384	Total in	1,071,422.5384
Out		Out	
Storage	0.0000	Storage	0.0000
Constant head	103,312.8906	Constant head	103,312.8906
Wells	0.0000	Wells	0.0000
Recharge	0.0000	Recharge	0.0000
ET	967,821.6875	ET	967,821.6875
River leakage	272.7555	River leakage	272.7555
Total out	1,071,407.3336	Total out	1,071,407.3336
In – out	15.2048	In – out	15.2048
Percent discrepancy	0.00	Percent discrepancy	0.00

RESULTS AND DISCUSSION

Simulated water balances

Based on the results of the simulation, the water balances were studied in the entire basin of the study area, to understand the functioning of the aquifers and their relationship with the Lake Nyasa. The aquifer is highly influenced by the diffuse recharge, which causes the water balances to change rapidly from positive to negative. The total water inflow (calculated as the sum of the net river/streambed recharge, the diffuse recharge and the lateral inflow) fluctuates greatly, due to climate variability. However, the total outflow (calculated as the sum of the evapotranspiration, outflow towards Lake Nyasa and the river leakage) remains practically stable. This fact is because the volume abstracted is nill or small compared to the total volume exchanged with basin, which is almost constant over time and is independent of the climate or of the aquifer exploitation (Table 2). Given that any surplus or deficit causes a change in the groundwater storage, and if it is assumed that the predevelopment pseudo-stationary state defines the equilibrium for the aquifer storage, the evolution can be observed of the accumulated deficits expected in the Lake Nyasa basin. However, The Lake Nyasa basin can exhibit the opposite behaviour. The total outflow term may be highly dependent on the total volume abstracted, while total inflows are less variable than in the case of the Lake Nyasa. Moreover, the inverse effect is expected if the water rights are not used, due to the importance of the streambed recharge (Table 3). The expected behaviour of these hydrogeological units, therefore, is clearly dependent upon the total abstraction.

Table 3. Model groundwater zone budget results between the stream gauge stations

	Inflow (m^3/day)	Outflow (m^3/day)
Recharge	1217	0
River leakage (baseflow)	0.025	1.87
Evapotranspiration	0	1149
From external zone	361	429

The net flow across the river boundary is calculated as follows:

$$\begin{aligned}\text{Net flow across river boundary} &= \text{Input river leakage into the aquifer} \\ &\quad - \text{Output river leakage out of the aquifer} \\ &= 0.025 - 1.87\ \text{m}^3/\text{day} \\ &= -1.85\ \text{m}^3/\text{day}\end{aligned}$$

A negative result indicates a net loss of water from the aquifer to the Kitiwaka River as all flow rates in MODFLOW are considered with respect to the aquifer. The recorded gain in net flow is 2.05 m^3/day while the model indicates a gain of 1.85 m^3/day. The difference (0.2 m^3/day) is small (~90% similarity) and well within the error of the model. The model result indicates that net loss of water from the aquifer may explain ~88% of the difference in stream flow between the two stream flow gauge stations. This is in agreement with the estimated baseflow component processed using baseflow Index [34] method (i.e. 85%).

Volumetric budget for each catchment

The water balance for each catchment (Table 4) was computed using the Zone-budget software package [44]. Zone-budget is a program that computes the water budget for a designated zone using results from the

MODFLOW-SURFACT groundwater flow model (Zone budget-version 3.01, updated year 2009). The Zone budget is a tool for measuring the flow from one part of the model to another. The zone budget was employed to create a baseflow measurement zone along Kitiwaka River, between the two stream gauging stations. This allows assessment of the gain or loss of stream flow. The measured and predicted baseflow values were then compared and the model parameters adjusted to improve the model calibration.

Table 4. Volumetric water budget for catchments

Inflow (1000 m³)				Outflow (1000 m³)			
Catchment	Recharge from rainfall	River leakage	From other catchment	ET	River leakage	Constant head (Lake Nyasa)	To other catchment
Kiwira	78,745	0.1	14,941	70,744	17	12,241	10,595
Lufirio	47,134	0.06	19,254	3205	33	17,795	16,723
Lumbria	86,265	0.05	9257	63,306	9.6	26,777	5380
Mbaka	30,993	0.05	16,389	27,823	15	6998	12,546
Mbawa	81,521	0.175	1968	63,468	2.1	17,539	2955
Mchuchuma	31,413	0.01	2464	27,091	0.9	4046	2899
Nkiwe	73,723	0.02	4332	62,559	21	7130	8403
Ruhuhu	509,090	0.13	10,406	501,830	128	10,786	6840
Rumakali	30,614	0.01	3798	20,775	0.03	0.05	13,627
Songwe	100,530	0.025	4743	98,151	47	0.08	7072

The two stream gauging stations located along the Kitiwaka River are indicated by a cross line symbol on Fig. 6. The measured stream flow at the downstream station was about 2.33 m^3/day and at the upstream station it was 0.28 m^3/day. The difference between these two numbers represents the river gain or water coming from the groundwater system and can be compared with the flow across the river boundary calculated by zone budget. To examine the amount of groundwater flow between the stream and the aquifer system, the river leakage under the input and output columns were extracted from the model output (Table 4).The projected population (Table 5), will likely have an impact on the estimated volumetric water budget.

Table 5. Projected population by catchment

(based on National Bureau of Statistics, Tanzania—2002 census data)

Catchment	Population			
	2012	2015	2025	2035
Kiwira	234,000	253,000	329,000	426,000
Lufirio	108,000	116,000	149,000	192,000
Lumbria	53,000	55,000	63,000	73,000
Mbaka	103,000	112,000	145,000	188,000
Mbawa	133,000	108,000	137,000	174,000
Mchuchuma	15,000	16,000	20,000	25,000
Nkiwe	38,000	39,000	47,000	55,000
Ruhuhu	401,500	430,000	544,500	689,000
Rumakali	15,000	16,000	17,000	19,000
Songwe	227,000	246,000	319,000	413,000
Grand total	1294,500	1,391,700	1,770,500	2,254,600

Annual safe yield/extraction limit

Based on the model the average net recharge and water in storage has been estimated (Table 6). In the absence of long term monitoring data to assess the performance of the aquifers the annual safe yield can be expressed as a percentage of the annual recharge [45]. Globally, if recharge can be assumed to be approximately 20% of precipitation, then deep percolation would be about 10% of recharge. Thus, a reasonably conservative estimate of sustainable yield would be 10% of the net recharge. The current concept of sustainable yield represents a compromise between theory and practice. In theory, a reasonably conservative estimate of sustainable yield would be about 10% of recharge. In practice, values higher than 10% may reflect the need to consider other factors besides conservation. Turner et al. [14] concluded that an aquifer system sustained by episodic recharge and the long-term gaining storage represents the maximum extractable volume.

To assess the potential groundwater resource for the basin the recharge has been estimated using the groundwater model. The estimated sustainable safe yields from the groundwater model for each catchment are compared to the estimated current domestic groundwater usage with a 20% contingency added for stock watering, small scale irrigation, commercial, institution (schools and Government buildings) and losses. Industrial, Mining and large scale agriculture are not included, they are considered socio-economic demands which may be developed after domestic uses are covered. It can be seen (Table 6) that currently the percentage of annual safe yield extracted ranges between 6.8 and 62%, averaging around 29%. This suggests there is room for socio-economic expansion of groundwater usage within all catchments.

Table 6. Estimated safe yield for each catchment and current groundwater usage

Catchment	Recharge volume (1000 m^3)	10% recharge (annual safe yield) (1000 m^3)	Estimated current groundwater usage (1000 m^3)	% of ASY currently used	Remaining groundwater available for development (1000 m^3)
Kiwira	78,745	7875	4865	61.8	3010
Lufirio	47,134	4713	2210	46.9	2503
Lumbria	86,265	8626	1089	12.6	7537
Mbaka	30,993	3099	212	6.8	2887
Mbawa	81,521	8152	2053	25.2	6099
Mchuchuma	31,413	3141	310	9.9	2831
Nkiwe	73,723	7372	768	10.4	6604
Ruhuhu	509,090	50,909	8655	17.0	42,254
Rumakali	30,614	3061	324	10.6	2737
Songwe	100,530	10,053	4879	48.5	5174
Total	1,070,028	107,001	27,273	25.5	79,728

The annual safe yield of the basin is taken as 10% of the total annual recharge and calculated to be 10.7×10^7 m^3. Current conservative estimates of groundwater use within the basin are around 27.3×10^5 m^3 which is around 30% of the basin annual safe yield. The current catchment groundwater usage ranges between 10 and 60% of the annual safe yield leaving between 90 and 40% of the annual safe yield depending on sub catchment available for future development [46]. Based on the current model result, estimate of groundwater recharge for the study area is about 1.07×10^6 m^3/day. Thus, the limits on aquifer pumping as a percentage of groundwater recharge range can be computed for the sustainable yield. Elsewhere, a value of total use of 50–70% of the sustainable yield was adopted.

Future scenario

Future predictions of groundwater usage and the resulting percentage of the annual safe yield used are provided in Table 7. In recent years, hydrological models, with various degrees of sophistication, have been used in various settings to evaluate water budgets for sustainable yield estimate under various climate or land-use scenarios (e.g. [11, 47]).

Table 7. Projected total groundwater usage, % of ASY and remaining available for development

Catchment	2015			2025			2035		
	Total GW usage	% ASY used	GW available for other uses	Total GW usage	% ASY used	GW available for other uses	Total GW usage	% ASY used	GW available for other uses
Kiwira	4865	61.8	3010	5767	73.2	2108	7373	93.6	502
Lufirio	2205	46.8	2508	2567	54.5	2146	3241	68.8	1472
Lumbria	1045	12.1	7581	1086	12.6	7540	1238	14.4	7388
Mbaka	2129	68.7	970	2511	81.0	588	3203	103.4	−104
Mbawa	2032	24.9	6120	2354	28.9	5798	2943	36.1	5209
Mchuchuma	304	9.7	2837	353	11.2	2788	447	14.2	2694
Nkiwe	745	10.1	6627	815	11.1	6557	931	12.6	6441
Ruhuhu	8559	16.8	42,350	9963	19.6	40,946	12,501	24.6	38,408
Rumakali	312	10.2	2749	318	10.4	2743	345	11.3	2716
Songwe	4856	48.3	5197	5755	57.2	4298	7379	73.4	2674
Total	27,048	25.3	79,953	31,489	29.4	75,512	39,601	37.0	67,400

Impacts on groundwater due to the population patterns may include increased abstraction for domestic, industrial or agricultural purposes, decreased recharge through land clearing and urbanisation, contamination from sewage and chemicals and localise increases in recharge due to storm water soakaway's in urban areas. Population concentrations in areas with limited access to gravity schemes, such as the south eastern part of Ruhuhu catchment near Songea, have resorted to groundwater for domestic water supplies.

These predictions do not account for possible reductions in the total annual recharge as a result of climate change [48]. It can be seen that by 2035 the percentage of annual safe yield extracted will increase to between 11.3 and 103%, averaging around 50%. This indicates that less water may be available for socio-economic development in the future. It also suggests that by 2035 Kiwira and Mbaka sub catchments will be nearly at or exceeding the annual safe yield leaving no room for socio-economic development, or the need to reduce existing socio-economic demands to meet domestic demands. The other option is to increase the percentage of total recharge allocated to the annual safe yield from 10 to 20% in these catchments.

Basin sustainability yield constraints

The current study shows that derivation of sustainable yield using conservation of mass principles leads to expressions for basin sustainable that can be readily determined using numerical modeling methods and selected on the basis of applied constraints. Bores are concentrated in areas of high population. For instance the village of Mlete, in Tanga Ward, Namtumbo District has 21 identified bores. The yield from each bore may not be high, but the combined yield of all 21 bores may be. Viewed on a large scale this is likely not an issue, however the combined yield of the bores may exceed the safe yield in a small area leading to localised drawdown.

The possibilities of severe, long-term droughts and climate change also should be considered. Because a climate stress on the hydrologic system is added to the existing or projected human-derived stress, droughts represent extreme hydrologic conditions that should be evaluated in any long-term management plan. Water reuse is one of the potential ways to complement the existing portable water sources. Practice of the water reuse for irrigation in Tanzania and similar region has been suggested by Kihila et al. [28]; Yihdego [49]. The challenges that need to be addressed, the benefits expected and global wastewater treatment options, capital and operation costs as well as the cost benefit analysis, implementation of water reuse will require among other things, proper defined policy and institutional framework, clear guidelines as well as more research and investment on wastewater treatment have been pointed out [28].

Transboundary considerations

When considering the groundwater potential and annual safe yield for the basin the potential for Transboundary impacts must be considered. Lake Nyasa is shared between three countries while the basin borders three countries, Zambia and Malawi along the north western boundary marked by the Songwe River and Mozambique on the southern boundary (Fig. 1). The Songwe River marks the border of Tanzania with Zambia and Malawi, the river is gaining having a high groundwater baseflow. On the Zambia/Malawi side are high hills and mountains which fall towards the river indicating the groundwater flow direction is also towards the river. Extraction of groundwater within the basin (Tanzania) is unlikely to impact groundwater across the border and the potential for transboundary impacts are considered very low. However over extraction of groundwater may impact baseflow to the river and therefore surface flows in the river which may impact users on both sides of the border. The southern basin boundary with Mozambique is an arbitrary line which doesn't follow topographic features, therefore there is a potential for transboundary impacts. However the length of border is short and in a mountainous area with high rainfall and low population limiting the potential for extraction. The potential for transboundary groundwater impacts are therefore considered low to very low.

CONCLUSIONS

To analysis the groundwater development based on sub-basin/sub-catchment, a three dimensional groundwater numerical modelling (MODFLOW-SURFACT coupled with Visual MODFLOW) were used to handle Vadose Zone and for greater numerical convergence stability. Groundwater modelling and water balance optimization were used for the sustainable yield estimation. The model has been calibrated to the hydraulic head and processed baseflow estimate, which adds confidence or none-uniqueness to the solution and hence as a basis good enough for prediction/scenario simulation. A water-budget model output was used in estimating recharge which allows for an evaluation of the future groundwater sustainability, with several sub-basin extraction limits/annual safe yields in the Lake Nyasa basin. The water balance error was 0% which could be considered as criteria (i.e. water balance stabilization), to constrain the estimate of sustainable extraction [13]. Based on the current model result, estimate of annual safe yield based on 10% of groundwater recharge for the study area is about 10.7×10^7 m^3. The current catchment groundwater usage ranges between 10 and 60% of the annual safe yield leaving between 90 and 40% of the annual safe yield, depending on sub catchment, available for future development. Future predictions of groundwater usage indicate that by 2035 the percentage of annual safe yield extracted will increase to between 11.3 and 103%. Sustainability indicators such as acceptable head can be examined based on climate and management decisions affecting water extraction and land use. The choice of the method employed in this study has proved successful for a sound understanding of the sub-basin groundwater resources sustainability among the stake holders, decision makers for an effective management. With this information, society can make better informed decisions about how to manage their ground water resources in a long-term context. Such analyses also ideally lead to the design and implementation of long-term hydrologic networks to monitor projected outcomes of the ground water development and to improve the ability to predict future system responses.

The basin size is relatively small and annual groundwater recharge volumes are low, assuming between 5 and 20% of rainfall and therefore sustainable groundwater extraction volumes will be low, and there is insufficient information to assess the component of deep circulation and hence the deep groundwater resource within the fractured rock systems. The possibility of transboundary issues are considered low due to the prevailing groundwater gradient in the areas of international borders. It is more likely groundwater extraction may lead to reduced baseflows especially in the Songwe River which may impact surface water users on both sides of the border. Basin groundwater quality is generally good due to the annual recharge of the shallow aquifers. Based on the limited available groundwater analysis all aquifer types in all catchments may be suitable for domestic and stock watering purposes subject to assessment of individual bores.

The effect of few management options and a simplified climatic-change scenario needs to be considered for the Lake Nyasa sub-basin aquifers together with a decrease in the net recharge (and possibly suggest Managed Aquifer Recharge/Aquifer Storage recovery), a change in river–aquifer interactions, and the total abstraction volumes identified as the main factors to address in management planning. Further efforts to constrain the hydrogeological parameters of the various geological units, including riverbed deposits, which could include targeted test pumping, could improve understanding of the system.

AUTHORS' CONTRIBUTIONS

YY performed numerical simulation and wrote most of the sections of the paper. AP analysed the field data and wrote some sections of the paper. All authors read and approved the final manuscript.

ACKNOWLEDGEMENTS

We acknowledge SMEC International and Tanzanian Ministry of Water. The study was funded by the International Development Association and the United Republic of Tanzania.

We would like to extend our thanks to Lake Nyasa Basin Water Board; Sokoine University of Agriculture, Tanzania; Drilling and Dams Construction Agency (DDCA), Dar es Salaam Office; Songea District Water Engineers Office, and Kyela District Water Engineers Office. The manuscript has benefitted from the reviewers' and editors' comments.

COMPETING INTERESTS

The authors declare that they have no competing interests.

REFERENCES

1. Al-Weshah R, Yihdego Y (2016) Modelling of strategically vital fresh water aquifers, Kuwait. Environ Earth Sci 75:1315.
2. Yihdego Y, Reta G, Becht R (2016) Human impact assessment through a transient numerical modeling on the UNESCO World Heritage Site, Lake Naivasha, Kenya. Environ Earth Sci.
3. Yihdego Y, Reta G, Becht R (2016) Hydrological analysis as a technical tool to support strategic and economic development: case of Lake Naivasha, Kenya. Water Environ J 30(1–2):40–48.
4. Yihdego Y, Drury L (2016) Mine water supply assessment and evaluation of the system response to the designed demand in a desert region, central Saudi Arabia. Environ Monit Assess J 188:619.
5. Ozdemir A (2011) GIS-based groundwater spring potential mapping in the Sultan mountains (Konya, Turkey) using frequency ratio, weights of evidence and logistic regression methods and their comparison. J Hydrol 411:290–308
6. Pourtaghi ZS, Pourghasemi HR (2014) GIS-based groundwater spring potential assessment and mapping in the Birjand Township, southern Khorasan Province, Iran. Hydrogeol J 22(3):643–662
7. Bredehoeft J (2005) The conceptualization model problem: surprise. Hydrogeol J 13:37–46
8. Alley WM, Leake SA (2004) The journey from safe yield to sustainability. Ground Water 42:12–16
9. Devlin JF, Sophocleous M (2005) The persistence of the water budget myth and its relationship to sustainability. Hydrogeol J 13:549–554
10. Sophocleous M (2000) From safe yield to sustainable development of water resources: the Kansas experience. J Hydrol 235:27–43
11. El-Kadi AI, Tillery S, Whittier RB, Hagedorn B, Mair A, Ha K, Koh G (2014) Assessing sustainability of groundwater resources on Jeju Island, South Korea, under climate change, drought, and increased usage. Hydrogeol J 22(3):625–642
12. Holman IP, Tascone D, Hess TM (2009) A comparison of stochastic and deterministic downscaling methods for modelling potential groundwater recharge under climate change in East Anglia, UK: implications for groundwater resource management. Hydrogeol J 17:1629–1641
13. Smith A, Walker G, Turner J (2010) Aquifer sustainability factor: a review of previous estimates. International Association of Hydrogeologists (AIH) and the Geological Society of Australia (GSA). EP104589
14. Turner RJ, Mansour MM, Dearden RÓ, Dochartaigh BÉ, Hughes AG (2015) Improved understanding of groundwater flow in complex superficial deposits using three-dimensional geological-framework and groundwater models: an example from Glasgow, Scotland (UK). Hydrogeol J 23:493–506
15. Whittier R, Rotzoll K, Dhal S, El-Kadi AI, Ray C, Chang D (2010) Groundwater source assessment program for the state of Hawaii, USA: methodology example application. Hydrogeol J 18:711–723
16. Conan C, de Marsily G, Bouraoui F, Bidoglio G (2003) A long-term hydrological modelling of the Upper Guadiana river basin (Spain). Phys Chem Earth 28:193–200

17. Bromley J, Cruces J, Acreman M, Martínez L, Llamas MR (2001) Problems of sustainable groundwater management in an area of over-exploitation: the upper Guadiana catchment, central Spain. Int J Water Resour Dev 17:379–396
18. Martínez-Santos P, Llamas MR, Martínez-Alfaro PE (2008) Vulnerability assessment of groundwater resources: a modelling-based approach to the Mancha Occidental aquifer, Spain. Environ Model Softw 23:1145–1162
19. Yihdego Y, Danis C, Paffard A (2015) 3-D numerical groundwater flow simulation for geological discontinuities in the Unkheltseg Basin, Mongolia. Environ Earth Sci J 73(8):4119–4133
20. Yihdego Y, Becht R (2013) Simulation of lake–aquifer interaction at Lake Naivasha, Kenya using a three-dimensional flow model with the high conductivity technique and a DEM with bathymetry. J Hydrol 503:111–122
21. Carrera J, Alcolea A, Medina A, Hidalgo J, Slooten LJ (2005) Inverse problem in hydrogeology. Hydrogeol J 13:206–222
22. Yihdego Y, Webb JA (2011) Modeling of bore hydrograph to determine the impact of climate and land use change in a temperate subhumid region of south-eastern Australia. Hydrogeology Journal 19(4):877–887
23. Yihdego Y, Webb JA (2015) Use of a conceptual hydrogeological model and a time variant water budget analysis to determine controls on salinity in Lake Burrumbeet in southeast Australia. Environ Earth Sci J 73(4):1587–1600
24. Yustres A, Navarro V, Asensio L, Miguel Candel M, García B (2013) Groundwater resources in the Upper Guadiana (Spain): a regional modelling analysis. Hydrogeol J 21:1129–1146
25. Orr S, Meystel AM (2005) Approaches to optimal aquifer management and intelligent control in a multi-resolutional decision support system. Hydrogeol J 13:223–246
26. Danish International Development Agency (DANIDA) (1988) Implementation of water master plan for Iringa, Ruvuma and Mbaya regions, for United Republic of Tanzania
27. Don Consult Ltd (2011) Provision of technical and facilitation consultancy services for rural water supply and sanitation sub-projects in Kyela district council, exploration drilling report. Tender No: LW/DC/F. 10/15/Vol. V/108, July 2008. Report for the United Republic of Tanzania, Prime Minister's Office Regional administration and Local Government, Water Sector Development Program
28. Kihila J, Mtei KM, Njau KN (2015) A review of the challenges and opportunities for water reuse in irrigation with a focus on its prospects in Tanzania. Int J Environ Eng 7(2):111–130.
29. SMEC (2013) Integrated water resources management and development plan—groundwater assessment. Ministry of Water, Dar es Salaam
30. Bootsma HA, Jorgensen SE (2006) Lake Malawi/Nyasa. Experience and lessons learned brief. http://Worldlakes.org. Accessed 25 March 2015
31. Kashaigili JJ (2010) Assessment of groundwater availability and its current and potential use and impacts in Tanzania, Report for International Water Management Institute, December 2010
32. Lake Nyasa Basin Water Board (LNBWB) (2010) Basin (Lake Nyasa) hydrological report, December 2010
33. Arnold JG, Allen PM, Muttiah R, Bernhardt G (1995) Automated baseflow separation and recession analysis techniques. Ground Water 33(6):1010–1018
34. Beck HE, Albert I, van Dijk JM, Miralles DG, de Jeu RAM, Bruijnzeel LA, McVica TR, Schellekens J (2013) Global patterns in baseflow index and recession based on stream flow observations from 3394 catchments. Water Resour Res 49(12):7843–7863
35. Smakhtin VU (2001) Estimating continuous monthly baseflow time series and their possible applications in the context of the ecological reserve. Water SA 27(2):213–218 **(ISSN 0378–4738)**

36. Howard AE (2011) A brief introduction to the geology and mining history of Tanzania. Report prepared for SIKA RESOURCES INC. Nebu Consulting LLC, Toronto
37. Hagedorn B, El-Kadi AI, Mair A, Whittier RB, Ha K (2011) Estimating recharge in fractured aquifers of a temperate humid to semiarid volcanic island (Jeju, Korea) from water table fluctuations, and Cl, CFC-12 and 3H chemistry. J Hydrol 409:650–662
38. HydroGeoLogic Inc (2002) MODFLOW-SURFACT: a comprehensive MODFLOW-based hydrologic modelling system: code documentation and user's manual. HydroGeoLogic Inc, Reston
39. Sanz D, Castaño S, Cassiraga E, Sahuquillo A, Gómez-Alday J, Peña S, Calera A (2011) Modeling aquifer–river interactions under the influence of groundwater abstraction in the Mancha Oriental System (SE Spain). Hydrogeol J 19:475–487
40. Fetter CW (2001) Applied hydrogeology, 4th edn. Prentice Hall, Englewood Cliffs
41. Barnett B, Townley LR, Post V, Evans RE, Hunt RJ, Peeters L, Richardson S, Werner AD, Knapton A, Boronkay A (2012) Australian groundwater modelling guidelines, waterlines report. National Water Commission, Canberra
42. Knowling MJ, Werner AD, Herckenrath D (2015) Quantifying climate and pumping contributions to aquifer depletion using a highly parameterized groundwater model: Uley South Basin (South Australia). J Hydrol 523:515–530
43. Middlemis H (2004) Benchmarking best practice for groundwater flow modelling. The Winston Churchill Memorial Trust of Australia
44. Harbaugh AW (1990) Zone budget. A computer program for calculating subregional water budgets using results from the US geological survey modular three-dimensional ground-water flow model: US geological survey open-file report 90–392
45. Kalf FRP, Woolley DR (2005) Applicability and methodology of determining sustainable yield in groundwater systems. Hydrogeol J 13:295–312
46. Mair A, Hagedorn B, Tillery S, El-Kadi AI, Westenbroek S, Ha K, Koh GW (2013) Temporal and spatial variability of groundwater recharge on Jeju Island, Korea. J Hydrol 501:213–226
47. Fornés JM, De La Hera Á, Llamas MR (2005) The silent revolution in groundwater intensive use and its influence in Spain. Water Policy 7:253–268
48. Yihdego Y, Webb JA (2016) Validation of a model with climatic and flow scenario analysis: case of Lake Burrumbeet in southeastern Australia. J Environ Monit Assess 188(308):1–14.
49. Yihdego Y (2015) Water reuse and recreational waters, urban water reuse handbook (UWRH), chapter 78. In: Eslamian S. (ed) Taylor and Francis, CRC Press, Boca Raton, pp 1029-1039. Print ISBN: 978-1-4822-2914-1. eBook ISBN: 978-1-4822-2915-8. CAT# K22608. https://www.crcpress.com/Urban-Water-Reuse-Handbook/Eslamian/p/book/9781482229141

CHAPTER 12

Two-Dimensional Seismic Interpretation and Petroleum System Modelling in the Hammerfest Basin, SW Barents Sea

Mohammedyasin MS*

Department of Civil Engineering, Debre Markos University, Debre Markos, Ethiopia

ABSTRACT

The Hammerfest Basin has undergone several structural and tectonic evolutions which affected its petroleum systems. Two-dimensional seismic interpretation and petroleum system modelling of the Hammerfest Basin was done in order to understand the petroleum system and effects of uplift and subsidence on hydrocarbon distribution, accumulations, maturity, migration and leakage. The central part of the Hammerfest Basin has been affected by Jurassic syn-rift faults forming a large graben structure in response to subsidence along the folded Asterias Fault Complex and Troms-Finnmark Fault Complex. The graben structure is characterized by gas chimney, bright spots and pockmarks. The modelling result shows that the Jurassic Hekkingen Formation is less mature (oil window) and that the Snadd and Kobbe Formations are mature to highly mature (gas window), respectively. The Stø and Tubåen reservoirs are enriched with oil and gas which migrated from the Snadd and Kobbe source rocks. The hydrocarbon accumulations show that the Stø Formation is the main reservoir in the Hammerfest Basin. Rapid burial during Permian and late Cretaceous was followed by uplift events. Three phases of erosion: the first two phases of erosion (75 to 60 Ma and 30 to 15 Ma) affected and eroded 200 m of Kveite Formation and 300 m of the Torsk Formation, respectively and the third phase of erosion (2.5 to 0.01 Ma) has eroded 100 m of Torsk Formation. The oil migration follows the deep seated faults along the Asterias Fault Complex, Troms-Finnmark Fault Complex and along the bounding faults of the graben structure at the center of the basin which leads to leakage. The migration of gas on the other hand, is vertically upward throughout the basin leaking out of the seal. The gas migration has been enhanced by the uplift and subsidence events which have been the main controlling factors of the hydrocarbon maturation and distribution within the Hammerfest Basin.

Keywords: Seismic interpretation; Petroleum system modelling; Erosion; Hammerfest basin; Source rock

INTRODUCTION

The Hammerfest Basin has undergone a long structural and tectonic evolution which has affected its petroleum systems [1]. Uplift and erosion in the Hammerfest Basin can have affected the integrity of the seals causing leakage of hydrocarbons [2-4]. This has increased the exploration risks leading to the high number of dry wells. In the eroded and uplifted areas, hydrocarbon generation has ceased locally due to a decrease in temperatures. The driving factors for the petroleum system of the Hammerfest Basin were uplift and erosion from the Paleocene til the Pleistocene which caused the depletion of hydrocarbon accumulations [5]. However, current investigations indicate that leakage is also a major exploration challenge in the basin [2,6-8]. Especially, petroleum system modelling and interpretation of the deeper horizons, including delineating the basement, have been another challenge. This paper will address the latter problem and provide information in the hydrocarbon potential of the study area through seismic interpretation and basin modelling. Basin modelling is a dynamic modelling of geologic processes include deposition, erosion, compaction, heat flow analysis, expulsion, phase dissolution, and hydrocarbon generation, accumulations, maturation and migration to know the sources, timing and prospectively of hydrocarbon. So, this work significantly help to (1) quantify the amount and timing of oil and gas generation in the basin, (2) estimate the hydrocarbon accumulation in the reservoirs, (3) reconstruct the possible hydrocarbon migration pathways,

(4) evaluate the possible leakage from the source rocks and reservoirs and (5) elucidate the effect of geologic events such as glaciation, uplift and subsidence on petroleum systems in the basin.

The SW Barents Sea margin, along the Senja Fracture Zone, developed during the Eocene opening of the Norwegian–Greenland Sea, first by continent–continent shear followed by continent–ocean shear, and has been passive since earliest Oligocene time. After breakup, the passive margin evolved in response to subsidence and sediment loading during the widening and deepening of the Norwegian- Greenland Sea. Sedimentation was modest until the Late Pliocene when the Northern Hemisphere Glaciation led to rapid progradation and greatly increased sedimentation rates forming huge, regional depocentres near the shelf edge offshore Mid–Norway and in front of bathymetric troughs in the northern North Sea and western Barents Sea [9]. Tectonic, paleoceanographic and paleoclimatic events that had a crucial influence on the hydrocarbon accumulation and distribution in the entire Barents Sea during Cenozoic and some of the major related consequences (1) cessation of petroleum generation, expulsion and migration from the source rocks, (2) phase changes, including the expansion of gas in reservoir, which resulted in the spilling of earlier trapped oil, (3) reactivation of faults and breaching of seals associated to the reduction of overburden and pressure fluctuations, and (4) leakage of hydrocarbons from the reservoir to the surface, with the possible formation of gas hydrates and pockmarks [4,8,10-13].

This work is therefore done to assess the petroleum system in the Hammerfest Basin through integrated approach of seismic interpretation and basin modelling. Fault and horizon interpretation, seismic-to-well tie, isochrone maps, velocity modelling and depth conversion were done to achieve this objective. The study area situated between latitude 70°50'-72°15'N and longitude 20-24°10'E and is bounded by the Finnmark Platform in the south and by the Loppa High and the Bjarmeland Platform in the north (**Figure 1**).

The evolution of the Hammerfest Basin

The Hammerfest Basin is a 150 km-long and 70 km-wide composite, asymmetric, elongated sedimentary basin, striking ENE/WSW. The basin is fault controlled and was probably established in the Late Carboniferous [14], although main subsidence occurred in the Triassic and the early Cretaceous. Basinal development largely culminated in the mid-Cretaceous, but highly condensed Upper Cretaceous and thin Lower Tertiary shales are also preserved in the basin, in spite of extensive late Tertiary uplift. There is no evidence of extensive late Palaeozoic evaporite deposition or of diapirism in the basin, in contrast to the Tromsø Basin to the west and the

Nordkapp Basin to the east. The internal structure of the basin is characterized by a central dome located along the basin axis and by a complex pattern of dominantly westerly and west-northwesterly trending faults; all of these features predominantly reflect late Jurassic tectonism [15].

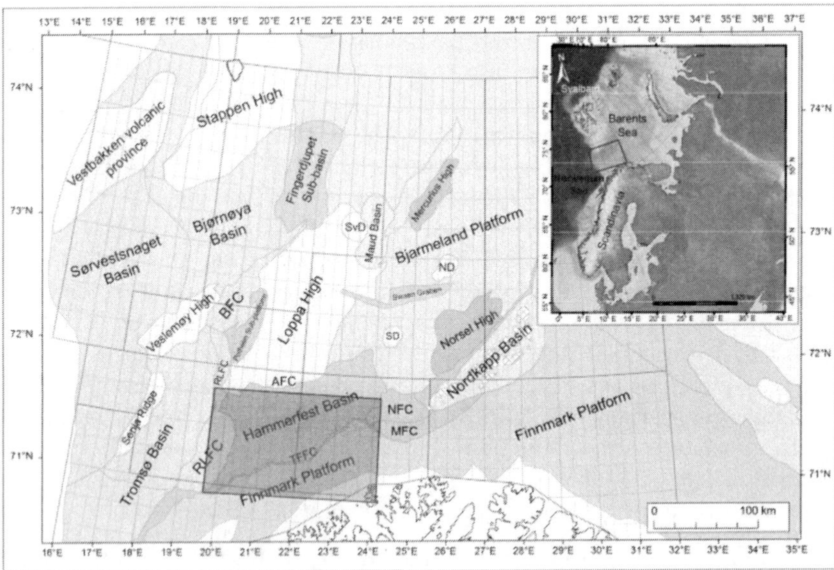

Figure 1: The red square shows the location of the Hammerfest Basin in the SW Barents Sea (AFC: Asterias Fault Complex; BFC: Bjørnøyrenna Fault Complex; MFC: Måsøy Fault Complex; ND: Nordvarg Dome; NFC: Nyslepp Fault Complex; RLFC: Ringvassøy–Loppa Fault Complex; SD: Samson Dome; SvD: Svalis Dome; SLHFC: Southern Loppa High Fault Complex; TFFC: Troms-Finnmark Fault Complex [12]. The inset shows the map of Northern Europe, with the approximate position of the study area.

Seismic data indicate the presence of a thick Permian sedimentary sequence in the eastern part of the Hammerfest Basin area. Possibly this area was part of the Nordkapp Basin in Paleozoic times, a basin developed synchronous to the Devonian to Carboniferous basins on Bjørnøya and Spitsbergen [16]. The tilting of the basin in late Carboniferous to early Permian times was caused by tensional strike slip movement with reactivation of underlying sediment fault trends. The depositional conditions during Permian were dominantly marine with general basin subsidence where the depocenters are in the northeastern and in the south-western parts of the present day Hammerfest Basin [1].

During the Triassic and early Jurassic the Hammerfest Basin was part of a regional basin. Its development as a separate depositional entity with flexuring and faulting of the northern flank and reactivation of rift was started during the Kimmeridgian rift phase and the northeastern Barents Sea from the western margin decoupled at this time [17]. After the main pulse of faulting, rapid subsidence followed by basin fill onlapping sequences on the flank of the basin was deposited [18,19]. The Middle and Upper Triassic sequences show an upward coarsening as result of the cyclic changes from open marine to the continentalenvironments. The intercalation of shaley continental and sandy shallow marine sediments during late Triassic to mid-Jurassic sequences was controlled by complex inter-play of tectonic subsidence, eustatic sealevel changes and local sediment input [1]. The Hammerfest Basin is bordered to the west by a marked faulted hinge zone (RLFC). There is a marked thickening of Upper Jurassic-Lower Cretaceous across the hinge zone [18].

Tertiary sedimentation in the Hammerfest Basin was started during the mid-Paleocene. Concurrent with the initiation of a marine environment, due to a transgression of the central parts of the basin, erosion of the Loppa High and Finnmark Platform was prevailed. In mid-late Paleocene time these were gradually transgressed, contemporaneously with a bathyal environment in the central part of the basin. A south-southwesterly progradation from the platform areas north-northeast of the Hammerfest Basin was initiated during the late Paleocene [1]. This, possibly combined with high biogenic activity, led to high accumulation rates in the basin areas. As a response to compressional uplift of the southwestern part of the Loppa High during the late Paleocene/early Eocene, south-southwesterly progradation was initiated proximal to the Loppa High. In the west-central part of the basin, subsidence led to a more westerly oriented progradation. Coeval sedimentation in the southern part of the Hammerfest Basin was dominated by low-energy deposition in a non-prograding environment. The sediment response is speculated as mirroring the latest rifting phase and the earliest phase of seafloor spreading (56 Ma) of the Norwegian-Greenland Sea, which gave rise to transpression along the western margin, and along the western part of the Hammerfest Basin, transtension (i.e., the Ringvassoy-Loppa Fault Complex). The Oligocene and Miocene were dominated by subsidence of the western Barents Sea margin, and the basins on the shelf including the Hammerfest Basin were alternately accumulating sediments and undergoing erosion [20]. The lithostratigraphy and summarized description of the Hammerfest Basin is presented on **Figure 2** and **Table 1**.

Table 1: Summary of lithostratigraphy of the Hammerfest Basin [22]. The lithologies in brackets are interbedded lithology. Gp: Group, Fm: Formation, Sst: sandstone, Cst:Claystone, Slt: Siltstone, Ms: mudstone, Lst: Limestone and Dlt: dolomite, E: early and L: late.

Name	Lithology	Depositional environment	Age
Nordland Gp	Cst and clay rich sandstone	Bathyal to glacial marine	L.Pliocene to Holocene
Torsk Fm	Cst, (Sst, Lst)	Open to deep marine shelf	L.Paleocene to Oligocene
Kveite Fm	shales and Cst, (Lst, Slt)	Deep open shelf	Cenomanian to Maastrichtian
Kviting Fm	calcareous Sst, (Ms, Cst, Lst)	Deep to shallow shelf	Campanian
Kolmule Fm	Cst and shale, (Slt, Lst, Dlt)	Open marine environments	Aptian to mid-Cenomanian
Kolje Fm	shale and claystone, Lst, Dlt	Distal open marine	E. Barremian to E Aptian
Knurr Fm	Cst, (Lst, Dlt)	Distal marine	Ryazanian to E.Barremian
Hekkingen Fm	shale and Cst, (Lst, Dlt, Slt, Sst)	Deep marine	E.Kimmeridgian to Ryazanian
Fuglen Fm	Mst, (Lst)	Highstand marine	L. Callovian to Oxfordian
Stø Fm	Sst, (shale, Slt)	Prograding coastal regimes	Pliensbachian to Bajocian
Nordmela Fm	Sst, (shale, Cst, coals)	Tidal flat to flood plain	Sinemurian to L.Pliensbachian
Tubåen Fm	Sst, (shale, coals)	Marginal marine	L Rhaetian to E.Hettangian
Fruholmen Fm	Mst, (Lst, shale)	Highstand marine	L. Callovian to Oxfordian
Snadd Fm	Shale, (Slt, Sst)	Distal marine	Ladinian to E. Norian
Kobbe Fm	shale, (Slt, Sst)	base clastic marginal marine	Anisian
Havert Fm	shale (Sst, Slt)	Marginal marine to open marine	Griesbachian to Dienerian
Ørret Fm	Sst, Slt, shale	Deltaic and lower coastal plain	Kungurian to Tatarian

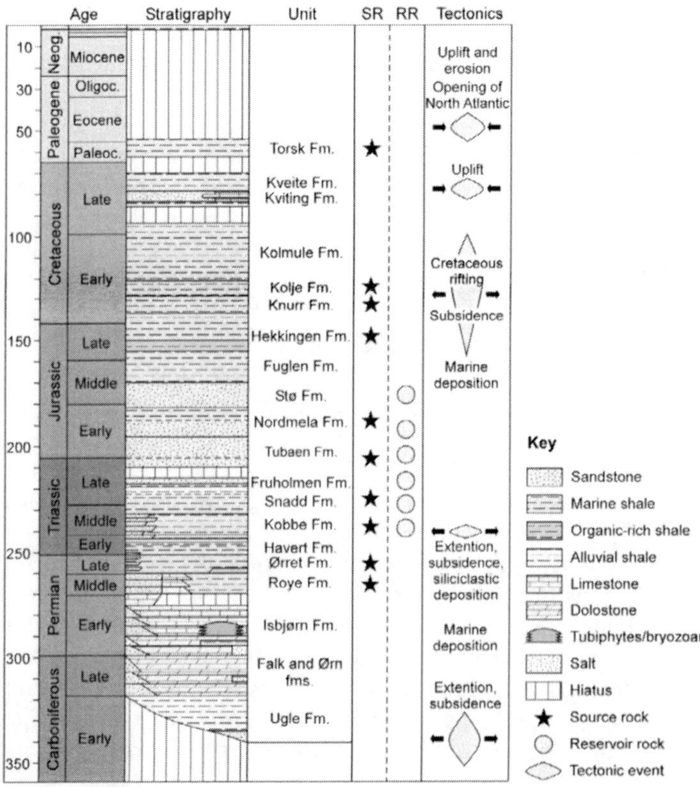

Figure 2: Lithostratigraphic chart of the Hammerfest Basin where potential source rocks (SR) and reservoirs (R) as well as the general overview of the geologic and tectonic history are composed from [4,5,21].

MATERIALS AND METHODS

Two-dimensional (2D) seismic data and eight wells were used in this study. In the first phase, Seismic interpretation of five dip lines were selected and seismic-to-well tie was done using Petrel software, where the seismic has best fit with synthetic seismogram. The depths also have a good agreement with known depths from the wells. The synthetic seismogram has been created in Petrel synthetic generation tool where all of the three wells show a good match with the seismic data. Additionally, these wells were correlated to one another. Faults were marked based on breaks of discontinuity of reflectors with visible displacement. The prominent strong and continuous train of wavelets running across the seismic sections was marked as horizon. On this basis, fourteen horizons were mapped. From these horizons one representative dip-line (BSSO1-104) was selected and depth converted after velocity modelling using 2 km/s for the basin (**Figure 3**). In the second phase, the depth-converted 2D seismic line was loaded into PetroMod software and faults and horizons were digitized. Then, layer splitting has been done to insert multi erosion events and to split the Jurassic and Triassic sequences into formations. The facies and erosion definition was done after age assignment and facies mapping as modified by Rodrigues Duran et al. [12] (**Table 2**).

Table 2: Facies and age definitions modified from [12]. Gp=Group, Fm=Formation, Fms=Formations, PSE=Petroleum system elements, OR=Overburden rock, UR=Underburden rock, SR=Source rock, SeR=Seal rock, RR=Reservoir rock, TOC=Total organic carbon, HI=Hydrogen index.

Name	TOC percentage (%)	HI (mgHC/gTOC)	PSE	SE Lithology
Nordland Gp (0.01-0 Ma)			OR	Siltstone (organic lean)
Torsk Fm (65.5-30 Ma)			OR	Shale (organic lean, typical)
Kveite-Kviting Fms (96.6-65.5 Ma)			OR	Siltstone (Organic lean)
Kolmule Fm (120-96.6 Ma)			OR	Shale (organic lean ,silt)
Kolje Fm (130-120 Ma)			OR	Shale (typical)
Knurr Fm (140.2-130 Ma)			OR	Shale (typical)
Hekkingen Fm (155.6-140.2 Ma)	10	300	SR	Shale (organic rich, 8% TOC)
Fuglen Fm (167.7-155.6 Ma)			SeR	Shale (organic lean, siliceous, typical)
Stø Fm (184-167.7 Ma)			RR	Sandstone (typical)
Nordmela Fm (196.5-184 Ma)			OR	Siltstone(Organic lean)
Tubåen Fm (200-196.5 Ma)			RR	Sandstone (clay poor)
Fruholmen Fm (210-200 Ma)			OR	Siltstone(Organic rich, 2-3% TOC)
Snadd Fm (237-210 Ma)	2	150	SR	Siltstone(Organic rich, 2-3% TOC)
Kobbe Fm (245-237 Ma)	3	200	SR	Siltstone(Organic rich, 2-3% TOC)
Klappmyss-Havert Fms (260-245 Ma)			UR	Shale (Organic lean, silty)
Ørret Fm (265-260 Ma)			UR	Shale (Organic lean, silty)
ILTR Fms			UR	Sandstone (typical)

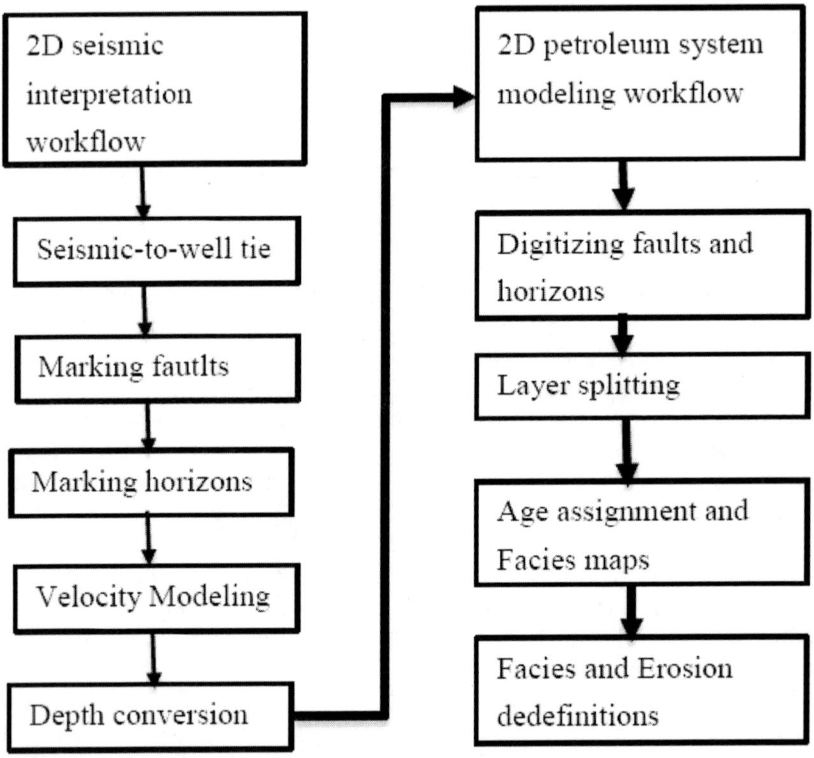

Figure 3: Workflow for 2D seismic interpretation and petroleum system modelling.

The paleowater depth ranges from 300 m to 500 m [23]. The water depth from different wells was used in this model and is below 400 m. The upper boundary condition of heat transfer in sedimentary basins is given by the temperatures either at the sub-aerial surface or at the sediment water interface, and is affected by the water depth, the paleogeographic position, and oceanic currents [24]. The Sediment Water Interface Temperature (SWIT) values in this model were assigned based on time latitude diagram which ranges from 5-25°C [24]. The interglacial and the present-day SWITs are 3 and 6°C, respectively [25,26]. The heat flow was automatically found from PetroMod. It shows increases from 41 mW/m^2 (2.83 Ma) to 55 mW/m^2 (130 Ma) and then decreases again to 47 mW/m^2 (196 Ma). Then, hybrid (darcy+flow path) migration method was used for simulation. The optimization of the model (relative difference with the reality) was 0.03. Hence, it is less than 1% and the model was built successfully.

RESULTS AND DISCUSSION

Seismic interpretation

Fourteen horizons were interpreted and named using the nomenclature of [14] and [27] (**Figure 4**). These include Lower Carboniferous (LC), Near Base of Permian (BP), Near Top of Permian (P1), Intra Lower Triassic (ILTR), Base Middle Triassic (BMTR), Intra-upper Triassic (IUTR), Base-upper Jurassic (BUJ), Intra Lower Cretaceous (?Hauterivian) (ILK1), Intra Lower Cretaceous (?Barremian) (ILK2), Intra Lower Cretaceous (ILK3), Base Tertiary (BT), Base Quaternary (BQ) and the Seabed (**Figure 5**). The bright reflectors shown locally in some parts are restricted laterally and different from the other strong reflectors. From IUTR to ILK1 the packages are thickening towards the discontinuities (AFC and TFFC) and thinning to the center of the basin in general. Some strong reflectors at the center of the basin between ILK1 and IUTR cross–cut the dip lines and are low amplitude and discontinuous on the strike lines.

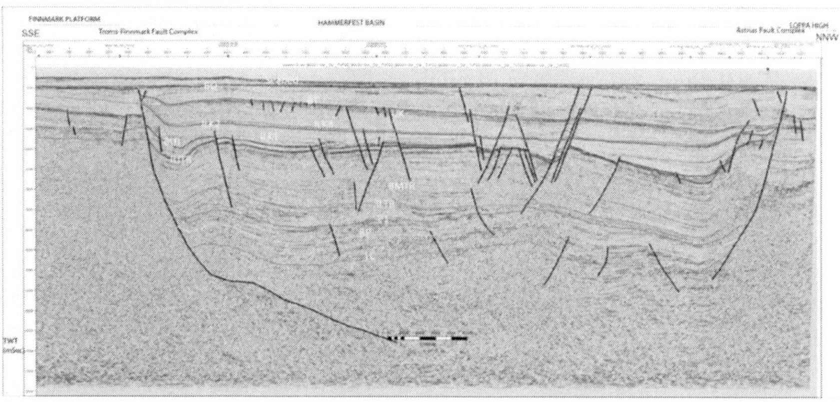

Figure 4: Faults and horizons interpretation of dip–line BSS01-104. 14 horizons and faults including the AFC and TFFC were interpreted.

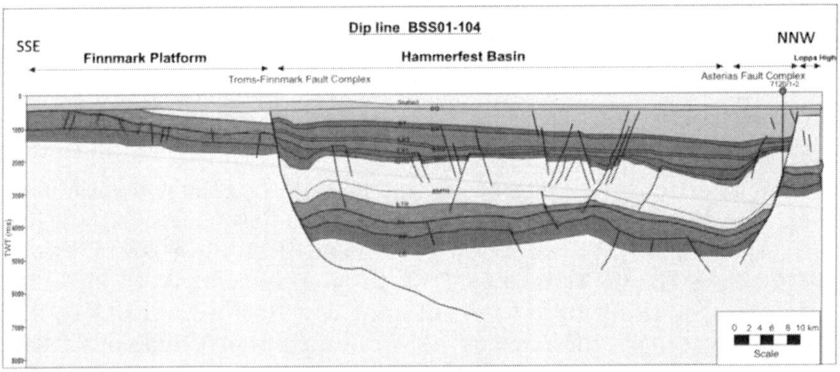

Figure 5: Stratigraphy of the Hammerfest Basin [28]. The colour code of the stratigraphy of the line from bottom to top is Carboniferous Formations (purple), Permian Formations (green),Triassic Formations (yellow), Jurassic Formations (red and pink), Cretaceous Formations (dark brown), Tertiary to Quaternary Formations (Light brown) and Quaternary and recent sediments (dark green).

The reflectors at the middle of the seismic section have different quality and architecture; some of them are high amplitude and continuous while the others are low amplitude and highly discontinuous. The discontinuous reflectors are more dominant along the Asterias Fault Complex (AFC), Troms-Finnmark Fault Complex (TFFC) and graben structure at the center of the basin. The reflectors along the active fault zones of the Asterias Fault Complex and the Troms–Finnmark Fault Complex are highly curved up following the master fault planes. The tectono-sedimentary evolution of the normal fault array can cause the sediments to bend along the fault complexes followed by onlapping sediments. Syn rift faults cutting IUTR, BUJ and ILK1, were recognized by thickening towards the faults. The discontinuities on the reflectors are faults which have reflection shadow and vertical displacements. The fault plane on the strike and dip lines has different nature of reflectance. All are normal faults dipping to NNW and SSE, dominantly extending from BMTR to BT.

The Troms-Finnmark Fault Complex changes its dip angle at depth and becomes gentle and steeper again (**Figures 4** and **5**). The deeper part of this fault was identified based on the structural relationship of faults on the dip and strike lines. It is dipping to the north on the dip– lines. The nature of the fault plane at this portion and further depths is not the same to its upper part. This part of the fault plane shows strong reflectors showing high rheological differences in acoustic impedance. It may encounter the basement at this depth as gravity studies indicate the basement is approximately at 9-10 km depth [29]. The subsidence along the AFC and

TFFC form synclines on both sides of the master faults caused an extension at the middle of the basin and subsequent subsidence leads to the formation of the graben structure.

Three possible phases of faulting have been identified based on repeated thickening of strata into the hanging wall of the basin near the faults. The first phase of rifting is the formation of the master faults e.g. Asterias Fault Complex and the Troms-Finnmark Fault Complex. It created accommodation space for the syn-rift sediments of Jurassic age. The second phase of rifting provides the Jurassic faults that extend to the intra-upper Triassic (IUTR) and Base of Middle Triassic (BMTR) as it can be evidenced from the folded strata in the AFC. The third phase of rifting can be the reactivation of the Jurassic faults which forms the graben structure at the middle of the Hammerfest basin. This continues up to Cretaceous and Tertiary formations. In this scenario, the formations above the Intra-lower Cretaceous (ILK1) are post-rift. The post rift formations are following the trend of syn-rift formations architecture at the base and become horizontal at the top. The deposition of the post rift significantly separated by erosional surface, at the Base Quaternary (BQ), since some of the Base Triassic to Intra-upper Triassic sediments are removed in the Finnmark Platform. The pre-rift formations are the Triassic and older formations. They have relatively uniform thickness throughout the basin. However, subsidence, uplift and fault-background collapse during the graben formation have affected the pre-rift sediments thickness along the flanks of the basin.

Stages of rifting from the reflectors architecture was used to identify the tectonic system tracts (**Figure 6**). S1 is the pre-rift stage to the Jurassic rifting phase and post rift to the Permian subsidence and rifting. The reflectors are parallel and uniform in thickness. The chaotic reflectors towards the master fault of Asterias Fault Complex can be coarse grained rock falls and talus. The rift initiation stage (S2) have wedge shape geometry where it is thicker towards the faults and thinner to its hanging wall by on lapping reflectors to the BUJ reflector. The accommodation space created by the rifting outpacing with subsidence has been filled with the Jurassic syn-rift sediments. After the rift climax stage (S4) the sedimentary layers start up- dipping towards the footwall shows the sediment supply is higher, the accommodation space formation by the rifting ceases. The post rift (S4) stage starts during Intra Late Cretaceous (ILK1). The tilting of the hanging wall and differential subsidence across the fault plane ceases and the rate of regional subsidence will decrease, although subsidence will continue due to the lithospheric thermal cooling effects.

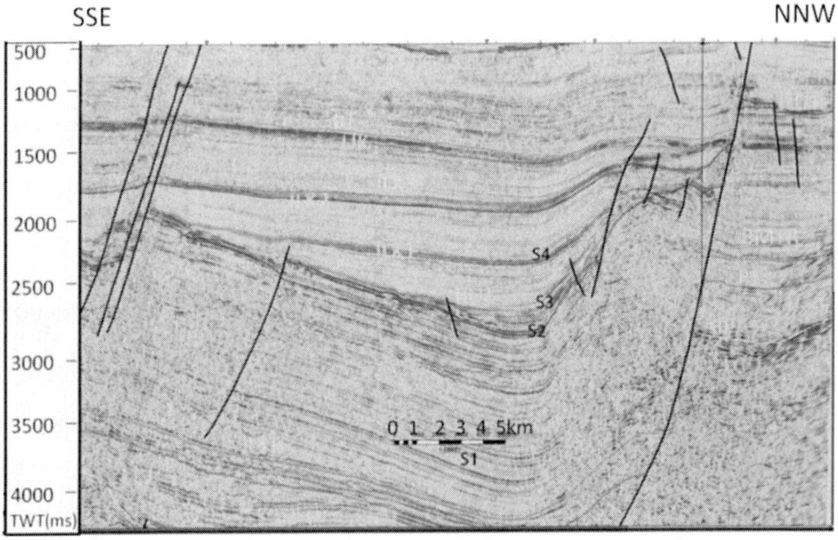

Figure 6: Stages of rifting and sedimentation based on the thickness variation and the architecture of the reflectors. S1: is the pre-rift stage, S2: the rift initiation stage, S3: is the rift climax and S4: post–rift stages are marking the tectonic and sedimentation evolution and their associated systems tracts.

Pockmarks are also visible along the seabed creating local depressions which are recognized in the seismic section [8,28] (**Figure 7**). They are dominantly seen above the fault graben structure. The graben is also characterized by the gas chimney and less clear reflectors from ILK1 to BQ. This indicates that pockmarks have a link with the gas leakage along the deep sited faults. Locally chaotic reflectors in this part can be good indicators for the subsidence and collapse of the graben structure. One can speculate the graben formation at the middle of the basin may be related with the subsidence of the footwall on both sides of the master faults of the Asterias Fault Complex and Troms-Finnmark Fault Complex. Deep seated faults and vents over the Snøhvit gas field are causing the hydrocarbon leakage which are manifest by pockmarks near and on the seabed [2,8,28]. The folded strata of the Asterias Fault Complex could suggest the nature of the sudden change of the dip of the master fault at depth and dragging of the hangingwall strata along the fault plane.

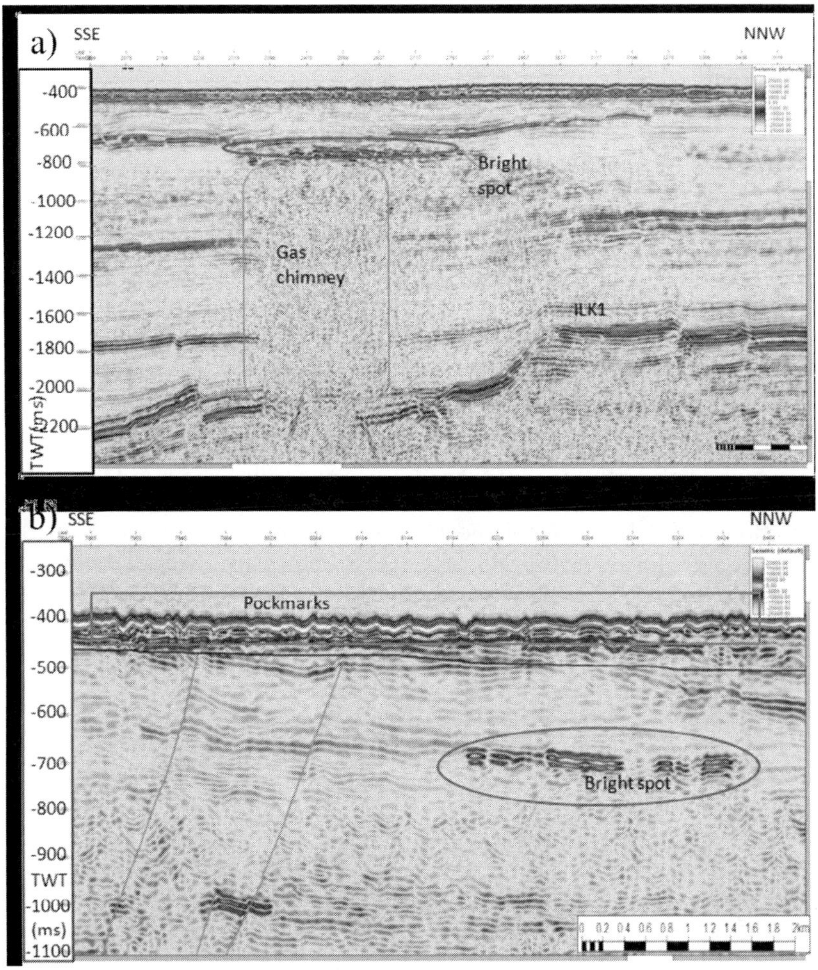

Figure 7: (a) Bright spot, gas chimney and (b) pockmarks.

2D petroleum system modeling

Heat flow calibration: The heat flow of this model ranges from low to average heat flow as confirmed in this model i.e., 41-55 mW/ m^2. The discrepancy in well 7120/12 between the observed temperature data and modelled temperature gradient suggests a recent perturbation in the basins thermal regime during the Cenozoic [3]. Thus, the calibration is not possible using a simple heat flow model. Which means as an increase in heat flux to match present-day temperatures forces the model to overestimate maturity. It indicates that the Hammerfest Basin is in a state of thermal disequilibrium, with the Snøhvit gas field is approximately 20°C above norm at the present day [3]. It is possible to infer that there was heat flow variation in the northern and southern parts of the basin. The thermal maturity difference in the northern and southern parts of the basin indicates that the heat flow modelling should consider the uplift/subsidence history. Many models have been tested with different erosion scenarios, fault properties and heat flows and has been observed that the accumulations also vary with these parameters.

Deposition and erosion: The rate of deposition in time and space is variable in the basin. There was rapid burial during Permian (Lopingian and Guadalupian) and late Cretaceous followed by uplifts. Three episodes of erosion (uplift) in the late Cretaceous, Oligocene– Miocene, due to tectonic uplift and Plio–Pleistocene glacial erosion events have been modelled in the basin affecting the hydrocarbon generation, accumulation, maturation and migration (**Figure 8**). This may be related to uplift and subsidence events. Different scenarios of erosion and uplift have been tested. Finally, three phases of erosion have been considered for modelling. 200 m of Kveite Formation and 300 m Torsk Formation have been eroded due to the tectonic uplift during 75-60 Ma and 30-15 Ma, respectively. The erosion event was continued during Pliocene-Pleistocene (2.5-0.01 Ma) causing the erosion of 100 m thickness of Torsk Formation (**Figure 9**).

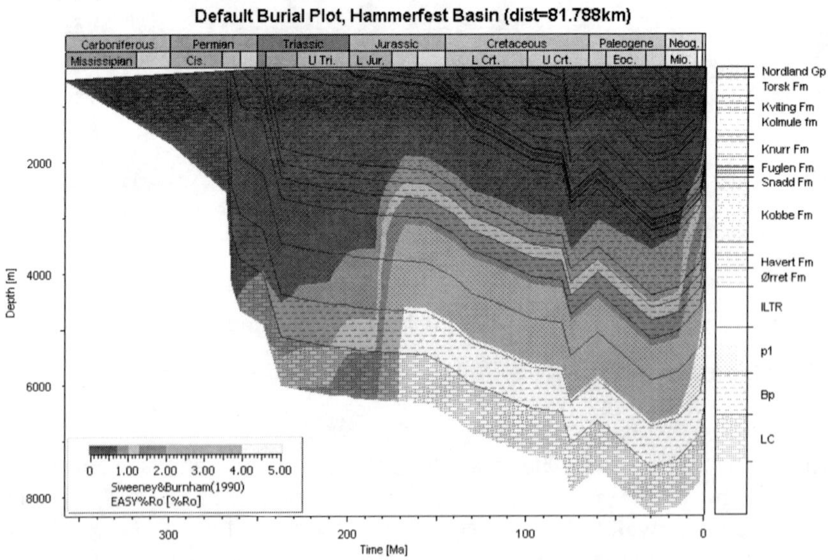

Figure 8: Burial history graph of the Hammerfest Basin superimposed with maturity (Ro) at 81 km.

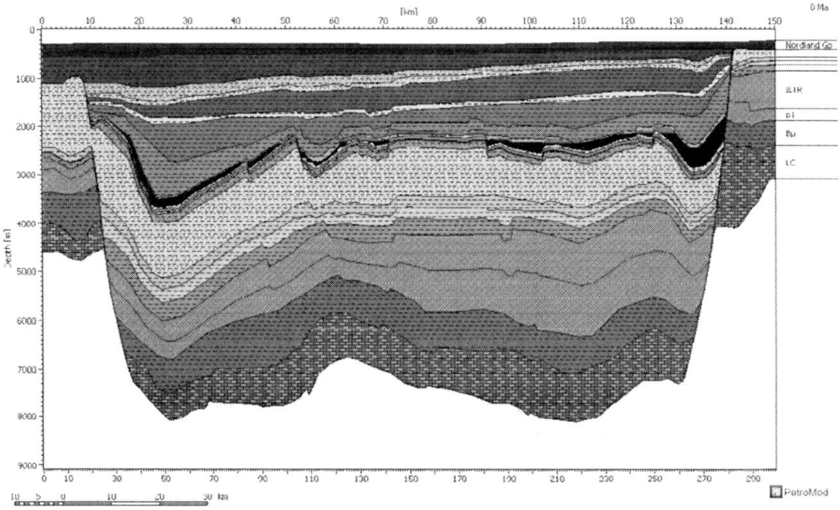

Figure 9: Present day model of Hammerfest Basin and hydrocarbon accumulations in the Stø and Tubåen reservoirs.

Hydrocarbon generation, accumulation and source rock maturity: Existence of appropriate source rock, and temperature and pressure condition is necessary for hydrocarbon generation in sedimentary basins. The hydrocarbon generation balance of the Hekkingen Formation is very high as compared to the Snadd and Kobbe source rocks but it is migrated upward by leakage. The accumulations in the reservoirs are from the Snadd and Kobbe Formations. On the other hand, this may also indicate that the Kobbe and Snadd Formations may be the major sources of gaseous hydrocarbon accumulation in the reservoirs. Then accumulations in the reservoirs become constant during that time. This is associated with tectonic uplift and erosion of the basin during Oligocene. It increases during Miocene and decreases again during Pliocene. The expansion of the gaseous phase causes spilling of oil out of the structures and affected primarily the liquid phase [12]. The hydrocarbon accumulation in each reservoir indicates that the Stø reservoir has a good reservoir potential and it is the main reservoir in the basin. The modelling result shows that the Tubåen reservoir has relatively small accumulations of hydrocarbon. The location of the accumulations has also been confirmed by well 7120/1-2 on the Aterias Fault Complex and 7121/7-2 at the middle of the basin. The Snadd and Kobbe Formations are thermally mature to highly mature which have entered the gas window, whereas the Hekkingen Formation is in the wet gas window to late oil window in the northern part and late oil to main oil windows in the southern part of the Hammerfest Basin (**Figure**

10 and **Table 3**). The maturation of the Hekkingen source rock has been interrupted during the uplift/subsidence phases. The increase in depth of burial and heat flow are also the main factors for maturation. The Permian Formations are highly to over mature.

Table 3: Oil and gas windows and vitrinite reflectance (VR) which is taken at the middle of the basin.

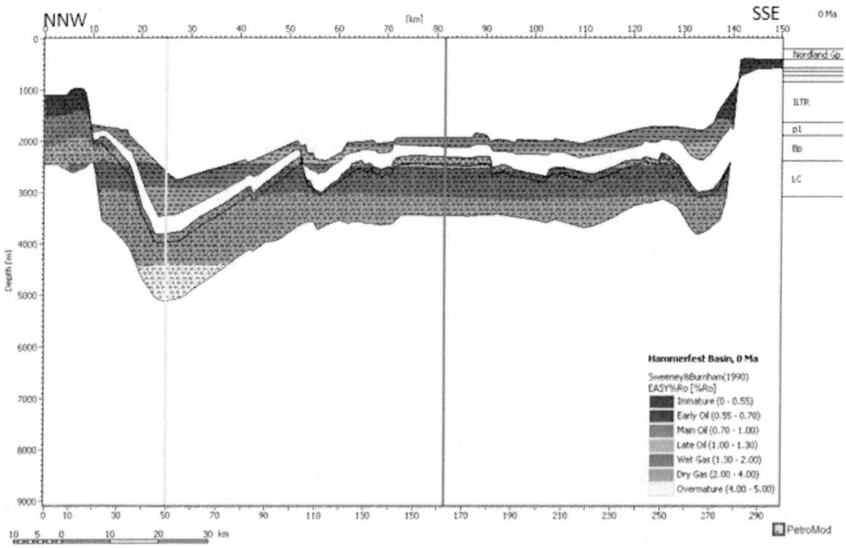

Figure 10: Source rock maturation variations in the southern and northern parts of the basin. Maturity increases toward the northern part of the basin.

Hydrocarbon migration and leakage: The migration pathway indicates upward and lateral migration across and within the formations (**Figures 11** and **12**). The upward migration is very dominant in the fault zones, along the Jurassic faults. The Hekkingen source rock has no downward migration pathways to the reservoirs. The closed fault type has more accumulation than the open type faults. But the migration and leakage were realized on both types of model have no big differences. Finally, the AFC and TFFC were defined as closed and the other faults are open which gives the best result in terms of accumulation. Different models have been tested in order to understand the dominant migration pathways. In most of the models the migration direction and leakage to the surface is related to the deep seated faults and the graben structure. Evidence from the seismic profiles includes fluid flow features such as gas chimnies and bright spots located at the crest of the graben. The oil migration follows the deep seated

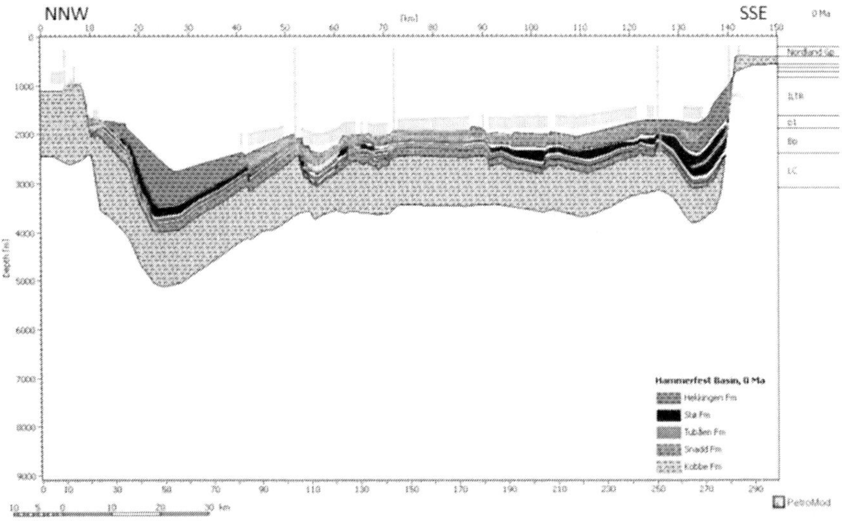

Figure 11: Oil upward migration from the source rocks into the reservoirs and the lateral migration within the source rocks. Further migration of oil out of the reservoir and leakage to the seabed follows the AFC, TFFC along the boundary faults of the graben structure.

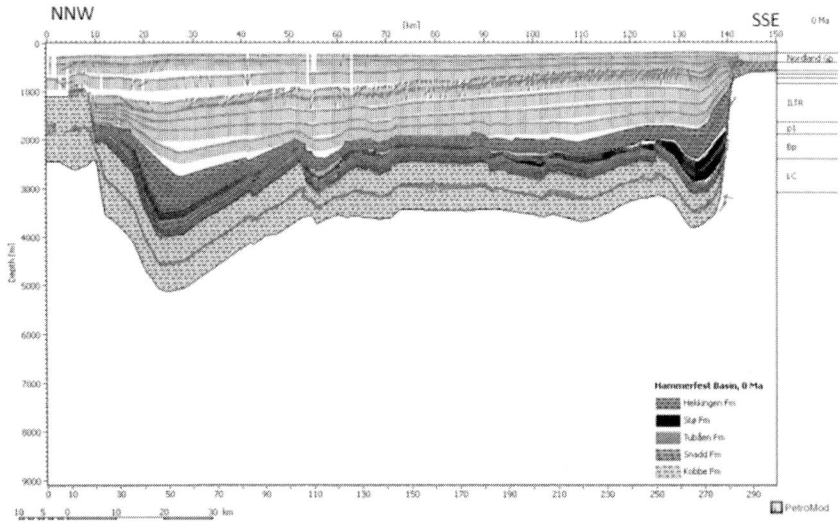

Figure 12: Gas upward migration from the source rocks into the reservoirs and the lateral migration within the source rocks. Further upward migration out of the reservoirs and leakage to the seabed throughout the basin is observed.

faults along the Asterias Fault Complex, Troms-Finnmark Fault Complex and along the bounding faults of the graben structure at the center of the basin which leads to leakage (**Figure 11**). The migration of gas on the other hand, is vertically upward throughout the basin leaking out of the seal (**Figure 11**). This could indicate that the seal is only effective for oil and not for the gas.

CONCLUSIONS

The hydrocarbons in the Hammerfest Basin are mainly gas and oil generated from the Hekkingen, Snadd and Kobbe Formations. The hydrocarbons generated from Snadd and Kobbe accumulated mainly in the Stø reservoir while some gas migrated into the Tubåen reservoir.

The Snadd and Kobbe Formations are thermally mature to highly mature and have reached the gas window. On the other hand, the Hekkingen Formation is less mature and it is in the wet gas window to late oil window in the northern part and late oil to main oil windows in the southern part of the Hammerfest Basin. The northern parts of the Hammerfest Basin have subsided more than the other parts. Hence, source rocks in the northern parts are more mature than the central and southern parts of the basin and showing lateral variations in maturation of source rocks.

Hydrocarbon migration from the Snadd and Kobbe source rocks into the Stø and Tubåen reservoirs is dominantly upwards assisted by the faults while the migration from Hekkingen source rock is upward to the Knurr Formation and the migration vector shows that it has no hydrocarbon contribution down to the reservoirs.

The first phase of erosion (75 to 60 Ma) erodes 200 m of Kveite Formation and the second (30 to 15 Ma) and the third erosion (2.5 to 0.01 Ma) phases erodes 300 m and 100 m of Torsk Formation, respectively. Thus, the uplift and subsidence events have been the main factors controlling the hydrocarbon maturation and distribution within the Hammerfest Basin.

ACKNOWLEDGEMENT

The author would like to acknowledge Dicky Harishidayat and Aamir Buriro for their technical help. Appreciation to the Norwegian Petroleum Directorate (NPD) for access to the seismic and borehole data used in this research and Schlumberger for provision of Petrel® for seismic interpretation and PetroMod® for petroleum system modelling.

REFERENCES

1. Berglund LT, Augustson JR,Gjelberg FJ, Ramberg-Moe H (1986) The evolution of the Hammerfest Basin. Habitat HydrocarbNorCont shelf p.319-338.
2. Ostanin I, Anka Z, di Primio R, Bernal A (2012) Hydrocarbon leakage above the Snøhvit gas field, Hammerfest Basin SW Barents Sea. Firs Brea 30: 55-60.
3. Cavanagh AJ, Di Primio R, Scheck-Wenderoth M, Horsfield B (2006) Severity and timing of Cenozoic exhumation in the southwestern Barents Sea. J GeolSoc London 163: 761-774.
4. Rodrigues Duran E, di Primio R, Anka Z, Stoddart D, Horsfield B (2013) Petroleum system analysis of the Hammerfest Basin (southwestern Barents Sea): Comparison of basinmodelling and geochemical data. Org Geochem 63: 105-121.
5. Ohm SE, Karlsen DA, Austin TJF(2008) Geochemically driven exploration models in uplifted areas: Example from the Norwegian Barents Sea. AmAssoc Pet Geol Bull 92: 1191-1223.
6. Hermanrud C, Halkjelsvik ME, Kristiansen K, Bernal A, Strömbäck AC (2012) Column Height Controls in the Western Hammerfest Basin.3rd EAGE International Conference on Fault and Top Seals, Fault and Top Seals.
7. Hermanrud C, Halkjelsvik ME, Kristiansen K, Bernal A, Strömbäck AC (2014) Petroleum column-height controls in the western Hammerfest Basin, Barents Sea. Pet Geosci 20: 227-240.
8. Ostanin I, Anka Z, di Primio R, Bernal A (2013) Hydrocarbon plumbing systems above the Snøhvit gas field: structural control and implications for thermogenic methane leakage in the Hammerfest Basin, SW Barents Sea. Mar Pet Geol 43: 127-146.
9. Faleide JI, Tsikalas F, Breivik AJ, Mjelde R, Ritzmann O, et al.(2008) Structure and evolution of the continental margin off Norway and the Barents Sea. Episodes 31: 82.
10. Chand S, Rise L, Ottesen D, Dolan MFJ, Bellec V, et al. (2009) Pockmark-like depressions near the Goliat hydrocarbon field, Barents Sea: morphology and genesis. Mar Pet Geol 26: 1035-1042.
11. Larsen RM, Fjaeran T, Skarpnes O (1993) Hydrocarbon potential of the Norwegian Barents Sea based on recent well results. ArctGeol Pet potential p.321-331.
12. Rodrigues Duran B, di Primio E, Anka R, Stoddart Z, Horsfield D (2013) 3D-basin modelling of the Hammerfest Basin (southwestern Barents Sea): A quantitative assessment of petroleum generation, migration and leakage. Mar Pet Geol 45: 281-303.
13. Chand S, Thorsnes T, Rise L, Brunstad H, Stoddart D, et al. (2012) Multiple episodes of fluid flow in the SW Barents Sea (Loppa High) evidenced by gas flares, pockmarks and gas hydrate accumulation. Earth Planet SciLett 331-332: 305-314.
14. Gabrielsen RH, Færseth RB, Jensen LN, Kalheim JE, Riis F (1990) Structural elements of the Norwegian Continental Shelf. Part 1: The Barents Sea region. NPD-Bulletin p.33.
15. Larsen RB (2002) The distribution of rare-earth elements in K-feldspar as indicator of petrogenetic processes in granitic pegmatites: Examples from two pegmatite fields in southern Norway. Can Mineral 40: 137-151.
16. RønnevikH, Jacosen H (1984) Structural highs and basins in the Western Barents Sea. PetrolGeol North Europe Marp.19-32.
17. Bjørkesett SS (2009) Late Paleozoic-Triassic evolution of the paleo- Loppa High, linked to tectonic events and depositional patterns. Master thesis, University of Oslo.
18. Rønnevik HP, Beskow H, Jacobsen B (2009) Structural and Stratigraphic evolution of the Barents Sea. Can Soc Pet Geol 8: 431-440.
19. Faleide JI, Gudlaugsson ST, Jacquart G (1984) Evolution of the western Barents Sea. Mar Pet Geol 1: 123-150.

20. KnutsenSM, Vorren TO (1991) Early Cenozoic sedimentation in the Hammerfest Basin. Mar Geol 101: 31-48.
21. Doré A (1995) Barents Sea geology, petroleum resources and commercial potential. Arctic 48: 207-221.
22. Norwegian Petroleum Directorate (2015) The Norwegian Petroleum Directorate Fact-Pages of Exploration Wellbores.
23. Lambeck K (1995) Constraints on the Late Weichselian ice sheet over the Barents Sea from observations of raised shorelines. QuatSci Rev 14: 1-16.
24. Wygrala BP (1989) Integrated Study of an Oil Field in the Southern Po Basin, Northern Italy.
25. Mienert SJ, Vanneste M, B¨unz S, Andreassen K, Haflidason H, et al. (2005) Ocean warming and gas hydrate stability on the mid-Norwegian margin at the Storegga Slide. Mar Pet Geol 22: 233-244.
26. Archer G, Martin D, Bu ett P, Brovkin B, Rahmstorf S (2004) The importance of ocean temperature to global biogeochemistry. Earth Planet SciLett 222: 333-348.
27. Breivik Y, Mjelde AJ, Grogan R, Shimamura P, Murai H, et al. (2005) Caledonide development offshore–onshore Svalbard based on ocean bottom seismometer, conventional seismic, and potential field data. Tectonophy 401: 79-117.
28. Mohammedyasin M, Lippard S, Omosanya K, Johansen S, Harishidayat D (2016) Deep-seated faults and hydrocarbon leakage in the Snøhvit Gas Field, Hammerfest Basin, Southwestern Barents Sea. Mar Pet Geol 77: 160-178.
29. Gernigon L, Brönner M, Roberts D, Olesen O, Nasuti A, et al. (2014) Crustal and basin evolution of the southwestern Barents Sea: from Caledonian orogeny to continental breakup. Tecton 33: 347-373.

CHAPTER 13

Investment in Deepwater Oil and Gas Exploration Projects: a Multi-Factor Analysis with a Real Options Model

Xin-Hua Qiu[1] • Zhen Wang[2] • Qing Xue[1]

[1].School of Business AdministrationChina University of PetroleumBeijingChina

[2].Academy of Chinese Energy StrategyChina University of PetroleumBeijingChina

ABSTRACT

Deepwater oil and gas projects embody high risks from geology and engineering aspects, which exert substantial influence on project valuation. But the uncertainties may be converted to additional value to the projects in the case of flexible management. Given the flexibility of project management, this paper extends the classical real options model to a multi-factor model which contains oil price, geology, and engineering uncertainties. It then gives an application example of the new model to evaluate deepwater oil and gas projects with a numerical analytical method. Compared with other methods and models, this multi-factor real options model contains more project information. It reflects the potential value deriving not only from oil price variation but also from geology and engineering uncertainties, which provides more accurate and reliable valuation information for decision makers.

Keywords: Investment decision Real options Multi-factor model Option pricing Deepwater oil and gas

1 INTRODUCTION

Deepwater petroleum investment has attracted much attention as offshore oil and gas resources are making up a large portion of worldwide energy potentials. However, due to the marine geographical environment, deepwater oil, and gas development projects contain higher geology and engineering risk than onshore or continental shelf projects. This situation increases the total amount of investment and the complexity of decision-making process. On one hand, the volatility of oil price causes more flexibility value for the deepwater projects which demand a longer duration for exploration. On the other hand, the technical risk of deepwater projects under development is much higher than that of onshore or continental shelf ones, and the effects of engineering and technological uncertainties on the value of deepwater projects are more significant. Under this background, the traditional theory of net present value cannot provide sufficient reliable reference for the decision making of deepwater oil and gas investment, because the value of flexibility from the uncertainties in oil price, engineering, and technology cannot be measured under the rigid assumptions. Therefore, the real options method based on uncertainty analysis is more suitable to evaluate deepwater oil and gas projects than the conventional ones.

The real options theory, originating from the financial option, regards the value from management flexibility as an option which could generate revenue. Myers (1977) analyzed the value of real options of additional investment opportunities for the first time. Since then, the flexibility value of real investment has received ongoing attention. Dixit and Pindyck (1994) summarized the research achievements of the real options theory and presented a systematic exposition of its construction and application. Firstly, they described the statistical characteristics of the uncertainty factor which influences the cash flow significantly. Secondly, they determined the functional relationship between uncertainty factors and the revenue and established the equation by non-arbitrage portfolio. Thirdly, they derive the equation for the real options model based on the assumptions and boundary conditions.

Since Brennan and Schwartz (1985) and Paddock et al. (1988) evaluated natural resources investment by adopting the real options method, research using the real options method has gradually increased. Dias (2004) presented an overview of real options models to evaluate investments in petroleum exploration and production projects. He pointed out that oil price was the only random variable in almost all the real option models and empirical studies, and other technical factors are assumed to be constant which could be obtained from engineers before the evaluation. In his review, a petroleum project was considered as a long-term investment and production process, during which the fluctuation of oil price could influence its economic value significantly. It could increase the flexibility value by adjusting production according to the oil price fluctuation. Due to the importance and financial attribute of oil price, simulation for the stochastic characteristics of oil price was a focus in real option research.

However, factors affecting the flexibility value of oil and gas projects are not limited to oil price, especially in the case of deepwater oil and gas exploration and development projects. The flexibility of geological understanding and engineering technology also has an important influence on the projects. The geological and technological information will be more accurate with increasing investment. The investors could make better decisions with the additional knowledge to realize flexibility value which may be ignored by the net present value method. In the theory of real options, the additional information and flexible management are valuable. If the flexible value of an underlying project is larger than its required investment, the project will be profitable. In order to evaluate the comprehensive flexibility value, a multi-factor real options model should be established with geological and technological factors in addition to the price factor.

Attempts to set up a multi-factor real options model have been made recently. Cortazar et al. (2001) added the information of geology and technology to the model of Brennan and Schwartz (1985) to evaluate a copper mine, but the study was not intensive. Cortazar et al. (2001) did not analyze the relationship between the information uncertainty and the flexibility value, and failed to describe the establishment or application of the model. The uncertainty factors and flexibility need further investigation. Fan and Zhu (2010) built a multi-factor real options model and applied it to an oil investment decision. However, the research did not consider the two important factors of geological and technological uncertainties. It introduced the exchange rate and resource tax rate to the model, none of which has a significant effect on the flexibility value for deepwater oil and gas projects. They proposed to adjust oil price to exchange rate and tax rate, and then substituted the volatility of oil price in the conventional real options model for integration of the three volatilities. However, the integration is meaningless because their integrated factor has no difference from a single factor in essence. Furthermore, the tax rate does not have a stochastic characteristic. Schmit et al. (2011) built a two factors real options model to estimate the influence of U.S. ethanol policy on plant investment decisions. Similarly, this research focused on the financial aspect only, which defined the two variables as revenue and cost, ignoring the important influence from technology and engineering.

Our multi-factor real options model and its application to deepwater projects will make several contributions. In the aspect of random factors, we analyze three of the most important factors: oil price, geological information, and engineering information, based on the characteristics of deepwater oil and gas projects. We also integrate the three factors based on the stochastic process theory. In the aspect of real options model, we extend the single-factor model with geological and technical factors to better describe the flexibility value of deepwater oil and gas projects on the basis of the integration model, because the partial differential equation for three factors is too complex to be solved. In the aspect of application, we provide an example to show the practical significance of key parameters and introduce the method of parameter assignment. We also apply the real options model to value a deepwater project under a typical production-sharing contract.

This paper is organized as follows: In the second section, we will describe and integrate the variables with stochastic process theory. In Sect. 3, the multi-factor real options model will be established based on the integration model and the non-arbitrage approach. In Sect. 4, we will discuss the parameter assignment. In Sect. 5, we will apply the model to a deep water oil and gas project and analyze the optimal investment decision. Section 6 is the conclusions.

2 THREE FACTORS AFFECTING THE VALUE OF FLEXIBILITY IN DEEPWATER OIL AND GAS PROJECTS

Considering the characteristics of marine geographical environments, the flexibility value of deepwater oil and gas projects is determined not only by the volatility of oil price but also by the uncertainty of geology and engineering technology. If the exploration and development scheme is adjusted based on the additional information, the economy value of project could be increased.

The flexibility value of geology conditions implies the unremitting objectives of minimization of investment and maximization of profit since the geology information updated as the project proceeds helps make the investment budget more precise. In the oil and gas industry, the investment in exploration gives investors priority over next stage's activities. So these investors have more prominent opportunities due to their information privilege. The flexibility of technology implies potential cost savings and production increases in the process of exploration and production with the uncertainties being gradually clarified and problems solved. On the other hand, under the background of whole block development, the flexibility of technology implies the value maximization for all projects located in the same area since investors could properly design the overall development program and share the facilities among different projects in that area. Besides, the flexibility value of oil price changes could never be neglected since deep water oil and gas development projects always take many years. Investors and their management team can adjust their actual production according to the price at the time under specific technology and engineering conditions, so as to realize the best economic value of oil and gas reserves.

Geological conditions, technologies, and oil price are the main factors that affect the real options value of deepwater oil and gas projects. This work aims to study these three factors first, and build a multi-factor real options model on such basis.

There are two requirements in simulation of random factors. Firstly, the model should be concise enough for practical use and could accurately render the dynamic characteristics of random factors since the purpose of simulation is to construct financial models and realize necessary computation rather than to make predictions on future situation. Secondly, many financial models are built on the basis of Ito's lemma by now, however the random factor models are also needed to follow Brownian movements as basic variables do (Itō 2010, 2011).

2.1 Oil price simulation

There has been a lot of research conducted on simulation and prediction of oil prices because of the importance of petroleum in the world economy and international relationships. It has been shown that the fluctuation of oil price follows a random walk process with sudden increases or jumps at certain periods. Dixit and Pindyck (1994) described the oil price with several models. They pointed out that the geometric Brownian motion (GBM) should be a foundational model and the mean reverting model could describe the stochastic characteristics of oil price more accurately since oil price fluctuated around the cost of oil production which was stable. But the difficulty and the cost of oil exploitation have increased rapidly with soaring demand during the past two decades, and this change has been reflected in the oil price fluctuation (see Fig. 1), so the mean reverting model is not as accurate to describe the characteristics of oil price.

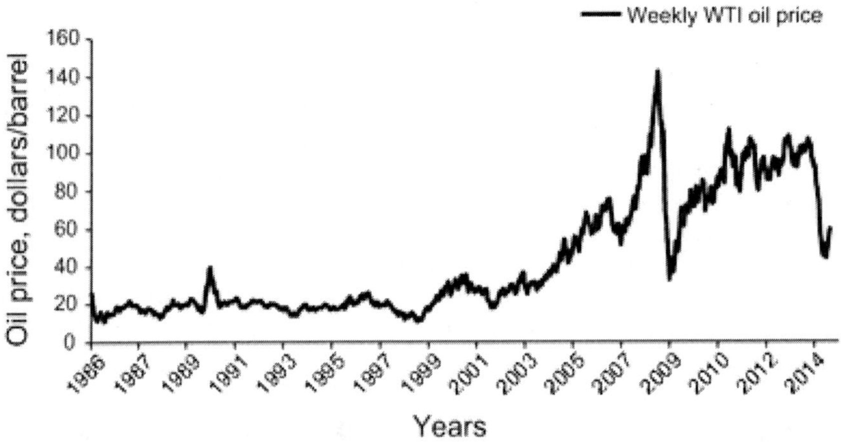

Fig. 1. The fluctuation of WTI oil price

Compared with the mean reverting model, GBM is more appropriate to embody oil price movements, as GBM conforms to both the stochastic characteristics of oil prices and the two requirements mentioned above.

Then, the simulation model of oil price will be:

$$\mathrm{d}P_t = \mu P_t \mathrm{d}t + \sigma_P P_t \mathrm{d}W_t, \tag{1}$$

where P_t is the oil price at time t, W_t is a Wiener process, μ and σ_P are constants.

Assuming an initial price of P_0, use the Ito stochastic integral to solve the equation and we will get the following:

$$P_t = P_0 \exp\left(\left(\mu - \frac{\sigma_P^2}{2}\right)t + \sigma_P W_t\right). \tag{2}$$

2.2 Simulation of technological and geological factors in engineering

Technological and geological factors in engineering can substantially affect the overall economic value of deepwater oil and gas exploration projects as economic factors on the product market do, and must be included in the evaluation model.

Deepwater exploration technology is rapidly evolving. Specialists and technicians are exploring better methods to describe geology, technological conditions, and risks in the seabed. Their ideas and models can be quite different, and most of them end up in describing various geological and technological factors in the form of probabilities. Therefore, probability theory can be used to study the geological and technological uncertainties in the evaluation model for deepwater oil and gas exploration projects. However, most technological models in engineering involve various parameters, which are complicated and confidential, and cannot be directly used in economic models. Thus the model with geological and technological uncertainties must be designed with comprehensive study of technological methods in engineering to meet the requirements of an evaluation model.

All the random variables in a stochastic system can be described by a stochastic process. According to research findings (Clapp and Stibolt 1991), engineering and technology data follow normal distribution with certain features, and can be described by a stochastic differential equation. Engineering technology covers a wide range of factors with correlations among some of them, and it is impractical to clarify each of them in an economic model. Thus geological and technological factors in engineering will be taken as one comprehensive factor in simulation and analysis.

We hereby define a one dimension geology technology factor G, which follows Brownian movement with zero drift and constant volatility.

$$dG = G\sigma_G dW_G, \tag{3}$$

where σ_G is volatility, dW_G is the standard Wiener increment.

2.3 Variables integration

The above analysis and treatment for oil prices and geological and technological factors with a stochastic model is to investigate how to put the uncertain technological factors separately and directly into the evaluation model, so as to provide precise evaluation results for deepwater oil and gas assets. However, it is over sophisticated to put both factors P and G into a proper model together with the application of Ito's lemma. Therefore factor P and factor G will be technically integrated into one factor to build a three-factor model (Itō 2010).

Both oil price (P) and geology technology (G) impact on the value of deepwater oil and gas projects, where P is mainly affected by market dynamics, and G affected by geological conditions under the sea floor and by achievements in technological development. Therefore assume factors P and G are independent, which means:

$$\mathrm{d}W_P \mathrm{d}W_G = 0 \tag{4}$$

and let

$$Z = F(P, G). \tag{5}$$

According to Ito's lemma, we substitute Eq. (1) and Eq. (3) into Eq. (5) and get:

$$\begin{aligned} \mathrm{d}Z = &\left(F_P P \mu + \frac{1}{2} F_{PP} P^2 \sigma_P^2 + \frac{1}{2} F_{GG} G^2 \sigma_G^2 \right) \mathrm{d}t \\ &+ F_P P \sigma_P \mathrm{d}W_P + F_G G \sigma_G \mathrm{d}W_G. \end{aligned} \tag{6}$$

In order to make Eq. (6) solvable and ensure the effectiveness of the evaluation model, we apply the principle of value additivity of stochastic process by Itō's (2010) and get:

$$Z = F(P, G) = PG. \tag{7}$$

Given that factors P and G following standard Wiener process, and they both are independent and unrelated incremental variables, Eq. (7) is correct according to Ito's theory and the two stochastic processes can be superimposed to get:

$$\frac{\mathrm{d}Z}{Z} = \mu \mathrm{d}t + \sigma_P \mathrm{d}W_P + \sigma_G \mathrm{d}W_G. \tag{8}$$

Hence the new variable Z embodies the same drift rate with oil price P but larger volatility than P:

$$\sigma_Z = \sqrt{\sigma_P^2 + \sigma_G^2}. \tag{9}$$

3 MODELING

3.1 Hypothesis in modeling

Actual economic problems are far more diversified and complicated. A series of hypothesis are established in the light of the problems' particularity and research targets for simulation and computation, so as to better describe and solve problems with mathematic modeling. The fundamental hypotheses in real options modeling for deepwater oil and gas exploration projects are as follows:

1. Oil price *P* follows GBM process, and its convenience yield is the function of oil price;
2. The geological technological variables follow Brownian movement;
3. Investment return *r* is known and constant;
4. The reproduction cost of the investment portfolio is negligible;
5. The real options value *V(Z, t)* in the form of variable *Z* and time *t* is second order differentiable, and follows Ito's lemma;
6. The compound option is perpetual since oil and gas exploration contracts last for many years.

3.2 Model establishment

The evaluation model is established on the basis of no-arbitrage portfolio theory. Assume *F(Z, τ)* is the price function of petroleum futures bought at time *t* with expiration time at *T*, where τ=T−tτ=T−t. According to Ito's lemma, the instant yield of the futures is:

$$\mathrm{d}F = \left(-F_\tau + \frac{1}{2}F_{ZZ}\sigma^2 Z^2\right)\mathrm{d}t + F_Z\mathrm{d}Z, \tag{10}$$

where F_Z and F_{ZZ} are first and second order partial derivatives.

And with Eq. (7) we have:

$$F_P = F_Z \times G \quad F_{PP} = F_{ZZ} \times G^2 \tag{11}$$

Then we generalize such an investment portfolio: An investor goes long on one unit of crude oil in the spot market and goes short on $(F_P)^{-1}$ unit of crude oil as underlying asset in the future market. Suppose no dividend is to be paid. According to Eqs. (10 and 11), the rate of return for this portfolio is:

$$\begin{aligned} &\frac{\mathrm{d}P}{P} + \frac{C(Z)\mathrm{d}t}{P} - (PF_P)^{-1}\mathrm{d}F \\ &\quad = (PF_P)^{-1}\left[F_P C(Z) - \frac{1}{2}F_{PP}\sigma^2 P^2 + F_\tau\right]\mathrm{d}t, \end{aligned} \tag{12}$$

where $C(Z)$ indicates convenience yield. Under the no-arbitrage principle of efficient market, the investment return of above portfolio equals the market return, which means:

$$\frac{1}{2}F_{PP}\sigma^2 P^2 + F_P(rP - C) - F_\tau = 0. \tag{13}$$

The boundary condition is:

$$F(P, G, 0) = P. \tag{14}$$

With Eq. (10), Eq. (11), and Eq. (8) we have:

$$\mathrm{d}F = F_P[P(\mu - r) + C]\mathrm{d}t + F_P P\sigma \mathrm{d}z. \tag{15}$$

Deepwater oil and gas exploration involves special risks and tremendous investments, and the economic value is mainly affected by oil price P, geology technology G, accumulative investment I, and time t. Taking V for the value of the petroleum asset, and with Eq. (7), we get:

$$V \equiv V(G, P, I, t) = V(Z, I, t). \tag{16}$$

With Ito's lemma Eq. (16) is changed into:

$$\mathrm{d}V = V_Z \mathrm{d}Z + V_I \mathrm{d}I + V_t \mathrm{d}t + \frac{1}{2}V_{ZZ}(\mathrm{d}Z)^2. \tag{17}$$

Let q be per unit investment, λ the average income tax rate, γ the rate of success in exploration, thus the after-tax cash flow of the exploration project will be:

$$\gamma V - q - \lambda V. \tag{18}$$

In order to get the partial differential equation of project value (V), we build another investment portfolio: buy one unit of oil asset and sell the same unit of oil futures, then the investment return will be:

$$\begin{aligned} & \mathrm{d}V + [\gamma V - q - \lambda V]\mathrm{d}t - (V_P/F_P)\mathrm{d}F \\ & = \frac{1}{2}\sigma^2 Z^2 V_{ZZ} - qV_I + V_t + (rP - C)V_P + [\gamma V - q - \lambda V]. \end{aligned} \tag{19}$$

In light of the no-arbitrage principle, the portfolio return is equal to the market return r_V. With Eq. (11) we have:

$$\begin{aligned} & \frac{1}{2}\sigma^2 Z^2 V_{ZZ} - qV_I + V_t + (rP - C)V_Z + q \\ & \quad - (r + \lambda - \gamma)V = 0. \end{aligned} \tag{20}$$

Taking deepwater oil and gas exploration projects as perpetual real options, then the operational period t is infinite. When an investor is operating one project, he also keeps seeking for other potential exploration blocks to ensure continuous cash inflow, which means t in the equation is not a variable ($V_t = 0$), and the value of real options is only related to its price and the geology technology uncertainties.

The value of a deepwater project is defined as $V(Z, I)$ under perpetual operation, then the maximum project value with optimal output level satisfies the following requirement:

$$\frac{1}{2}V_{ZZ}Z^2\sigma_Z^2 + (rZ - C)V_Z + qV_I - q - (r + \lambda + \gamma)V = 0. \tag{21}$$

The boundary conditions are:

$$\begin{aligned} & V(0, I) = 0 \\ & V_Z(0, I) = 0 \\ & \lim_{Z\to\infty} V_{ZZ}(Z, I) = 0. \end{aligned} \tag{22}$$

Taking Eq. (21) together with Eq. (22) to establish the multi-factor real options model under conditions of uncertainty, and it is generally difficult to obtain analytical solutions for this model. A numerical simulation method is adopted instead for this model.

4 VARIABLE SIMULATION AND PARAMETER ANALYSIS

4.1 Variable simulation

In Sect. 2 the oil price, geological factor and technological factor are described by using stochastic differential equation (SDE), and the SDE is solved to get the key parameters of the equation. Also, the three factors are integrated together to build up the integration model. Figure 2 shows the simulation results of the factors based on the SDE and the Gauss fitting result.

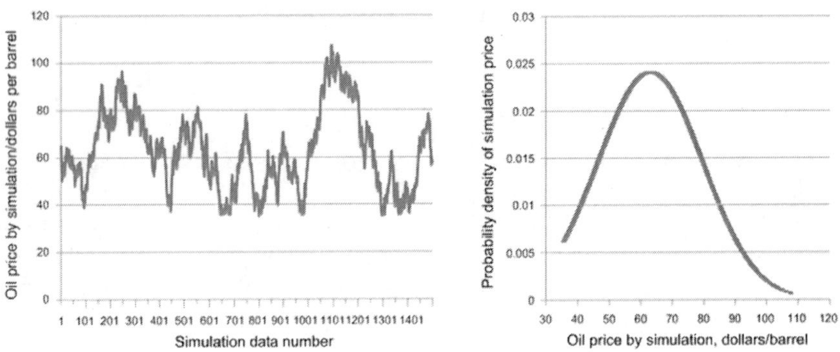

Fig. 2. The stochastic simulation and Gauss fitting of oil price

We collect 1500 WTI (West Texas Intermediate) weekly oil price data and fit these data for comparison analysis, as shown in Fig. 3.

Figures 2 and 3 indicate that the simulated movements and actual oil price movements share the same characteristics of stochastic process if excluding unpredictable sudden jumps caused by political or economic emergencies. Therefore, the above simulation model and the given values acquired successfully to reflect the characteristics of real oil price movements, and the basic form of the model conforms to a Wiener process, thus it can be used in financial models for oil and gas asset evaluation.

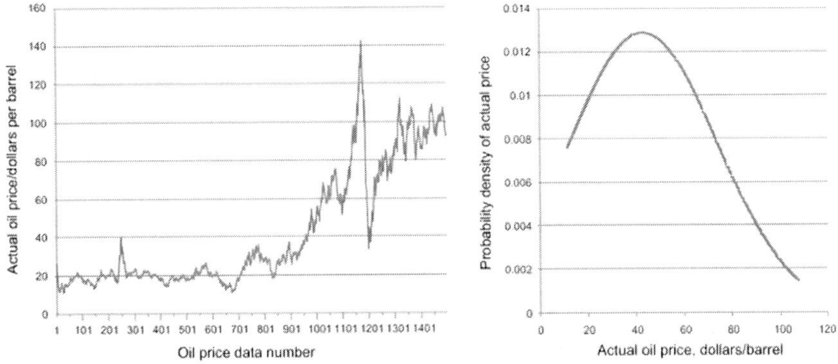

Fig. 3. The plot and Gauss fitting of WTI oil price

Simulation results of the geology technology factor *G* and the integrated three factors are shown in Figs. 4 and 5.

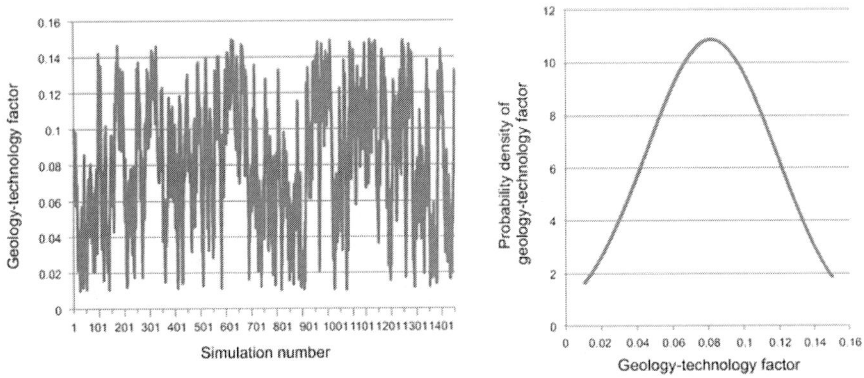

Fig. 4. The stochastic simulation and Gauss fitting of geology technology factor

According to Figs. 2, 3, 4 and 5, the integration made in this work meets the hypothesis of Ito's lemma, and can be the basis of the parametric estimate in the multi-factor real options model. In addition, the fitting results of integrated and single factors are different from each other, which also prove that the study for multi-factor real options model makes sense.

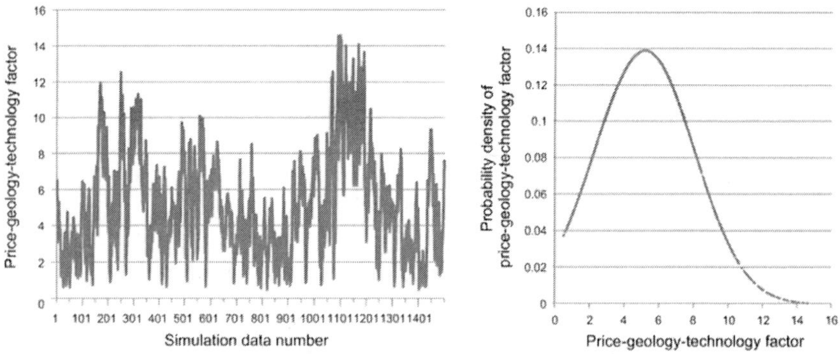

Fig. 5. The stochastic simulation and Gauss fitting of oil price-geology technology factor

4.2 Parameter analysis

Wang and Li (2010) analyzed the parameters of the real options model and demonstrated the significant influence of the parameters to the valuation result. In order to make the multi-factor real options model applicable, accurate and understandable, study of the parameters value must be first conducted. Deepwater oil and gas exploration and development projects involve interests of many parties, which are stipulated in complicated contract clauses. The study of convenience yield is computed based on these clauses, which is analyzed in the research by Liu et al. (2012).

4.2.1 Investment rate and its influence on project value

In deepwater oil and gas exploration projects, investors are confronted with many uncertain factors, which on the other hand provide them with great flexibility in project management: they can expand or hold down the investment volume according to updated geological and technological conditions, and increase or reduce their production in consistence with market dynamics. Whereas there is one thing in common for most projects: some of the investment is irreversible despite succeeding investment policies as the initial investment is sunk or at least partially sunk. The investment in exploration is totally irreversible no matter if it succeeded or failed in finding recoverable resources.

The investment rate in exploration indicates the capital expenditure in search of recoverable oil and gas resources, and it is a crucial factor for total investment returns in the real options evaluation model. When oil companies increase their investment in exploration, they will acquire information about recoverable reserves with higher volume and better accuracy, and they will also have more confidence to realize greater economical value of the project. Thus it is reasonable to assume that the increase in exploration investment adds value to the project. To be more simplified, we suppose the exploration investment and project value are in positive linear correlation.

4.2.2 Success rate in exploration

Success rate in exploration changes with many factors such as location, reservoir conditions, exploration engineering and equipment, and oil companies have different success rates in their exploration works in different areas and different blocks. It is fairly difficult to calculate or predict with limited geology and engineering information, while an alternative is to use empirical values from projects with similar location, water depth and other parameters.

4.2.3 Convenience yield rate

Convenience yield rate is a sensitive parameter in a real options evaluation model. Gibson and Schwartz (1990) suggested that convenience yield was the value-added cash flow naturally derived from products, and it belonged to the holders of such products rather than holders of derivative contracts; Convenience yield depended on the volume of inventory: products of less inventory and higher spot price could achieve higher convenience yield, and vice versa. Paddock et al. (1988) concluded that investment return of developed reserves consisted of two parts: operating profits from production sales and capital gains from intrinsic value growth of remaining reserves. Thus let convenience yield be:

$$C_t = \frac{\omega[\Pi_t - V_{bt}]}{V_{bt}}, \tag{23}$$

where ω is the decline rate of production in percentage; Π_t is after-tax profit of oil sales; V_{bt} is oil value of developed reserves per barrel.

4.2.4 Market investment return

We can adopt the return of investment used in discounted cash flow analysis and adjust it according to the characteristics of assets in the same region.

5 EVALUATION AND DECISION ANALYSIS

An overseas deepwater asset located in West Africa is an example for this multi-factor real options model. This project is under a production-sharing contract (PSC). At the initial stage of investment, the oil price was at a relatively low level but moving upward, so the exploration scheme was conservatively designed for modest oil production since only pre-exploration had been conducted, and detailed exploration data were not available. At the initial decision point, the discounted cash flow analysis did not show too much promise even though adjacent oil blocks showed promising economic returns. So the decision making of this project does not completely refer to DCF analysis results.

Parameters in the real options model are calculated (see Table 1) according to the contract and the analysis in Sect. 4.

Table 1. Parameter estimation

Market price of developed reserves V_b, \$/bbl	15.38
After-tax earnings Π, \$/bbl	19.01
Total cost D, \$/bbl	13.04
Exploration cost E, \$/bbl	2.96
Influence of investment changes to project value V_I	1.9
Success rate in exploration γ	0.2
Production decline rate ω, %	6.25
Investment rate q, %	23.00
Market investment return r, %	10.00
Convenience yield C, %	1.50

Assume that oil price is at USD65$ with volatility of 0.18 and factor *G* equals 0.1 with volatility of 0.25, then the value of this project with geology technology uncertainties amounts to USD13.53$ per barrel. In a more pessimistic scenario with higher geology technology risks, assuming factor *G* to be 0.07, the evaluation result from the real options model suggests a project value of USD6.48$ per barrel.

While a single-factor real options model is often applied to evaluate the flexibility value with uncertainties in oil price, this study adopts both multi-factor real options model and net present value method to better demonstrate the uncertainties. The comparison results are shown in Table 2.

Table 2. Results of three evaluation methods

Evaluation method	NPV	Single-factor real options model	Multi-factor real options model with high G rate	Multi-factor real options model with low G rate
Result, USD/barrel	5.76	7.38	13.53	6.48
Deviation from NPV result	–	28.13 %	135 %	12.5 %

The results in Table 2 show that the multi-factor real options model can better reflect the flexibility value. Besides, projects in adjacent blocks are endowed with excellent geological conditions. They have exhibited great potential and gained much higher economic returns than their initial evaluation results, which provide evidence for the effectiveness of multi-factor real options model to some extent.

On the other hand, the results under higher geology technology risks scenario by a multi-factor real options model are much more conservative than those given by the single-factor model, thereby it confirms that multi-factor model is more capable in reflecting the impact of engineering factors on projects' flexibility value while single-factor model only focuses on the impact of oil prices but ignores the impact of geology and technology.

6 CONCLUSIONS

The management of flexibility value is not only embodied in the feasibility to elevate the economic value of reserves by adjusting production to oil prices. It's also shown in the flexibility to design exploration schemes according to the uncertainty of geological information and technology, especially for deepwater oil and gas exploration projects. We analyzed several influential factors for project value with reference to the characteristics of real options in deepwater projects. We established a multi-factor real options model under uncertain conditions for project evaluation, and employed the idea of multi-uncertainty factors integration to make the model practical.

The model has been successfully applied to a real deepwater project. The evaluation result shows that the multi-factor real options model could be more accurate than the single-factor model. This multi-factor model gives investors more reliable theoretical supports to make reasonable decisions. Our sample project has been operated for more than five years, and the real practice has also showed that the estimated value with multi-factor real options model is a better approximation to reality. So the multi-factor real options model could be a good reference approach for investment decisions about deepwater oil and gas projects.

ACKNOWLEDGMENTS

The authors would like to thank the experts for their helpful discussion and suggestions on the 18th Annual International Conference on Real Options. This paper is supported from the National Science and Technology Major Project under Grant No. 2011ZX05030.

REFERENCES

1. Brennan MJ, Schwartz ES. Evaluating natural resource investments. J Bus. 1985;58(2):135–57.
2. Clapp R, Stibolt R. Useful measures of exploration performance. J Pet Technol. 1991;10:1252–7 Cortazar G, Schwartz ES, Casassus J. Optimal exploration investments under price and geological-technical uncertainty: a real options model. R&D Manag. 2001;31:181–9.
3. Dias MA. Valuation of exploration and production assets: an overview of real options models. J Pet Sci Eng. 2004;44:93–114.
4. Dixit AK, Pindyck RS. Investment under uncertainty. Princeton: Princeton University Press; 1994.
5. Fan Y, Zhu L. A real options based model and its application to China's overseas oil investment decisions. Energy Econ. 2010;32:627–37.
6. Gibson R, Schwartz ES. Stochastic convenience yield and the pricing of oil contingent claims. J Financ. 1990;45(3):959–76.
7. Itō K. Stochastic process. Beijing: Postal and Telecom Press; 2010 **(IN CHINESE)**.
8. Itō K. Probability theory. Beijing: Postal and Telecom Press; 2011 **(IN CHINESE)**.
9. Liu M, Wang Z, Zhao L, et al. Production sharing contract: an analysis based on an oil price stochastic process. Pet Sci. 2012;9(3):408–15.
10. Myers SC. Determinants of corporate borrowing. J Financ Econ. 1977;5(2):147–75.
11. Schmit TM, Luo J, Conrad JM. Estimating the influence of U.S. ethanol policy on plant investment decisions: a real options analysis with two stochastic variables. Energy Econ. 2011;33:1194–205.
12. Paddock JL, Seigel DR, Smith JL. Option valuation of claims on real assets: the case of offshore petroleum. Q J Econ. 1988;103(3):479–508.
13. Wang Z, Li L. Valuation of the flexibility in decision-making for revamping installations: a case from fertilizer plants. Pet Sci. 2010;7(3):428–34.

CHAPTER 14

Correlations between Formation Properties and Induced Seismicity during High Pressure Injection into Granitic Rock

Mark W. McClure[a,], Roland N. Horne[b]

[a] *Department of Petroleum Geosystems Engineering, University of Texas at Austin, 200 E. Dean Keeton, C0300, Austin, TX 78704, United States*

[b] *Energy Resources Engineering, Stanford University, 367 Panama Street, Green Earth Sciences 065, Stanford, CA 94305-4007, United States*

ABSTRACT

We reviewed published results from six projects where hydraulic stimulation was performed in granitic rock. At each project, fractures in the formation were well-oriented to slip at the injection pressures used during stimulation. In all but one case, thousands of cubic meters of water were injected, and in every case, flow rates on the order of tens of liters per second were used. Despite these similarities, there was a large variation in the severity of induced seismicity that occurred in response to injection. At the three projects where induced seismicity was significant, observations at the wellbore showed evidence of well-developed brittle fault zones. At the three projects where induced seismicity was less significant, observations at the wellbore indicated only crack-like features and did not suggest significant fault development. These results suggest that assessments of the degree of fault development at the wellbore may be useful for predicting induced seismicity hazard. We cannot rule out that the differences were caused by variations in frictional properties that were unrelated to the degree of fault development (and it is possible that there is a relationship between these two parameters). The projects with more significant seismicity tended to be deeper, and if this is a meaningful correlation, it is unclear whether depth influenced seismic hazard through the degree of fault development, frictional properties, or some other variable. The results of this paper are not conclusive, but they suggest that there may be significant opportunity for future research on identifying geological conditions that increase induced seismicity hazard.

Keywords: Induced seismicity; Geothermal; Water injection; Fault development; Rate and state friction Gadget timed out while loading

1. INTRODUCTION

Induced seismicity is an issue of growing importance for the exploitation of geothermal energy (Majer *et al.*, 2007; Majer *et al.*, 2011; Cladouhos *et al.*, 2010 ; Evans *et al.*, 2012), wastewater disposal (Frohlich *et al.*, 2011; Kim, 2013 ; Zhang *et al.*, 2013), CO_2sequestration (Zoback and Gorelick, 2012), hydrocarbon production (Suckale, 2009), and other activities (McGarr *et al.*, 2002 ; Hitzman, 2012). Appropriate management of induced seismicity requires estimation of induced seismic hazard.

Some methodologies for estimating induced seismic hazard are purely statistical (Bommer *et al.*, 2006 ; Bachmann *et al.*, 2011) or hybrid statistical/fluid flow models (Shapiro *et al.*, 2007 ; Gischig and Wiemer, 2013). However, these methods are site-specific and must be conditioned by performing the activity for which seismic hazard needs to be assessed. Low natural seismicity does not necessarily indicate that induced seismic hazard will be low, though there is some correlation (Evans et al., 2012).

In addition to purely statistical methods, numerical simulation has been used for induced seismic hazard analysis. This may involve kinematic modeling of deformation to estimate induced stress on neighboring faults (Segall, 1989; Hunt and Morelli, 2006 ; Vörös and Baisch, 2009) or dynamic modeling that couples fluid flow, stresses induced by deformation, and friction evolution (Baisch *et al.*, 2010 ; McClure and Horne, 2011). However, model results are dependent on assumptions and input parameters that may be challenging to estimate. For example, change in stress on a fault may be estimated, but it is unclear how this should be related quantitatively to increased hazard (due to uncertainties such as the fault stress state and frictional properties).

All approaches to induced seismicity hazard analysis, whether purely statistical, physically based, or a hybrid of both, could benefit from methodologies that relate geophysical and geological observations to hazard (e.g., Davis and Frohlich, 1993). McGarr (1976) predicted that induced seismic moment release should be proportional to the volume of fluid injected, and this has been borne out by subsequent experience (Rutledge *et al.*, 2004; Bommer *et al.*, 2006; Hunt and Morelli, 2006 ; Baisch and Vörös, 2009). However, the constant of proportionality between injection volume and moment release varies over orders of magnitude between different locations. For example, there are over 35,000 wells in the United States that have been hydraulically fractured in unconventional shale resources, and there are only a handful of confirmed instances of an induced event felt at the surface (page 76 of Hitzman, 2012). Induced seismicity does appear to be associated with oil and gas activities in many cases, but this is apparently due to long term fluid injection (Frohlich et al., 2011) or extraction (van Eijs et al., 2006). On the other hand, hydraulic stimulation for exploitation of geothermal resources in crystalline rock has routinely induced seismic events felt at the surface (Majer *et al.*, 2007; Majer *et al.*, 2011; Cladouhos *et al.*, 2010 ; Evans *et al.*, 2012). Even among geothermal hydraulic fracturing projects, a huge diversity in induced seismicity hazard has been observed (Table 1; Kaieda et al., 2010). More work is needed to explain how geological conditions cause these large variations in induced seismicity.

Table 1. Summary of experiences with induced seismicity at six EGS projects. Supporting references are given in the text below. The assessments of "volume of fluid injected" are not necessarily complete because in some projects a variety of different injection operations were carried out over many years. We have not made an extensive effort to document all of the injections performed at these projects.

	Depth range	Maximum magnitude	Temperature	Degree of fault development	Volume of fluid injected during stimulation
Basel	4.6–5.0 km	3.4	190 °C at 5.0 km	High	11,570 m^3
Cooper Basin (Habanero 1)	4.1–4.4 km	3.7	250 °C at 4.4 km	High	20,000 m^3 in 2003, 25,000 m^3 in 2005
Fjällbacka	0.5 km	~0.2	16 °C at 0.5 km	Low	400 m^3 in Fjb1 and 36 m^3 in Fjb3
Ogachi (OGC-1)	0.99–1.0 km	−1.0 except a 2.0 outlier	230 °C at 1.0 km	Low	10,140 t (approximately 9200 m^3)
Rosemanowes	1.7–2.65 km	0.16	100 °C at 2.6 km	Low	100,000 m^3 over two months in RH11 and RH12 (1982) and 5700 m^3 in RH15 (1985)
Soultz (shallow)	2.8–3.4 km	1.9	150 °C at 3.4 km	High	Two stimulations of 20,000 m^3 each
Soultz (deep)	4.5–5 km	2.9	200 °C at 5.0 km	High	Three wells stimulated at volumes between 20,000 and 35,000 m^3

It is universally accepted that hydraulic fracturing in an oil and gas settings causes the initiation and propagation of new fractures (Economides and Nolte, 2000). However, during injection, fluid could leakoff into existing faults and cause slip (and potentially seismicity). In EGS, it is typically believed that injection predominantly causes induced slip on preexisting fractures (Cladouhos et al., 2011), though some authors have argued that there is probably more new fracture propagation than is commonly believed (McClure, 2012; Jung, 2013 ; McClure and Horne, submitted for publication).

For significant induced seismicity to occur: (1) faults must be oriented properly with respect to the prevailing stress field so that they slip in response to imposed changes in stress and/or fluid pressure, (2) faults must have appropriate frictional properties so that slip occurs rapidly enough to generate seismicity, and (3) faults must be large enough to host significant events.

The first requirement, appropriately oriented faults, can be understood in the context of Coulomb theory (Chapter 2 of Hitzman, 2012). It has been argued that, in general, the crust is in a state of failure equilibrium (Townend and Zoback, 2000), suggesting that nearly everywhere in the subsurface, faults are present that will slip in response to an increase in fluid pressure.

The second requirement, rapid slip, can be explained in the context of results from laboratory friction experiments. Experiments have shown that the tendency for faults to slip seismically (rapidly) or aseismically (gradually) depends on the frictional properties of the minerals contacting in the fracture walls. Rock type, temperature, and other factors affect the tendency for fractures to slip seismically or aseismically (Dieterich, 2007).

There is a significant untapped opportunity to apply the results from friction experiments to help relate induced seismic hazard to lithology and depth. For example, differences in frictional properties may be the reason why hydraulic fracturing in granite has led to much greater induced seismic hazard than hydraulic fracturing in sedimentary formations. Assessments of induced seismic hazard from fluid injection in sedimentary formations (e.g. CO_2 sequestration or wastewater disposal) would benefit from efforts to identify lithologies where fault frictional properties are most favorable for seismic slip. Zhang et al. (2013) identified an apparent correlation between induced seismicity hazard and injection into basal aquifers.

The third requirement is that faults must be large enough to host significant-sized events. Seismic imaging and stratigraphic study can be used to identify major faults in layered formations. But in nonlayered formations such as crystalline basement rock, these techniques are limited because of the lack of seismic reflectors and discernible stratigraphic offsets. A very thick fault zone is required to generate a visible reflection at significant depth in crystalline rock. Faults in the basement may not extend into the overlying sediments. Even in layered sedimentary formations, hidden faults may be capable of hosting significant induced seismicity.

In this paper, we investigate whether wellbore observations could be used to estimate induced seismic hazard by assessing the degree to which large, brittle faults are present in a formation. Wells only sample the formation locally and may not intersect the most seismically important faults. However, formations that contain large faults are likely to contain abundant faults at all levels of development, and the overall degree of fault development in the formation should be observable at the wellbore. This theory is supported by the general observation that induced seismicity typically follows a Gutenberg–Richter distribution (Baisch *et al.*, 2009; Baisch *et al.*, 2010 ; Bachmann *et al.*, 2011), that large faults are surrounded by sizable damage zones, and that fracture size distributions are usually found to obey a power law or exponential distribution (Chapter 3 of Scholz, 2002). The mechanistic reason is that faults develop from accumulated deformation over time, starting with small, isolated cracks, which eventually link up and develop into large, continuous features (Chapter 3 of Segall and Pollard, 1983; Scholz, 2002 ; Mutlu and Pollard, 2008). We refer to formations that have significantly developed faults that have linked up and formed larger features as having a high "degree of fault development."

To test whether the degree of fault development may be correlated to induced seismic hazard, we reviewed six projects where hydraulic stimulation (high rate fluid injection) was performed in granitic rock for the exploitation of geothermal energy: the projects at Cooper Basin, Australia; Soultz, France; Ogachi, Japan; Rosemanowes, United Kingdom; Basel, Switzerland; and Fjällbacka, Sweden. Projects using hydraulic fracturing to develop geothermal energy reservoirs are sometimes called "Enhanced Geothermal Systems,", or EGS. Some well-known EGS projects in granite, such as the projects at Fenton Hill, USA and Hijiori, Japan, were not included in this study because we were unable to find references that would permit an assessment of the degree of fault development. The Rosemanowes and Fjällbacka projects were performed for research purposes and targeted lower temperature reservoirs than would be typical for geothermal exploitation.

To control for the possible effect of lithology on the frictional properties of the faults, only projects in granite were included. To minimize the potential effect of fault orientation, we included only hydraulic stimulation tests where large injection pressures were used. The bottomhole fluid pressure likely reached or exceeded the minimum principal stress at every project considered, except possibly Basel, where estimates of the minimum principal stress are not available (McClure and Horne, submitted for publication). At these elevated pressures, faults with a wide variety of orientations are capable of slip (though not all orientations). In the "Fault orientation" section, we review literature that shows that fractures were present that were capable of slipping at the injection pressure at all six projects. A source of uncertainty is that pressure decreases with distance away from the wellbore during injection.

Fracture observations from a linear well are biased because wells are more likely to intersect faults perpendicular to their orientation than parallel to their orientation (Mauldon and Mauldon, 1997). This may particularly be an issue for vertical wells in strike-slip faulting regimes, where faults are expected to be vertical. Of the projects reviewed in this paper, Cooper Basin was in a reverse faulting regime (Baisch et al., 2006), Fjällbacka was in a reverse/strike-slip faulting regime (Wallroth et al., 1999), Soultz was in a normal/strike-slip regime (Valley and Evans, 2007), Basel, Ogachi, and Rosemanowes were in strike-slip faulting regimes (Pine and Batchelor, 1984 ; Häring *et al.*, 2008), and different references specify Ogachi as either strike-slip (Kiho et al., 1999) or reverse faulting (Ito, 2003). The wells at all these projects were nearly vertical, except at Rosemanowes, where the wells were deviated around 30° (Pine and Batchelor, 1984). Despite the potential for sampling bias, the approach in this paper assumes that wellbore observations will be adequate to characterize the degree of fault development in the formation. This approach relies on the idea that there was enough natural variability in fault orientation to provide an adequate sample of the formation from a borehole, and also on the idea that a formation containing large faults (which may not be intersected by the well) will also contain numerous smaller features that provide evidence of brittle fault development.

The six projects could be placed in two categories. At three of the projects, cataclasite and evidence of well-developed faults were observed at the wellbore, and induced seismicity was significant. At the other three projects, cataclasite and well-developed faults were not observed at the wellbore, and induced seismicity was much less significant. These results seem to indicate that the degree of fault development is a good indicator for induced seismic hazard, as long as the faults are well-oriented for slip in the ambient stress field and have frictional properties favorable for seismicity.

The results cannot be considered conclusive, especially because only six projects were included in the study. It cannot be ruled out that variation in frictional properties and/or depth accounted for the variation in seismicity, nor can their effects be fully disentangled because they may be interrelated. Even though the results are not conclusive, the results of this study suggest that formation frictional properties and degree of fault development could be useful in seismic hazard assessment and that these concepts merit further study.

2. METHODOLOGY

Six EGS projects in granitic rock were reviewed. Many projects involved multiple wells and stimulations. Other than at Fjällbacka, thousands or tens of thousands of cubic meters of fresh water were injected into the wells during stimulation. At Fjällbacka, injection volume was on the order of hundreds of cubic meters. In all cases, injection was performed into uncased sections of the wellbore at rates in excess of ten liters per second. Maximum magnitudes at each project are quoted directly from literature sources.

The degree of fault development was assessed from the thickness of observed faults and/or the presence of cataclasite and ultracataclasite. In projects with a "high" degree of fault development, fault zone features with thicknesses of meters were observed, or alternatively, drilling cuttings or core showed the presence of cataclasite and ultracataclasite, consistent with the classical concept of a fault with a damage zone (Faulkner et al., 2010). In projects with a "low" degree of fault development, only crack-like features were observed with thickness no greater than a few millimeters. These categorizations are coarse, but this was intentional due to the lack of precision inherent to assembling results from many different sources.

Maximum magnitudes are reported for the period during stimulation or shortly after. At Rosemanowes and Fjällbacka, larger events occurred during long-term unbalanced circulation (more fluid injected that produced) between injector/producer pairs. Long-term unbalanced circulation is a different process from stimulation, involving the net injection of a greater volume of fluid over a larger period of time and contained in a larger spatial region. Therefore, the induced seismic hazard would reasonably be expected to be greater in these cases, and so for consistency, induced events during long-term unbalanced circulation were not included in the assessments of maximum magnitude that are reported in Table 1.

3. RESULTS

Table 1 summarizes the results. In the following sections, the specific details of each project are reviewed.

3.1. Basel

One well was stimulated in an openhole section from 4629 m to 5000 m with 11,570 m^3 of water. The temperature at the bottom of the hole was around 190 °C. When the well was being drilled, drill cuttings were recovered containing ultracataclasite and cataclasite from five discrete zones (Kaeser *et al.*, 2007 ; Häring *et al.*, 2008). The fault thickness could not be directly observed with image logs due to enlarged borehole diameter (Häring et al., 2008), but the presence of cataclasite indicates that well-developed faults were present. Mazurek (1998) described wells that were continuously cored through the crystalline basement of northern Switzerland within distances of 20–100 km from the Basel site. In these wells, thick well-developed faults were abundant. We classified the degree of fault development as high. The maximum magnitude was 3.4 (Evans et al., 2012).

3.2. Cooper Basin

The well Habanero 1 was completed openhole in granite from 4135 m to 4421 m and stimulated twice, in 2003 and 2005. In 2003, over 20,000 m^3 was injected (Baisch et al., 2006), and in 2005, around 25,000 m^3 was injected (Baisch et al., 2009). The temperature at the bottom of the hole was around 250 °C (Baisch *et al.*, 2006 ; Baisch *et al.*, 2009). In wellbore imaging logs, a major fault zone was identified with a core with thickness of a few meters, surrounded by subsidiary fracturing within ten meters of the core [*Doone Wyborn, personal communication*]. Flow from the wellbore localized at this major fault. A broad, horizontal region of microseismicity spread from the well at the same depth as the fault (the minimum principal stress is vertical), and a well drilled subsequently several hundred meters away intersected a large fault at the same depth as the microseismic cloud (Baisch et al., 2006). We classified the degree of fault development as high. Maximum magnitude was 3.7 (Asanuma et al., 2005).

3.3. Fjällbacka

Two wells were stimulated from openhole sections around 500 m depth. Fjb1 was stimulated with 400 m^3 of water (Wallroth et al., 1999), and Fjb3 was stimulated with 36 m^3 (Evans et al., 2012). Temperature at 0.5 km was 16 °C (Wallroth et al., 1999). One of the wells was continuously cored, and Eliasson et al. (1990) performed a detailed analysis of the mineralogy of infilling of the fractures found in the wellbore core. Hydrothermally altered fracture zones were found, but there was no report of cataclasite.

We classified the degree of fault development as low. The maximum observed magnitude during or after stimulation was – 0.2. During unbalanced circulation between the wells, when more fluid was injected than produced, there was an event that was felt only at the project site, but no estimate of magnitude is available (Evans et al., 2012).

3.4. Ogachi

OGC-1 was drilled and stimulated from an openhole section from 990 to 1000 m with 10,140 t (approximately 9200 m^3) (Kaieda et al., 2010). Hundreds of meters of OGC-1 and an adjacent well, OGC-2, were continuously cored. Temperature was around 230 °C at 1 km (Kitano et al., 2000). Faults with gouge less than two millimeters were abundant, but faults were not observed with thickness greater than 5 mm (Ito, 2003). Hydrothermal breccias were observed, but Ito (2003) specifically noted that these features were associated with fluidization, not shear deformation, because they contained neither cataclasites nor slickensides. A major flowing zone in the well OGC-2 was found to occur along an andesite dike, not a brittle fault structure. We classified the degree of fault development as low.

Suzuki and Kaieda (2000) discussed the results of a seismic reflection survey and a magneto-telluric survey that were performed at the Ogachi project. Their interpretation was that there is a large fault roughly 500 m from the Ogachi wells. For several reasons, we did not include this feature in our assessment of the degree of fault development. First, the fault was not observed directly and was inferred based on assumptions made in interpretation and processing. The interpretation was especially challenging because the formation is crystalline and not layered. Second, even if a structural offset exists, the seismic interpretation did not give any detail about its inner structure — whether it is a brittle fault structure, a ductile shear zone, or some other feature. Because the inferred fault is unconfirmed by direct measurements and no information is available about its inner structure, we feel this report is not enough to change our assessment, particularly because such a detailed core analysis is available from Ito (2003).

The largest event was magnitude 2.0, but the next largest event was – 1.0 (Kaieda et al., 2010). This seismic behavior was unusual compared to other projects, where Gutenberg–Richter distributions of magnitude versus frequency were observed (Baisch *et al.*, 2009; Baisch *et al.*, 2010; Bachmann *et al.*, 2011). Other than the outlier event, the magnitude-frequency distribution at Ogachi fell along a Gutenberg–Richter distribution with a b value of 1.1 [*Hideshi Kaieda, personal communication*].

3.5. Rosemanowes

Three wells were stimulated in openhole sections at different depths between 1.7 and 2.65 km. Temperature at 2.6 km was around 100 °C (Richards et al., 1994). An analysis of wellbore imaging logs at the Rosemanowes wells discusses only thin fractures, not thicker fault features (Pearson et al., 1989). Fracture mapping from surface and mineshafts for the project discuss only jointing, not thicker fault features (Whittle, 1989; Randall *et al.*, 1990). Heath (1985) studied wells that were continuously cored to 700 m depth at a separate site around 10 km from the Rosemanowes project and reported faults with displacements of no more than a few millimeters. We classified the degree of fault development as low. The maximum magnitude observed during or immediately after stimulation was 0.16, but magnitude 1.7 and magnitude 2.0 events were reported during long term, unbalanced circulation between an injector and producer (when more fluid was being injected than produced) (Evans et al., 2012). In 1982, 100,000 m^3 of water was injected into RH11 and RH12 over two months. In 1985, 5700 m^3 of water was injected into RH15 (Evans et al., 2012).

3.6. Soultz (shallow)

One well was stimulated from an openhole section in granite from 2.8 to 3.4 km depth. Two stimulations were performed with 20,000 m^3 each (Evans et al., 2005). Temperature at 3.4 km was around 150 °C (Genter et al., 2010). GPK1 was extensively logged with wellbore imaging logs, and an adjacent well, EPS1, was continuously cored to 2230 m. Thick fault zones (up to 27.5 m thick) with the pattern of cataclasite and breccia in a core surrounded by intensely fractured and hydrothermally altered surrounding zones were observed in the core (Genter et al., 2000). We classified the degree of fault development as high. Maximum magnitude was 1.9 (Evans et al., 2012).

3.7. Soultz (deep)

Three wells were stimulated in openhole sections in granite from roughly 4.5 to 5 km. Temperature at 5 km was around 200 °C (Genter et al., 2010). Consistent with the shallower Soultz reservoir, the degree of fault development was high (Dezayes et al., 2010). Maximum magnitude was 2.9. The wells were each stimulated with volumes from 20,000 m^3 to 37,000 m^3 (Evans et al., 2012).

4. DISCUSSION

4.1. Seismicity

Seismicity was significantly greater at the three projects where evidence of well-developed faults was found, Cooper Basin, Soultz, and Basel, than at the three where it was not, Rosemanowes, Fjällbacka, and Ogachi.

At the less seismically active projects, there were a handful of outlier events. At Ogachi, there was a report of a single magnitude 2.0 even though the next largest event was reported to be magnitude – 1.0. At Rosemanowes and Fjällbacka, the maximum magnitudes during stimulation were very low, 0.16 and – 0.2. But there were a few reports of larger events (magnitude 1.7 and 2.0 events at Rosemanowes and an event felt at the project site at Fjällbacka) during long term unbalanced circulation. Despite these outlier events, the overall intensity of seismicity at these projects was much less than at the more active projects. These outlier events do not fall along a typical Gutenberg–Richter distribution of magnitudes, as was observed at Cooper Basin (Baisch et al., 2009), Soultz (Baisch et al., 2010), and Basel (Bachmann et al., 2011).

4.2. Degree of fault development

The correlation between the degree of fault development and seismicity could be explained in terms of slip surface continuity or by considering the overall formation deformation that occurs in response to a unit of fluid injected.

Earthquake ruptures have been observed to terminate when they reach fault stepovers (Wesnousky, 2006). Because it is difficult for ruptures to propagate from one fracture to another, large slip events should be prevented from occurring in formations with low fault development, where large, continuous fractures are less likely to be present.

However, the difference in seismicity between the projects cannot be explained only from slip surface continuity. The overall amount of seismic moment release was orders of magnitude greater for the projects with a high degree of fault development. The projects with much lower maximum magnitudes would have required a huge number of small events (with an unusually high b value) to have matched the seismic moment release from the projects with large maximum magnitudes and typical magnitude–frequency distributions. This was not observed.

Formations with a higher degree of fault development may be expected to experience more overall shear deformation in response to injection.

Preexisting flaws deform much more easily than intact rock, which is why stiffness and rock strength decrease with increasing length scale (Heuze et al., 1990). As a simple example, injecting fluid into a fault cut in a granite block loaded with shear stress could lead to significant shear deformation. Injecting fluid into an intact granite block loaded under the same conditions would lead to much less shear deformation (unless the intact rock failed in compression or tension). Analogously, highly faulted formations should experience more shear deformation in response to injection than formation that are sparsely fractured. All other factors held equal, the amount of shear deformation caused by injection should be greater for formations with greater fault development. The scaling of this relationship deserves further study. It is possible that for formations with sufficiently developed faults, further fault development does not increase the amount of deformation that will occur in response to injection. van Eijs et al. (2006) performed a statistical investigation across geologically similar formations in the Netherlands and found a correlation between induced seismicity hazard and "fault density" in the formation as estimated from interpretations of seismic reflection surveys.

4.3. Fault orientation

A Coulomb analysis can be used to demonstrate that fault orientation in the ambient stress state strongly affects the tendency for a fault to slip in response to injection (Chapter 2 of Hitzman, 2012). In the EGS projects reviewed in this paper, injection pressure reached the minimum principal stress in almost every case (see review in McClure and Horne (submitted for publication)). With such high fluid pressure, critical stress analysis may be less useful for assessing hazard because many or most fractures in the formation are likely to be able to slip. Of course, there will always be a fluid pressure gradient away from the well during injection, so fluid pressure elevation during injection will decrease as a function of distance.

Analyses performed at Soultz (Evans, 2005 ; Meller *et al.*, 2012), Fjällbacka (Jupe et al., 1992), Rosemanowes (Pine and Batchelor, 1984), Ogachi (Kiho *et al.*, 1999 ; Moriya *et al.*, 2000), and Basel (Häring *et al.*, 2008 ; Delacou *et al.*, 2009) indicate that abundant fractures were present at these projects that were oriented appropriately to slip at the injection pressure. We do not have access to statistics on fracture orientation from the wellbore imaging logs at Cooper Basin, but it is certain that natural fractures slipped in response to injection at this site because there were large magnitude events and the microseismic cloud was oriented subhorizontally in a reverse faulting stress regime (Baisch et al., 2006).

In different applications, where downhole fluid pressure may be more modestly elevated, fracture orientation and stress analysis may be more useful for estimating induced seismic hazard.

4.4. Depth and temperature

In our dataset, there was a correlation between depth and induced magnitude. This was most apparent at the Soultz project, where the maximum magnitude during the stimulation of the reservoir at 3 km was lower than in the reservoir at 5 km.

One explanation could be that because stress scales linearly with depth, stress drop also scales with depth. In fact, there is some evidence that stress drop scales with depth for shallower earthquakes. Shearer et al. (2006) reviewed earthquakes in Southern California and found that the median stress drop increased from around 0.6 MPa near the surface to around 2.2 MPa at 8 km. However, the total seismic moment release and maximum magnitude at the sites reviewed in this paper varied by orders of magnitude, more than be explained solely from stress drop.

Depth might relate to seismicity in other ways. A rate and state analysis suggests that the critical patch size for unstable slip decreases for greater normal stress (Dieterich, 2007). This suggests that aseismic slip is more likely under the lower stress conditions at shallower depth, all other factors held equal.

Temperature may play a role. Blanpied et al. (1995) performed laboratory experiments on wet fractures in granite which showed that stable slip should occur above 350 °C (higher temperature than any project reviewed in this project). There were mixed results in the temperature range of 250 °C–350 °C, mostly stable slip at room temperature, and consistently unstable slip between 100 °C and 250 °C. The Fjällbacka project was within the lower temperature range where stable slip appeared to be favored (somewhere below 100 °C), and the Rosemanowes project was in the neighborhood of the (not clearly delineated) lower transition temperature between stable and unstable sliding. The Basel and Soultz projects were within the temperature range consistent with unstable slip. The Cooper Basin and Ogachi projects were approaching the upper temperature range where there were mixed results indicating the possibility of stable slip. Application of these laboratory experiments should be considered highly imprecise because variations in granite mineralogy, fault gouge, hydrothermal alteration, and other factors could lead to considerable differences between the experiments and in-situ conditions.

A final possibility is that deeper formations are more likely to contain well-developed faults because they are under greater stress (though their depth is unlikely to have been constant over geological time), and perhaps deformation at greater stress is more likely to contribute to rock failure and fault zone development. To fully disentangle this issue, more data points would be needed, either shallow projects with well-developed faults or deep projects in formations without well-developed faults.

4.5. Seismic and aseismic slip

Because all projects reviewed were in granitic rock, we hoped to control for variations in the frictional properties of the fractures in the formations. However, we cannot independently verify that the frictional properties of the fractures in these formations were the same. The frictional properties of the faults could be related to the variation in composition of the different granites, different hydrothermal alterations and fracture infilling mineralogies, or different temperatures and depths. It is also possible that well-developed cataclasite fault cores in granite have frictional properties that are especially favorable for seismic slip. These issues could only be addressed with careful studies specifically directed at addressing the frictional properties of fractures in the individual formations. Variance of frictional properties of the formation has received little attention in the induced seismicity literature, though Zoback et al. (2012) discussed this possibility with respect to variations in microseismicity during hydraulic fracturing of shale.

Several authors have argued that aseismic deformation must have taken place at the shallow Soultz reservoir (Cornet *et al.*, 1997 ; Evans, 1998). Caliper logs indicated that faults had experienced greater shear offsets greater than could be explained from the total seismic moment release. It seems plausible that the greater maximum magnitude observed in the deeper Soultz reservoirs could be related to the frictional properties of the faults at different depths. This would suggest that even among formations with well-developed faults, variations of frictional properties will strongly affect induced seismicity hazard.

4.6. Spatial continuity

If our hypotheses are correct that the degree of fault development can be considered a formation property, it can be estimated from wellbore observations, and it can be related to seismic hazard, it would be useful to understand the spatial continuity of "degree of fault development." During hydraulic fracturing, the region of fluid pressure increase is relatively localized around the wellbore. But for long term injection (e.g., wastewater disposal, CO_2 sequestration, or unbalanced circulation in an EGS injector/producer pair) the region of pressure perturbation could spread significant distances from the wellbore. In these cases, assessments of the degree of fault development from observations made at the wellbore may be less reliable.

Analyses have shown induced seismicity at individual sites following predictable statistical behavior, such as the linear relationship between injection volume and moment release (Rutledge *et al.*, 2004; Bommer *et al.*, 2006; Shapiro *et al.*, 2007 ; Bachmann *et al.*, 2011). But since these statistical relationships are highly variable between sites, determining the region where statistical stationarity can be assumed becomes an important issue. If a region of stress and/or pressure perturbation is growing over time, it will become increasingly likely that the pressure perturbation will reach regions of the subsurface that behave statistically differently. These transitions may not necessarily be easy to predict. Variations in lithology could potentially cause breaks in stationarity, as could faults that have formed along preexisting interfaces between lithologies. Outlier events were observed at Rosemanowes and Ogachi, and one possible explanation for these events is that the fluid pressure perturbation reached more distant faults that had statistically different behavior than the faults sampled near the wellbore.

As discussed in Section 3.4, a detailed core analysis at Ogachi indicated a low degree of fault development (Ito, 2003), but a seismic reflection survey was interpreted to indicate that a large fault exists 500 m from the wells. One interpretation could be that the inferred fault caused the outlier event at Ogachi and that it behaves statistically differently than the formation immediately surrounding the wells.

4.7. Injection volume

There was no apparent correlation between injection volume and maximum induced magnitude. At a particular site, increasing injection volume would be expected to increase overall seismic moment release and maximum magnitude. However, when comparing between projects, no correlation is apparent, evidently because site-specific factors overwhelmed the effect due to injection volume.

5. CONCLUSIONS

In this paper, we reviewed six projects where hydraulic stimulation was performed in granitic rock. Even though similar injection volumes were used (except Fjällbacka, where a lower volume was used) and the formations contained fractures well-oriented to slip at the injection pressure, there was striking variability in the magnitude of induced earthquakes. With the limited data, it is impossible to draw firm conclusions. However, there was a correlation between the degree of fault development observed at the well and the induced seismic deformation. We cannot rule out that variations in the frictional properties of the faults in the formations caused the observed variability (or that frictional properties are related to the degree of fault development). Comparison with laboratory experiments suggests that temperature may play a role, though the effect on the projects discussed in this paper is unclear. The three projects with the greatest seismicity and degree of fault zone development were also the deepest projects. From the six projects in this comparison, the relationships between depth, degree of fault development, temperature, and induced seismicity cannot be disentangled.

To our knowledge, prediction of induced seismic hazard by characterizing the degree of fault development based on wellbore observations has not previously been proposed in the literature. At many projects, measurements are not taken that would make such characterizations possible. Wellbore imaging logs, mud logs, and (if practical) wellbore core can be used to characterize the degree of fault development observed at the wellbore. Imaging logs can identify large features, and core and mud logs can be used to identify the presence of cataclasite. The results of this study suggest that this approach deserves further consideration, and operators should attempt to characterize the degree of fault development whenever feasible. Characterizing the degree of fault zone development may have other uses, such as helping understand flow and fracture network connectivity in a reservoir. The variability in seismicity between the sites reviewed in this paper may also be related to frictional behavior, another concept that has rarely been discussed in the induced seismicity literature but which has significant promise to improve induced seismicity hazard assessments.

ACKNOWLEDGMENTS

Part of this work was performed with support from the Precourt Institute for Energy at Stanford University. Several individuals generously helped us track down information about various projects. Thank you very much to Ingrid Stober, Albert Genter, Eva Schill, Steve Hickman, Doone Wyborn, and Günter Zimmermann. We discussed several issues with Nick Davatzes and thank you to him for his very stimulating comments. Thank you very much to Art McGarr and four other anonymous reviewers for their very helpful comments, which significantly improved the paper.

REFERENCES

1. Asanuma, H., Soma, N., Kaieda, H., Kumano, Y., Izumi, T., Tezuka, K., Niitsuma, H., Wyborn, D., 2005. Microseismic monitoring of hydraulic stimulation at the Australian HDR project in Cooper Basin, Paper Presented at the World Geothermal Congress, Antalya, Turkey (http://www.geothermal-energy.org/pdf/IGAstandard/WGC/2005/1615.pdf).
2. Bachmann, C.E., Wiemer, S., Woessner, J., Hainzl, S., 2011. Statistical analysis of the induced Basel 2006 earthquake sequence: introducing a probability-based monitoring approach for Enhanced Geothermal Systems. Geophys. J. Int. 186 (2), 793–807.
3. Baisch, S., Vörös, R., 2009. Deep heat mining Basel — seismic risk analysis: AP 3000, induced seismicity. Q-Con Report (http://www.wsu.bs.ch/geothermie.htm).
4. Baisch, S., Weidler, R., Vörös, R., Wyborn, D., de Graaf, L., 2006. Induced seismicity during the stimulation of a geothermal HFR reservoir in the Cooper Basin, Australia. Bull. Seismol. Soc. Am. 96 (6), 2242–2256.
5. Baisch, S., Vörös, R., Weidler, R., Wyborn, D., 2009. Investigation of fault mechanisms during geothermal reservoir stimulation experiments in the Cooper Basin, Australia. Bull. Seismol. Soc. Am. 99 (1), 148–158.
6. Baisch, S., Vörös, R., Rothert, E., Stang, H., Jung, R., Schellschmidt, R., 2010. A numerical model for fluid injection induced seismicity at Soultz-sous-Forêts. Int. J. Rock Mech. Min. Sci. 47 (3), 405–413.
7. Blanpied, M.L., Lockner, D.A., Byerlee, J.D., 1995. Frictional slip of granite at hydrothermal conditions. J. Geophys. Res. 100 (B7), 13045–13064.
8. Bommer, J.J., Oates, S., Cepeda, J.M., Lindholm, C., Bird, J., Torres, R., Marroquín, G., Rivas, J., 2006. Control of hazard due to seismicity induced by a Hot Fractured Rock geothermal project. Eng. Geol. 83 (4), 287–306.
9. Cladouhos, T., Petty, S., Foulger, G., Julian, B., Fehler, M., 2010. Injection induced seismicity and geothermal energy. Trans. Geotherm. Resour. Counc. 34, 1213–1220.
10. Cladouhos, T.T., Clyne, M., Nichols, M., Petty, S., Osborn, W.L., Nofziger, L., 2011. Newberry volcano EGS demonstration stimulation modeling. Trans. Geotherm. Resour. Counc. 35, 317–322.
11. Cornet, F.H., Helm, J., Poitrenaud, H., Echecopar, A., 1997. Seismic and aseismic slips induced by large-scale fluid injections. Pure Appl. Geophys. 150 (3), 563–583.
12. Davis, S.D., Frohlich, C., 1993. Did (or will) fluid injection cause earthquakes? — criteria for a rational assessment. Seismol. Res. Lett. 64 (3–4), 207–224.
13. Delacou, B., Sartori, M., Philippossian, F., Savary, J., Grosjean, G., 2009. AP2000 Report — Structural Model of Potential Seismogenic Faults,
14. Deep Heat Mining Basel — Seismic Risk Analysis, SERIANEX. Dezayes, C., Genter, A., Valley, B., 2010. Structure of the low permeable naturally fractured geothermal reservoir at Soultz. Compt. Rendus Geosci. 342 (7–8), 517–530.
15. Dieterich, J.H., 2007. 4.04 — Applications of Rate- and State-Dependent Friction to Models of Fault Slip and Earthquake Occurrence. In: Schubert, G. (Ed.), Treatise on Geophysics. Elsevier, Amsterdam, pp. 107–129.
16. Economides, M.J., Nolte, K.G., 2000. Reservoir Stimulation, 3rd ed. John Wiley, Chichester, England, New York.
17. Eliasson, T.M., Tullborg, E.-L., Landström, O., 1990. Fracture filling mineralogy and geochemistry at the Swedish HDR research site, Paper Presented at the Hot Dry Rock: Proceedings Camborne School of Mines International Hot Dry Rock Conference. Robertson Scientific Publications.
18. Evans, K., 1998. Does Significant Aseismic Slip Occur on Fractures in HDR Systems Under Stimulation Conditions, Paper Presented at the 4th International HDR Forum, Strasbourg, Germany.

19. Evans, K.F., 2005. Permeability creation and damage due to massive fluid injections into granite at 3.5 km at Soultz: 2. Critical stress and fracture strength. J. Geophys. Res. 110 (B4).
20. Evans, K.F., Moriya, H., Niitsuma, H., Jones, R.H., Phillips, W.S., Genter, A., Sausse, J., Jung, R., Baria, R., 2005. Microseismicity and permeability enhancement of hydrogeologic structures during massive fluid injections into granite at 3 km depth at the Soultz HDR site. Geophys. J. Int. 160 (1), 389–412.
21. Evans, K.F., Zappone, A., Kraft, T., Deichmann, N., Moia, F., 2012. A survey of the induced seismic responses to fluid injection in geothermal and CO2 reservoirs in Europe. Geothermics 41, 30–54.
22. Faulkner, D.R., Jackson, C.A.L., Lunn, R.J., Schlische, R.W., Shipton, Z.K., Wibberley, C.A.J., Withjack, M.O., 2010. A review of recent developments concerning the structure, mechanics and fluid flow properties of fault zones. J. Struct. Geol. 32 (11), 1557–1575.
23. Frohlich, C., Hayward, C., Stump, B., Potter, E., 2011. The Dallas–Fort Worth earthquake sequence: october 2008 through may 2009. Bull. Seismol. Soc. Am. 101 (1), 327–340.
24. Genter, A., Traineau, H., Ledésert, B., Bourgine, B., Gentier, S., 2000. Over 10 years of geological investigations within the HDR Soultz project, France, Paper Presented at the World Geothermal Congress, Kyushu–Tohoku, Japan (http://www.geothermal-energy.org/pdf/IGAstandard/WGC/2000/R0710.PDF).
25. Genter, A., Evans, K., Cuenot, N., Fritsch, D., Sanjuan, B., 2010. Contribution of the exploration of deep crystalline fractured reservoir of Soultz to the knowledge of enhanced geothermal systems (EGS). Compt. Rendus Geosci. 342 (7–8), 502–516.
26. Gischig, V.S., Wiemer, S., 2013. A stochastic model for induced seismicity based on nonlinear pressure diffusion and irreversible permeability enhancement. Geophys. J. Int. 194 (2), 1229–1249.
27. Häring, M.O., Schanz, U., Ladner, F., Dyer, B.C., 2008. Characterisation of the Basel 1 enhanced geothermal system. Geothermics 37 (5), 469–495.
28. Heath, M.J., 1985. Geological control of fracture permeability in the Carnmenellis granite, Cornwall: implications for radionuclide migration. Mineral. Mag. 49 (351), 233–244.
29. Heuze, F.E., Shaffer, R.J., Ingraffea, A.R., Nilson, R.H., 1990. Propagation of fluid—driven fractures in jointed rock. Part 1—development and validation of methods of analysis. Int. J. Rock Mech. Min. Sci. Geomech. Abstr. 27 (4), 243–254.
30. Hitzman, M.W., 2012. Induced Seismicity Potential in Energy Technologies. National Research Council of the National Academies.
31. Hunt, S.P., Morelli, C., 2006. Cooper Basin HDR Hazard Evaluation: Predictive Modeling of Local Stress Changes due to HFR Geothermal Energy Operations in South Australia. South Australia Department of Primary, Industries and Resources.
32. Ito, H., 2003. Inferred role of natural fractures, veins, and breccias in development of the artificial geothermal reservoir at the Ogachi Hot Dry Rock site, Japan. J. Geophys. Res. 108 (B9).
33. Jung, R., 2013. EGS — goodbye or back to the future, in effective and sustainable hydraulic fracturing. In: Bunger, A.P., McLennan, J., Jeffrey, R. (Eds.), InTech, pp. 95–121. http://
34. Jupe, A.J., Green, A.S.P., Wallroth, T., 1992. Induced microseismicity and reservoir growth at the Fjällbacka Hot Dry Rocks project, Sweden. Int. J. Rock Mech. Min. Sci. Geomech. Abstr. 29 (4), 343–354.
35. Kaeser, B., Kalt, A., Borel, J., 2007. The Crystalline Basement Drilled at the Basel-1 Geothermal Site, a Preliminary Petrological–Geochemical Study. Institut de Géologie et d'Hydrogéologie. Université de Neuchâtel, Switzerland.
36. Kaieda, H., Sasaki, S., Wyborn, D., 2010. Comparison of characteristics of microearthquakes observed during hydraulic stimulation operations in Ogachi, Hijiori, and Cooper Basin HDR projects, Paper Presented at the World Geothermal

Congress, Bali, Indonesia (http://www.geothermal-energy.org/pdf/IGAstandard /WGC/2010/ 3157.pdf).
37. Kiho, K., Kitano, K., Kaieda, H., Ito, H., Shin, K., Eguchi, Y., Suenaga, H., 1999. Overview of the Ogachi HDR project. Trans. Geotherm. Resour. Counc. 23.
38. Kim, W.-Y., 2013. Induced seismicity associated with fluid injection into a deep well in Youngstown, Ohio. J. Geophys. Res. 118.
39. Kitano, K., Hori, Y., Kaieda, H., 2000. Outline of the Ogachi HDR project and character of the reservoirs, Paper Presented at the World Geothermal Congress, Kyushu–Tohoku, Japan (http://www.geothermal-energy.org/pdf/IGAstandard /WGC/2000/ R0435. PDF).
40. Majer, E.L., Baria, R., Stark, M., Oates, S., Bommer, J., Smith, B., Asanuma, H., 2007. Induced seismicity associated with Enhanced Geothermal Systems. Geothermics 36 (3), 185–222.
41. Majer, E., Nelson, J., Robertson-Tai, A., Savy, J., Wong, I., 2011. Protocol for addressing induced seismicity associated with Enhanced Geothermal Systems. http://www1. eere.energy.gov/geothermal/pdfs/egs-is-protocol-final-draft-20110531.pdf.
42. Mauldon, M., Mauldon, J.G., 1997. Fracture sampling on a cylinder: from scanlines to boreholes and tunnels. Rock Mech. Rock. Eng. 30 (3), 129–144
43. Mazurek, M., 1998. Technical Report 93-12: Geology of the Crystalline Basement of Northern Switzerland and Derivation of Geological Input Data for Safety Assessment Models. National Cooperative for the Disposal of Radioactive Waste, Switzerland (http://www. nagra.ch/documents/database/dokumente/%24default/ Default% 20Folder/ Publikationen/NTBs%201991-1993/e_ntb93-12.pdf).
44. McClure, M.W., 2012. Modeling and Characterization of Hydraulic Stimulation and Induced Seismicity in Geothermal and Shale gas Reservoirs. PhD Thesis Stanford University, Stanford, California.
45. McClure, M.W., Horne, R.N., 2011. Investigation of injection-induced seismicity using a coupled fluid flow and rate/state friction model. Geophysics 76 (6), WC181–WC198.
46. McClure, M.W., Horne, R.N., 2014. An investigation of stimulation mechanism in Enhanced Geothermal Systems. Int. J. Rock Mech. Min. Sci. (submitted for publication).
47. McGarr, A., 1976. Seismic moments and volume changes. J. Geophys. Res. 81 (8), 1487–1494.
48. McGarr, A., Simpson, D., Seeber, L., 2002. 40. Case histories of induced and triggered seismicity. In: Lee, W.H.K., Kanamori, H. (Eds.), International Geophysics. Academic Press, pp. 647–661.
49. Meller, C., Kohl, T., Gaucher, E., Genter, A., 2012. Approach for Determination of the Failure Probability of Fractures at the Soultz-Sous-Forêts EGS Project, Paper Presented at the Thirty-Seventh Workshop on Geothermal Reservoir Engineering, Stanford, CA.
50. Moriya, H., Niitsuma, H., Kaieda, H., 2000. Reevaluation of Reservoir Structure at Ogachi HDR Field by Precise Source Location of AE Multiplet, Paper Presented at the World Geothermal Congress. Ogachi, Japan.
51. Mutlu, O., Pollard, D.D., 2008. On the patterns of wing cracks along an outcrop scale flaw: a numerical modeling approach using complementarity. J. Geophys. Res. 113 (B6).
52. Pearson, R.A., Moore, P.W., Harker, C., Lanyon, G.W., 1989. Logging results, section 1 — proprietary logging results. In: Parker, R.H. (Ed.), Hot Dry Rock Geothermal Energy: Phase 2B Final Report of the Camborne School of Mines Project. Pergamon Press, Oxford, New York, pp. 423–558.
53. Pine, R.J., Batchelor, A.S., 1984. Downward migration of shearing in jointed rock during hydraulic injections. Int. J. Rock Mech. Min. Sci. Geomech. Abstr. 21 (5), 249–263.

54. Randall, M.M., Lanyon, G.W., Nicolls, J., Willis-Richards, J., 1990. Evaluation of jointing in the Carnmenellis granite, Paper Presented at the Hot Dry Rock: Proceedings Camborne School of Mines International Hot Dry Rock Conference. Robertson Scientific Publications.
55. Richards, H.G., Parker, R.H., Green, A.S.P., Jones, R.H., Nicholls, J.D.M., Nicol, D.A.C., Randall, M.M., Richards, S., Stewart, R.C., Willis-Richards, J., 1994. The performance and characteristics of the experimental Hot Dry Rock geothermal reservoir at Rosemanowes, Cornwall (1985–1988). Geothermics 23 (2), 73–109.
56. Rutledge, J.T., Phillips, W.S., Mayerhofer, M.J., 2004. Faulting induced by forced fluid injection and fluid flow forced by faulting: an interpretation of hydraulic-fracture microseismicity, Carthage Cotton Valley Gas Field, Texas. Bull. Seismol. Soc. Am. 94 (5), 1817–1830.
57. Scholz, C.H., 2002. The Mechanics of Earthquakes and Faulting. Cambridge University Press, Cambridge, UK. Segall, P., 1989. Earthquakes triggered by fluid extraction. Geology 17 (10), 942–946.
58. Segall, P., Pollard, D.D., 1983. Nucleation and growth of strike slip faults in granite. J. Geophys. Res. 88 (B1), 555–568.
59. Shapiro, S.A., Dinske, C., Kummerow, J., 2007. Probability of a given-magnitude earthquake induced by a fluid injection. Geophys. Res. Lett. 34 (22
60. Shearer, P.M., Prieto, G., Hauksson, E., 2006. Comprehensive analysis of earthquake source spectra in southern California. J. Geophys. Res. 111.
61. Suckale, J., 2009. Induced seismicity in hydrocarbon fields. Adv. Geophys. 51 (2), 55–106.
62. Suzuki, K., Kaieda, H., 2000. Geological structure around the Ogachi Hot Dry Rock test site using seismic reflection and CSAMT surveys, paper presented at the World Geothermal Congress. Kyushu - Tohoku, Japan.
63. Townend, J., Zoback, M.D., 2000. How faulting keeps the crust strong. Geology 28 (5), 399–402.
64. Valley, B., Evans, K., 2007. Stress state at Soultz-sous-Forêts to 5 km depth from wellbore failure and hydraulic observations, Paper Presented at the Thirty-Second Workshop on Geothermal Reservoir Engineering. Stanford University (http://www.geothermal-energy.org/pdf/IGAstandard/SGW/2007/valley.pdf).
65. van Eijs, R.M.H.E., Mulders, F.M.M., Nepveu, M., Kenter, C.J., Scheffers, B.C., 2006. Correlation between hydrocarbon reservoir properties and induced seismicity in the Netherlands. Eng. Geol. 84 (3–4).
66. Vörös, R., Baisch, S., 2009. Deep Heat Mining Basel — Seismic Risk Analysis: AP 4000, Triggered Seismicity. Q-Con Report (http://www.wsu.bs.ch/geothermie.htm). Wallroth, T., Eliasson, T., Sundquist, U., 1999. Hot Dry Rock research experiments at Fjällbacka, Sweden. Geothermics 28 (4–5), 617–625.
67. Wesnousky, S.G., 2006. Predicting the endpoints of earthquake ruptures. Nature 444 (7117), 358–360.
68. Whittle, R., 1989. The granites of South West England, section 2 — jointing in the Carnmenellis granite. In: Parker, R.H. (Ed.), Hot Dry Rock Geothermal Energy: Phase 2B Final Report of the Camborne School of Mines Project. Pergamon Press, Oxford, New York, pp. 119–182.
69. Zhang, Y., Person, M., Rupp, J., Ellett, K., Celia, M.A., Gable, C.W., Bowen, B., Evans, J., Bandilla, K., Mozley, P., Dewers, T., Elliot, T., 2013. Hydrogeologic controls on induced seismicity in crystalline basement rocks due to fluid injection into basal reservoirs. Groundwater 51 (4), 525–538.
70. Zoback, M.D., Gorelick, S.M., 2012. Earthquake triggering and large-scale geologic storage of carbon dioxide. Proc. Natl. Acad. Sci. 109 (26), 10164–10168.
71. Zoback, M., Kohli, A., Das, I., McClure, M., 2012. The importance of slow slip on faults during hydraulic fracturing stimulation of shale gas reservoirs. SPE 155476, Paper Presented at the SPE Americas Unconventional Resources Conference, Pittsburgh, Pennsylvania USA.

CHAPTER 15

An Automatic Deconvolution Method for Modified Gaussian Model using the Exchange Monte Carlo Method: Application to Reflectance Spectra of Synthetic Clinopyroxene

Peng K Hong[1*], Hideaki Miyamoto[1-3], Takafumi Niihara[1,4], Seiji Sugita[2], Kenji Nagata[5], James M Dohm[1] and Masato Okada[5]

[1]The University Museum, The University of Tokyo, 7-3-1 Hongo, Bunkyo-ku, Tokyo 113-0033, Japan

[2]Department of Earth and Planetary Science, The University of Tokyo, 7-3-1 Hongo, Bunkyo-ku, Tokyo 113-0033, Japan

[3]Planetary Science Institute, 1700 East Fort Lowell, Tucson, AZ 85719-2395, USA

[4]Lunar and Planetary Institute, Universities Space Research Association, 3600 Bay Area Boulevard, Houston, TX 77058, USA

[5]Department of Complexity Science and Engineering, The University of Tokyo, 5-1-5 Kashiwanoha, Kashiwa, Chiba 277-8561, Japan

ABSTRACT

Deconvolution analysis of reflectance spectra has been a useful method to infer mineral composition and crystal structure. Many of the recent deconvolution analyses of reflectance spectra of major rock-forming minerals, such as olivine and pyroxene, have been based on a modified Gaussian Model (MGM). The numerical algorithm of the widely used MGM, however, utilizes the steepest descent method, which has a local minima problem. With inaccurate initial parameters, the steepest descent method converges into a local minimum, thus the analyzer must manually adjust initial parameters and calculate the model repeatedly to obtain the desired solution. In order to avoid the local minimum problem, we utilized Bayesian spectral deconvolution with the exchange Monte Carlo method, which is an improved algorithm of the Markov chain Monte Carlo method, aimed to both avoid local minima traps and remove the arbitrariness originated from initial parameters. We applied the model to visible to near infrared reflectance spectra of 31 synthetic clinopyroxene samples with wide ranging Mg, Fe and Ca compositions (solid solution). We obtained results consistent with the previous studies based on conventional MGM analyses, suggesting that the exchange Monte Carlo method can yield results consistent with the conventional MGM analyses purely based on the observed data. We also find that the center wavelengths of 1 μm absorption bands of high-Ca pyroxene samples have a linear dependence on Fe/Mg component. Both 1 μm and 2 μm absorption bands seem to follow approximation lines in the three-dimensional spaces of center wavelengths, Ca and Fe components. The successful application of the exchange Monte Carlo method to a wide range of clinopyroxenes would have a potential to expand the applicability of MGM to a variety of space/ground-based observations, especially when we cannot rely on prior information of the mineralogy.

Keywords: Pyroxene; Synthetic mineral; Reflection spectroscopy; Spectral deconvolution; Modified Gaussian model; Monte Carlo method

INTRODUCTION

Remote sensing of reflectance spectra of Earth and other planetary bodies can be useful for identifying mineral distribution on their surfaces, especially in remote regions that are exceedingly challenging to perform field-based investigation, and those planetary surfaces yet to have in situ observation, mapping, characterization, sampling, and analyses [1-8]. Many different factors, however, can influence the surface spectra, such as various alteration and weathering processes, and observational conditions [2,3,5,9-12]. Because a reflectance spectrum is a complex non-linear mixture of the above mentioned factors [13-16], it is highly challenging to segregate each factor and extract the true mineral spectra, based solely on remotely observed reflectance spectra, and thus confidence in the resulting signatures should be gained by comparing with the reference spectra obtained by field or laboratory measurements [17]. Though challenging, investigating the spectral change due to the variation of elemental composition should not be avoided, since it is one of the ultimate goals of remote sensing of reflectance spectra of planetary surfaces [2,18]. Compared to terrestrial surfaces, those of extraterrestrial bodies such as the Moon and asteroids are not covered by liquid water and vegetation, and have negligible to no atmosphere, and thus may be considered to be the best places to observe the true nature of mineral spectra [19-25]. Yet, there are many factors to contaminate reflectance spectra of such planetary surfaces such as regolith particles, space weathering, and horizontal and vertical mixing by impact cratering, thus identifying the variation of mineral distribution on extraterrestrial bodies is still difficult [10,22,26-30]. Therefore, studying the spectral change using simple pure minerals that compose Earth and planetary surfaces is a critical foundational step to analyze reflectance spectra and is a prerequisite of reflectance spectroscopy, in light of continued application to planetary surfaces [8,31-35].

Deconvolution of reflectance spectra has been a common procedure for interpreting the experimental data of minerals and observation data of terrestrial and extraterrestrial surfaces. Among the most recognized spectral deconvolution methods in planetary science is the modified Gaussian model (MGM) [36,37]. In their paper, Sunshine et al. [36] showed that for 1 μm absorption band of orthopyroxene, Gaussian functions in the wavelength space fit better than Gaussians in the frequency space. The MGM express reflectance spectrum by the following function:

$$\log R = C(x) + \sum_{k=1}^{k} s_k . \exp\left(\left\{\frac{-(x-\mu_k)^2}{2\sigma_k^{\ 2}}\right\}\right)$$

where R is reflectance, x is wavelength, and sk, μk and σk are the strength (amplitude), center (mean) and width (standard deviation) for kth Gaussian function, respectively. K is the number of Gaussian functions. The function C(x) is continuum of reflectance spectra:

$$C(x) = C_0 + \frac{C_1}{x}$$

where C0 and C1 are respectively intercept and slope of the continuum in frequency space.

The technical difficulty of applying the modified Gaussian model (MGM) is due to a local minima problem. The numerical algorithm of the widely used MGM utilizes the steepest descent method [36,38] or total inversion algorithm [37,39], both of which are not guaranteed to converge into a global solution. The analyzer can find an optimal solution only when the appropriate initial parameters are provided, based on preliminary knowledge of mineralogy [40,41]. With inaccurate initial parameters, however, these gradient descent methods converge into local minima, and thus the analyzer must manually adjust initial parameters and calculate the model iteratively to obtain the desired solution [16,42]. This would be a significant obstacle when one needs to automatically analyze large spectral databases obtained by space missions. In addition, preliminary knowledge of mineralogy may not always be available for space/ground-based observations, especially when the reflectance spectra are the only useful obtained data for interpreting the mineralogy of target bodies. Given the recent rapid increase of reflectance spectral data, automation of deconvolution analysis without requiring preliminary information on mineralogy is warranted. Makarewicz et al. [43] and Parente et al. [42] recently developed an algorithm to select initial band parameters automatically, based on inflection points of the derivatives of observed spectra. Although their algorithm does not depend on prior information, many spurious local minima and inflection points due to noise lead the authors to apply a smoothing filter, yielding arbitrariness on their analyses.

In order to overcome the local minima problem, the exchange Monte Carlo method, also known as parallel tempering [44], has been widely applied in the fields of physics, chemistry, biology, engineering and materials science [45]. Nagata et al. [46] developed a Bayesian spectral deconvolution model combined with the exchange Monte Carlo method with application to visible to near-infrared (Vis/NIR) reflectance spectra of fayalite and forsterite. This method is an improved algorithm of the Markov chain Monte Carlo method, aimed to avoid local minima traps [47] and to remove the arbitrariness originated from initial parameters. In order to solve the local minima problem, the simulated annealing scheme [48] has been incorporated into the model of Nagata et al. [46]. This algorithm introduces a pseudo-temperature and attempts to find the global minimum by heating and cooling the system. Nagata et al. [46] showed that the method can deconvolve reflectance spectral data of fayalite and forsterite into a few Gaussians with a continuum, purely based on the observed data, without requiring preliminary information of the band structure of olivine absorptions. In this paper, we report the applicability of the exchange Monte Carlo method to more complex rock-forming minerals (i.e., clinopyroxene). As described below, since the behavior of pyroxene spectra with the change of chemical composition is relatively well understood, the use of reflectance spectra of pyroxene minerals is suitable for testing the new spectral deconvolution method. Clinopyroxene (Cpx), with its general formula being (M2)(M1) (SiAl)2O6, is one of the most important rock-forming mineral groups due to both its rich abundance on solid bodies in the solar system and distinguished absorption features [20,34,49]. Cpx includes a wide range solid solution of Mg, Fe and Ca compositions and has two crystal structures of C2/c and P21/c [50], which could reflect various physical and chemical processes inside planetary bodies, such as the thermal history of magma [51]. Due to its wide range of chemical compositions, reflectance spectra of Cpx minerals vary significantly, and are generally grouped into three types: type-A, B and A/B [20,52,53]. Three major absorption bands are observed in type-B spectra, centered around 1.0, 1.2, and 2 μm, attributed to spin-allowed crystal field transitions of Fe cations in the octahedral (M1 and M2) sites [20]. These band centers are known to vary due to total iron and calcium content [19,52-55]. On the other hand, type-A spectra lack a strong 2 μm band, interpreted as a low Fe2+ content in the M2 site [53]. Type-A/B spectra are intermediate between type-A and B, although the boundaries are not well defined. MGM analyses have been performed to natural Cpx [36,53], synthetic Cpx [56], and mixtures of orthopyroxene-clinopyroxene [37,41].

METHODS

Exchange Monte Carlo method

The technical details of the exchange Monte Carlo method is described elsewhere [46], thus we briefly summarize the key parameters of the model. In our study, the hyperparameters for Gamma and Gauss distributions used to yield probability densities of the parameters are those identified in Nagata et al. [46]: ηa = 3.0, λa = 2.0, ν0 = 1.25, ξ0 = 2.5, ηb = 5.0 and λb = 0.04. The total number of temperatures L in our study was 80, and the inverse temperature βl given by:

$$\beta_l = \begin{cases} 0 & (l = 1) \\ 1.25^{l-L} & (l \neq 1) \end{cases}$$

Figure 1 shows typical examples of the evolution of root mean square (RMS) during the exchange Monte Carlo calculations for sample 088 using 4 Gaussian functions. Rapid decreases of RMS for the lowest temperature can be observed at about Monte Carlo steps = 300, 500, 700, 1500, and 6000, due to the parameter exchange between the lowest and middle temperatures. From Figure 1, it can be observed that the RMSs of higher temperatures are generally larger than those of lower temperatures, showing attempts to find better global minimum with wider fluctuations. On the other hand, for lower temperatures, each iteration attempts to find local minimum within a parameter range narrower than higher temperatures. After the Monte Carlo step exceeds 104, although the model still attempts to find better solution with certain fluctuations of parameters, the RMSs remain almost constant, and the expectation values of parameters converge to the best solution. Similar to Nagata et al. [46], the Monte Carlo calculation was iterated through 100,000 steps for the burn-in period and 20,000 steps for the expectation value calculations. Errors of band parameters are estimated from 2σ based on ten runs using a different series of random numbers.

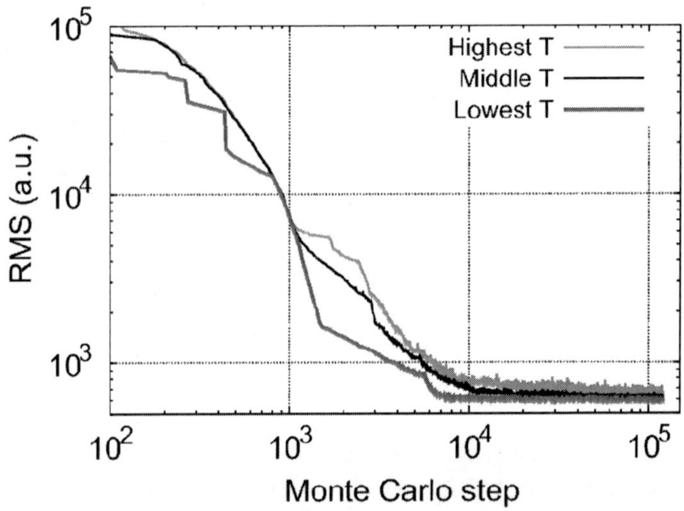

FIGURE 1: Evolution of root mean square (RMS) with every 10 Monte Carlo steps during the exchange Monte Carlo calculation for sample 088 with 4 Gaussian functions. Significant reduction of RMS for the lowest temperature can be observed approximately at Monte Carlo steps: 300, 500, 700, 1500, and 6000, which shows that the parameter exchange between the lowest and middle temperatures have occurred. The first 100,000 steps were used for the burn-in period and the last 20,000 steps for the expectation value calculations.

The number of Gaussian functions, K, is an important parameter for deconvolution analysis. The model usually improves with more Gaussians. Though, too many Gaussians may cause overfitting, and the solution can be physically unrealistic. We performed spectral deconvolution using a various number of Gaussian functions. For one spectrum, we varied K from 3 to 10, thus the total number of free parameters ranges from 11 to 32, including the intercept and slope of the continuum. In order to select an optimal K for the deconvolution, we calculate the free energy, or stochastic complexity [57,58], which is an evaluation function for the model section problem [46]. With increasing K, we find that the free energy decreases, and when K is higher than about 5, it converges and fluctuates around a low value. Thus, we chose minimum Ks from the region where the free energies converge and interpret them as optimal Ks, listed in Table 1.

TABLE 1: Compositions of synthetic clinopyroxene samples and results of spectral deconvolution calculated using the exchange Monte Carlo method. Errors are estimated from 2σ based on ten runs using different series of random numbers.

Sample ID	Mineral[a]	Composition[b]				1 µm band		1.2 µm band		
		En	Fs	Wo	Center (µm)	FWHM (µm)	Strength	Center (µm)	FWHM (µm)	Strength
9	Pigeonite	43	47	10	0.956 ± 0.0014	0.223 ± 0.004	-1.32 ± 0.03	1.229 ± 0.008	0.30 ± 0.03	-0.24 ± 0.03
11	Pigeonite	36	50	14	0.9579 ± 0.0003	0.2342 ± 0.0007	-2.040 ± 0.004	1.243 ± 0.0011	0.230 ± 0.003	-0.416 ± 0.0013
53	Pigeonite	23	70	8	0.9646 ± 0.0003	0.233 ± 0.0011	-2.042 ± 0.004	1.258 ± 0.0013	0.244 ± 0.003	-0.458 ± 0.005
88	Pigeonite	0	90	10	0.9801 ± 0.0004	0.2114 ± 0.0007	-1.334 ± 0.003	1.245 ± 0.002	0.351 ± 0.003	-0.3592 ± 0.0008
50	Augite	19	58	23	0.9608 ± 0.0003	0.235 ± 0.0013	-2.307 ± 0.007	1.278 ± 0.002	0.265 ± 0.003	-0.510 ± 0.005
51	Augite	39	34	27	0.991 ± 0.002	0.20 ± 0.010	-1.6 ± 0.11	1.26 ± 0.014	0.34 ± 0.07	-0.35 ± 0.08
54	Augite	6	70	23	0.988 ± 0.0013	0.250 ± 0.008	-2.50 ± 0.02	1.290 ± 0.007	0.30 ± 0.010	-0.70 ± 0.02
55	Augite	18	56	26	0.998 ± 0.002	0.201 ± 0.005	-1.9 ± 0.12	1.261 ± 0.003	0.331 ± 0.005	-0.475 ± 0.008
56	Augite	18	60	22	0.9715 ± 0.0004	0.241 ± 0.002	-1.32 ± 0.014	1.260 ± 0.002	0.257 ± 0.006	-0.32 ± 0.011
57	Augite	36	39	25	0.9915 ± 0.0004	0.1998 ± 0.0008	-1.414 ± 0.004	1.269 ± 0.004	0.28 ± 0.02	-0.182 ± 0.005
58	Augite	28	45	27	1.001 ± 0.004	0.198 ± 0.007	-1.3 ± 0.14	1.26 ± 0.011	0.28 ± 0.02	-0.29 ± 0.03
66	Augite	15	48	36	1.0128 ± 0.0003	0.1804 ± 0.0005	-1.43 ± 0.02	1.227 ± 0.004	0.386 ± 0.009	-0.443 ± 0.009
67	Augite	52	9	39	1.016 ± 0.003	0.17 ± 0.011	-1.0 ± 0.14	1.3 ± 0.13	0.3 ± 0.2	-0.2 ± 0.10
68	Augite	29	33	38	1.008 ± 0.008	0.20 ± 0.011	-1.8 ± 0.3	1.26 ± 0.010	0.33 ± 0.014	-0.38 ± 0.012
73	Augite	36	25	39	1.0099 ± 0.0003	0.163 ± 0.0010	-1.083 ± 0.007	1.12 ± 0.011	0.53 ± 0.014	-0.22 ± 0.010
74	Augite	24	37	39	1.0108 ± 0.0005	0.181 ± 0.002	-1.32 ± 0.03	1.204 ± 0.006	0.41 ± 0.010	-0.41 ± 0.013
85	Augite	0	61	39	1.024 ± 0.002	0.165 ± 0.007	-0.91 ± 0.05	1.23 ± 0.05	0.4 ± 0.13	-0.2 ± 0.13
87	Augite	0	71	29	1.000 ± 0.004	0.200 ± 0.007	-2.10 ± 0.06	1.245 ± 0.005	0.409 ± 0.006	-0.707 ± 0.009
33	Diopside	42	8	49	1.04 ± 0.03	0.5 ± 0.3	-0.3 ± 0.14	1.5 ± 0.11	0.8 ± 0.2	-0.1 ± 0.10
36	Diopside	27	24	49	1.03 ± 0.03	0.6 ± 0.5	-0.7 ± 0.3			
39	Diopside	29	22	49	1.04 ± 0.013	0.5 ± 0.4	-0.7 ± 0.3			
43	Diopside	45	6	49	1.08 ± 0.010	0.51 ± 0.04	-0.44 ± 0.07			
75	Diopside	46	9	45	1.03 ± 0.02	0.16 ± 0.02	-0.7 ± 0.13	1.2 ± 0.10	0.3 ± 0.2	-0.18 ± 0.07
77	Diopside	52	3	45	1.0193 ± 0.0002	0.161 ± 0.001	-0.701 ± 0.003	1.106 ± 0.0012	0.507 ± 0.004	-0.208 ± 0.002
79	Diopside	38	15	47	1.024 ± 0.0013	0.165 ± 0.003	-0.93 ± 0.03	1.05 ± 0.02	0.62 ± 0.03	-0.51 ± 0.04
37	Hedenbergite	16	35	49	1.04 ± 0.02	0.29 ± 0.09	-0.8 ± 0.4	1.3 ± 0.2	0.2 ± 0.3	-0.2 ± 0.3
70	Hedenbergite	14	41	45	1.028 ± 0.007	0.17 ± 0.010	-1.1 ± 0.10	1.23 ± 0.05	0.36 ± 0.07	-0.4 ± 0.2
71	Hedenbergite	23	31	46	1.025 ± 0.002	0.171 ± 0.002	-1.06 ± 0.02	1.24 ± 0.03	0.2 ± 0.10	-0.1 ± 0.11
76	Hedenbergite	18	35	46	1.03 ± 0.011	0.19 ± 0.04	-0.6 ± 0.2	1.2 ± 0.10	0.4 ± 0.11	-0.20 ± 0.07
82	Hedenbergite	1	50	49	1.06 ± 0.02	0.23 ± 0.09	-1.0 ± 0.3			
83	Hedenbergite	0	49	51	1.07 ± 0.02	0.25 ± 0.04	-0.4 ± 0.14	1.21 ± 0.02	0.26 ± 0.05	-0.7 ± 0.2

Sample ID	2 μm band			Continuum		
	Center (μm)	FWHM (μm)	Strength	C_0	C_1	K^c
9	2.046 ± 0.004	0.69 ± 0.02	-0.66 ± 0.02	-1.38 ± 0.02	-9.69E-3 ± 0.05	4
11	2.1228 ± 0.0003	0.6938 ± 0.0009	-1.171 ± 0.0014	-0.5536 ± 0.00010	-1.51E-4 ± 0.0007	4
53	2.1798 ± 0.0006	0.712 ± 0.002	-1.133 ± 0.002	-0.3729 ± 0.0002	-1.86E-3 ± 0.005	4
88	2.2022 ± 0.0002	0.6354 ± 0.0007	-0.565 ± 0.0010	-0.8344 ± 0.0002	-5.94E-2 ± 0.0010	4
50	2.2620 ± 0.0006	0.728 ± 0.002	-1.263 ± 0.003	-0.3828 ± 0.00014	-2.20E-4 ± 0.006	4
51	2.29 ± 0.04	0.73 ± 0.12	-0.90 ± 0.07	-0.393 ± 0.006	-4.85E-3 ± 0.12	5
54	2.2860 ± 0.0009	0.706 ± 0.004	-1.321 ± 0.006	-0.556 ± 0.003	-3.05E-3 ± 0.010	5
55	2.3060 ± 0.0005	0.731 ± 0.004	-1.072 ± 0.004	-0.264 ± 0.0014	-3.85E-3 ± 0.009	5
56	2.190 ± 0.0013	0.751 ± 0.009	-0.769 ± 0.008	-0.258 ± 0.004	-2.89E-3 ± 0.02	4
57	2.264 ± 0.0010	0.59 ± 0.011	-0.58 ± 0.010	-1.00 ± 0.02	-6.46E-2 ± 0.014	4
58	2.285 ± 0.0010	0.76 ± 0.02	-0.79 ± 0.02	-0.19 ± 0.02	-2.38E-2 ± 0.03	5
66	2.3139 ± 0.0009	0.665 ± 0.003	-0.804 ± 0.004	-0.24 ± 0.013	-4.77E-2 ± 0.02	5
67	2.328 ± 0.003	0.52 ± 0.02	-0.50 ± 0.02	-0.45 ± 0.05	-9.43E-2 ± 0.09	5
68	2.328 ± 0.0012	0.652 ± 0.008	-0.92 ± 0.011	-0.33 ± 0.03	-5.26E-2 ± 0.06	6
73	2.3078 ± 0.0006	0.494 ± 0.002	-0.427 ± 0.002	-0.952 ± 0.009	-4.66E-2 ± 0.02	4
74	2.312 ± 0.0013	0.614 ± 0.005	-0.682 ± 0.009	-0.50 ± 0.03	-6.63E-2 ± 0.05	5
85	2.290 ± 0.004	0.54 ± 0.02	-0.34 ± 0.02	-0.58 ± 0.05	-1.21E-1 ± 0.09	6
87	2.2882 ± 0.0005	0.638 ± 0.002	-1.056 ± 0.002	-0.4928 ± 0.0002	-7.09E-4 ± 0.002	5
33	2.33 ± 0.07	0.8 ± 0.3	-0.2 ± 0.10	-0.17 ± 0.07	-3.79E-2 ± 0.04	6
36	2.33 ± 0.06	0.2 ± 0.2	-0.07 ± 0.09	-0.86 ± 0.04	-4.20E-3 ± 0.11	6
39	2.30 ± 0.011	0.39 ± 0.03	-0.17 ± 0.010	-0.52 ± 0.07	-4.34E-2 ± 0.14	5
43	2.39 ± 0.013	0.26 ± 0.06	-0.111 ± 0.009	-0.29 ± 0.03	-2.62E-2 ± 0.04	5
75	2.33 ± 0.02	0.59 ± 0.06	-0.44 ± 0.03	-0.29 ± 0.05	-7.15E-2 ± 0.05	6
77	2.310 ± 0.0012	0.574 ± 0.004	-0.338 ± 0.002	-0.344 ± 0.002	-4.53E-4 ± 0.003	4
79	2.297 ± 0.002	0.47 ± 0.02	-0.42 ± 0.02	-1.01 ± 0.03	-8.65E-2 ± 0.09	5
37	2.3 ± 0.3	0.3 ± 0.7	-0.08 ± 0.09	-0.74 ± 0.05	-1.20E-1 ± 0.11	7
70	2.302 ± 0.002	0.58 ± 0.02	-0.63 ± 0.02	-0.32 ± 0.04	-3.74E-2 ± 0.09	6
71	2.302 ± 0.002	0.52 ± 0.013	-0.47 ± 0.010	-0.72 ± 0.03	-1.29E-1 ± 0.07	5
76	2.25 ± 0.02	0.36 ± 0.05	-0.11 ± 0.05	-1.55 ± 0.05	-2.05E-1 ± 0.2	5
82	2.297 ± 0.003	0.39 ± 0.05	-0.16 ± 0.02	-0.40 ± 0.04	-2.36E-2 ± 0.06	7
83	2.35 ± 0.02	0.4 ± 0.4	-0.05 ± 0.02	-0.48 ± 0.04	-9.79E-3 ± 0.06	10

[a]Based on Morimoto et al. [50].
[b]Data from Klima et al. [56]. En: Enstatite; Fs: Ferrosilite; Wo: Wollastonite.
[c]Number of Gaussian function.

Spectral data

We used the visible to near-infrared spectra of synthetic clinopyroxene samples with a wide compositional range collected at the KECK/NASA Reflectance Experiment Laboratory (RELAB) at Brown University [59,60]. The sample IDs for the RELAB catalogue is summarized in Table 1. In order to compare our results with conventional MGM analyses, we collected 31 reflectance spectra that have been analyzed by a previous study [56]. The method of synthesis is detailed in Turnock et al. [61]. Individual synthetic pyroxene grains typically have 15-25 μm in size, however, these grains formed clumps [56]. Thus, samples are crushed and sieved at <45 μm. The spectra were measured at 5 nm intervals over the wavelength range of 0.3-2.6 μm. The incidence and emission angles were 30° and 0°, respectively [56,59,60]. We performed spectral deconvolution only over the wavelength range of 0.4-2.6 μm, since the standard deviations become larger near the shorter wavelengths [46]. The chemical compositions of the samples are measured with electron microprobe by Klima et al. [56]. The compositions of individual pyroxene samples are indicated in molar ratio with endmember compositions of enstatite (En: Mg2Si2O6), ferrosilite (Fs: Fe2Si2O6) and wollastonite (Wo: Ca2Si2O6) and plot on a pyroxene quadrilateral (Table 1 and Figure 2). Minor compositions typical for natural CPx samples, such as Cr, Mn, Al and Fe3+, are not observed [56]. Only the mineral structure of sample 088 is reported to be P21/c [62], while the mineral structures for the remaining samples are not available. Based on the nomenclature of clinopyroxene [50], we assumed low-Ca pyroxene specimens with Wo < 20 to be pigeonite, high-Ca pyroxene with Wo > 45 and En > 25 diopside, and high-Ca pyroxene with Wo > 45 and Fs > 25 hedenbergite. The remaining samples with intermediate Wo contents are assumed to be augite. Many of the samples locate in the "forbidden zone" where pyroxene exists merely as a metastable state under standard temperature and pressure [51]. These samples, however, were synthesized under a high-pressure up to 22.5 kbar [56], and thus are stable even under standard temperature and pressure for a geological timescale [63]. All of the reflectance spectra of synthetic pyroxene samples are shown in Figure 3. Spectra of pigeonite and augite are shown in Figures 3A-3C. The band minima of 1 μm absorptions for pigeonite and augite locate slightly shorter than 1 μm, similar to synthetic orthopyroxene [64]. Their band minima of 2 μm absorptions, caused by spin-allowed crystal field transition of Fe2+ in the M2 site [52,65], locate at longer than 2 μm similar to Ferich low-Ca pyroxene (orthopyroxene). Pigeonite and augite also show distinctive 1.2 μm absorption bands, which is attributed to spin-allowed crystal field transition of Fe2+ in the M1 site [66]. In Figures 3A- 3C, the spectra of pigeonite and augite specimens, which are assigned to type-B spectra [52], show distinctive 1 and 2 μm absorption bands. On the other hand, some of the high-Ca pyroxene (i.e., diopside and hedenbergite) lack distinctive 2 μm absorption bands, as shown in Figures 3D and 3E. They are assigned to type-A pyroxene, in which the M2 sites are saturated cations other than Fe2+, such as Ca2+ [52]. The spectrum of sample 083 has a broad 1 μm absorption band, probably due to a composite absorption by M1 bands near 1 and 1.2 μm [56]. This type of spectrum, exemplified by sample 083, may not be a suitable subject for MGM analyses, since the shape of 1 μm absorption band is far from Gaussian. However, since our objective in this paper is to test the applicability of the exchange Monte Carlo method and to compare the results with those of conventional MGM analyses, we included sample 083 in our analysis. Fitting additional small absorption bands centered near 0.50 and 0.55 μm in type-B spectra, which are associated with spin-forbidden crystal field transitions [66], is beyond the scope of this paper, since our focus was to evaluate the applicability of the model to two major absorption bands of clinopyroxene centered about 1 and 2 μm.

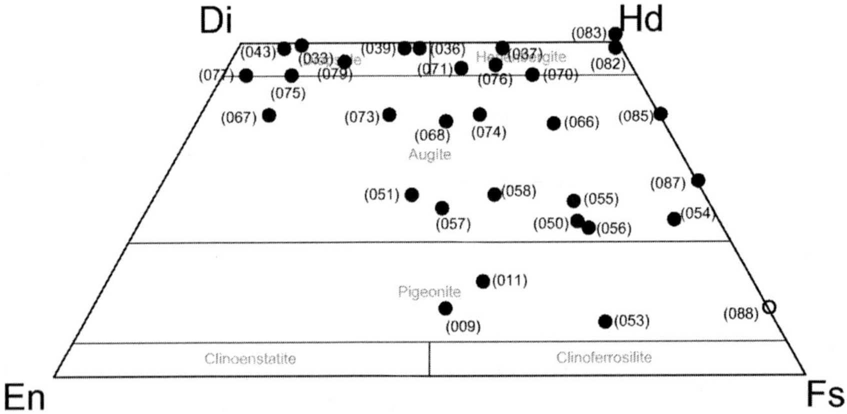

FIGURE 2: Clinopyroxene quadrilateral with the sample names. Compositions are measured with an electron probe microanalyzer (EPMA) by Klima et al. [56]. EPMA data for sample 088 is not available, thus it is plotted as an open circle based on the initial composition of the synthesis. The nomenclature of clinopyroxene is based on Morimoto et al. [50].

RESULTS

Deconvolution with the exchange Monte Carlo method

Deconvolution results of pyroxene spectra are shown in Figure 4. The best optimized parameters for 1, 1.2, and 2 μm bands are summarized in Table 1. Errors estimated from 2σ based on ten runs using different series of random numbers are also shown. Most of the spectra are best fitted by 4 to 6 Gaussians with appropriate continuum. For example, sample 053 is fitted with 4 Gaussians with their centers locating at 0.091, 0.965, 1.258, and 2.180 μm and a nearly constant continuum. For pigeonite and augite, our results show that the spectra are mainly fitted with 3 Gaussians with their centers locating at about 1, 1.2, and 2 μm, and with one Gaussian in the wavelength region ranging from ultraviolet to visible and with a continuum being nearly constant or gradually decreasing toward shorter wavelengths (Figures 5a-5d). Each Gaussian function could be physically interpreted to represent spin-allowed crystal field transitions with 1 μm for Fe2+ in the M1 and M2 sites, 1.2 μm for Fe2+ in the M1 site and 2 μm for Fe2+ in the M2 site (Figures 6a-6d). A Gaussian in the wavelength region ranging from ultraviolet to visible could represent oxygen-metal charge transfers which are centered within the ultraviolet region [34] (Figure 7). Although additional small bands centered near 0.7 μm may be necessary to improve the fitting such as shown in sample 054, we find that the additional band does not significantly affect the band center for 1, 1.2, and 2 μm absorptions. Our deconvolution results for pigeonite and augite are consistent with those by Klima et al. [56].

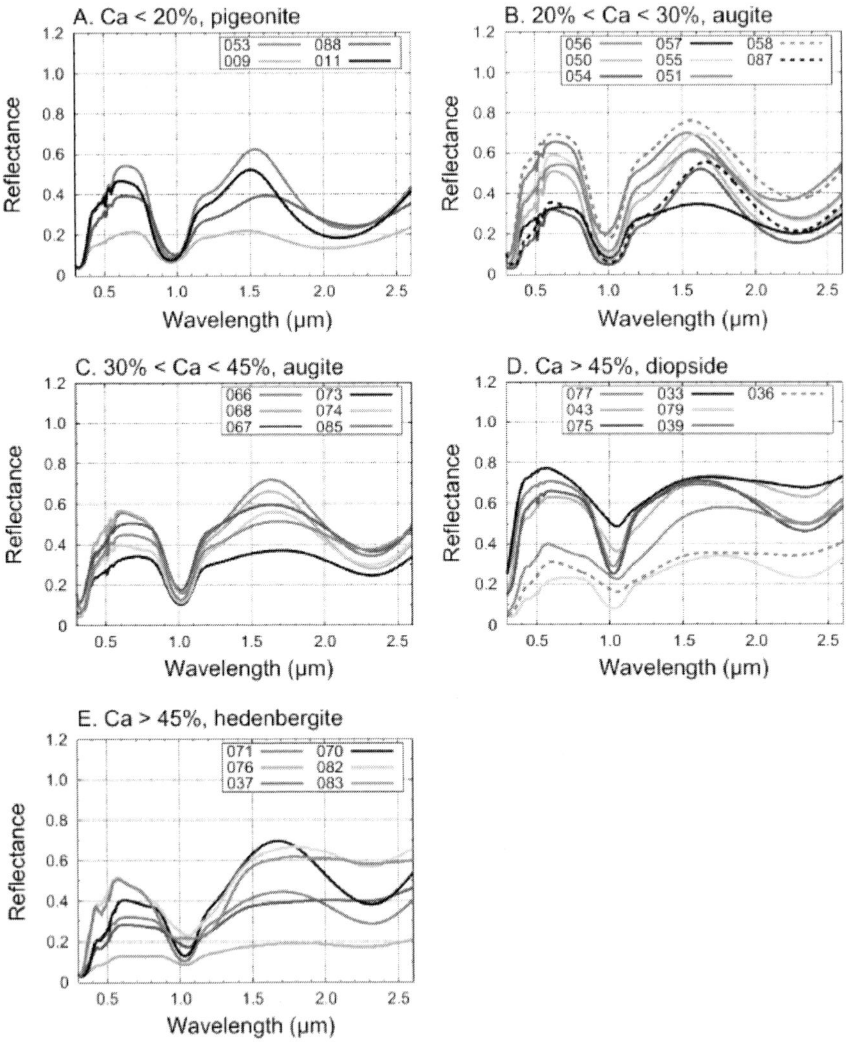

FIGURE 3: Visible to near infrared reflectance spectra of synthetic clinopyroxene. (3A) Pigeonite, (3B) Low-Ca augite, (3C) High-Ca augite, (3D) Diopside, (3E) Hedenbergite.

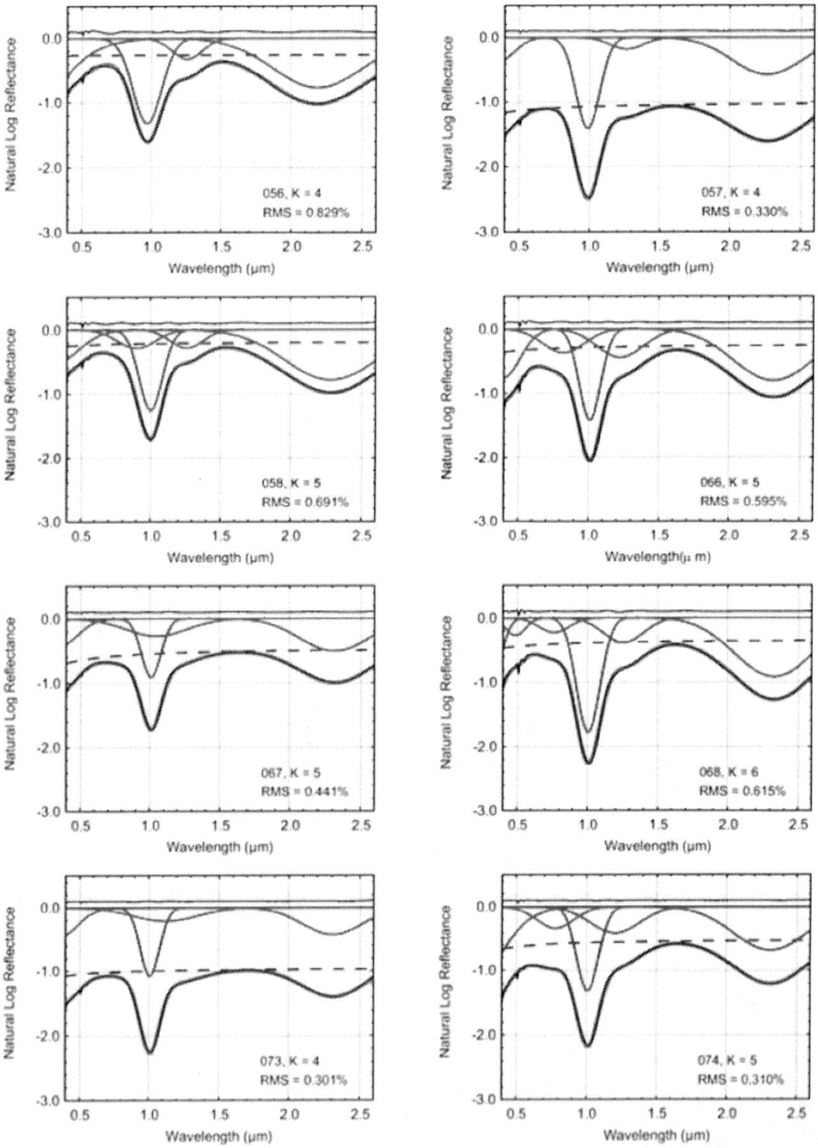

FIGURE 4: Spectral deconvolution results of synthetic Cpx samples using the exchange Montel Carlo method. Each Gaussian (blue solid lines) and continuum functions (black broken lines) compared to the synthetic spectra using the exchange Monte Carlo calculations (red lines) which approximate the original spectra (solid black lines). The residual errors between the modeled and the actual spectra are shown as solid black lines at top in each figure (offset +0.1 for clarity).

Deconvolution analyses for type-A spectra including some of the diopside and hedenbergite are not as straightforward as type B, since the 1 μm bands are too narrow or too wide for a single Gaussian to fit, and some of the 1 and 2 μm bands are asymmetrical. The non-Gaussian shape of type-A spectra results in larger errors of band parameters when compared to type-B spectra (Table 1). Also, more Gaussian functions are needed to fit type-A spectra, reflecting the complex shape of the spectra. For example, samples 037 and 082 were fitted with 7 Gaussians, while sample 083 with 10 Gaussians of which 4 Gaussians are centered below 0.5 μm. Comparing our fitting results of sample 082 and 083 with those reported by Klima et al. [56], we find that the exchange Monte Carlo method yields more symmetrical configurations for Gaussians in 1 μm band. We also find that our optimal deconvolution results for samples 082 and 083 do not require the very weak absorption bands which were assigned as M2 absorptions in Klima et al. [56]. These M2 absorptions in Klima et al. [56] are significantly weak compared with the strongest M1 absorption in 1 μm band, and the M2 absorption near 1 μm was mostly covered with the M2 absorptions. For type-A spectra, it is difficult to resolve the discrepancy between the results obtained by Klima et al. [56] and this study, however, since both of the modeling results can reproduce the observed spectra almost equally well. Nevertheless, the large errors of band parameters for type-A spectra indicate that caution should be taken when performing spectral deconvolution for type-A spectra using MGM. Thus, being able to estimate errors in the fitting results based on random initial parameters is also an advantage of the exchange Monte Carlo method for assessing the statistical robustness of fitting results, which has not been performed by previous MGM analyses.

Band shift as functions of Ca, Fe, and Mg Contents

The center positions of Gaussian functions corresponding to 1, 1.2, and 2 μm absorptions are summarized in Table 1. Position shifts of 1 μm band as functions of Ca, Fe, and Mg components (solid solution) are shown in Figure 5, while position shifts of 2 μm band are shown in Figure 6. Our results are found to be consistent with those of Klima et al. [56], suggesting that the exchange Monte Carlo method developed in this study is able to extract results similar to those obtained by conventional MGM analysis.

The band centers of 1 μm absorption have a linear dependence on Wo (Ca/(Mg+Fe+Ca) molar%) contents, moving to longer wavelengths with increasing Wo. For diopside and hedenbergite with Wo $\approx$ 50, the band centers of 1 μm absorption largely scatter within the range of 1.03-1.08 μm. Pigeonite also shows a large variance as a function of Wo component, however, the band centers of 1 μm absorptions have a linear dependence on En (Mg/(Mg+Fe+Ca) molar%) or Fs (Fe/(Mg+Fe+Ca) molar%) components. The band centers of 1 μm absorption of augite scatter widely as a function of Fs content, and distinct dependence on Fs content were not observed. For diopside and hedenbergite samples, with the exception of sample 043, the band centers of 1 μm absorptions seem to have a linear dependence on Fs content, moving to longer wavelengths with increasing Fs content. En content seems to have no obvious influence on the band center of 1 μm absorption, except for pigeonite. We find that pigeonite, augite and diopside-hedenbergite can be separated with the use of 1 μm band position as a function of Fs or En content, as shown in Figure 5.

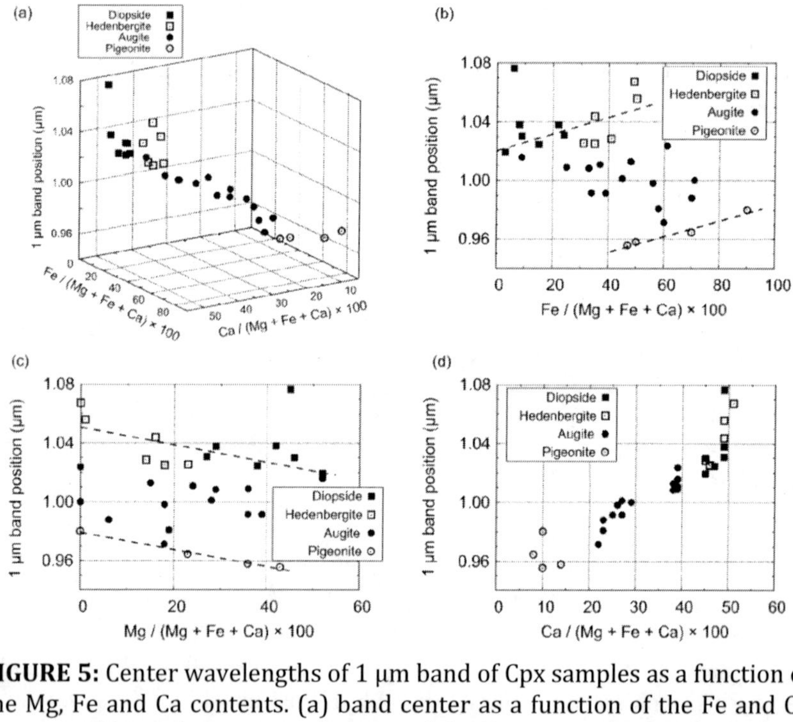

FIGURE 5: Center wavelengths of 1 μm band of Cpx samples as a function of the Mg, Fe and Ca contents. (a) band center as a function of the Fe and Ca contents. (b) band center as a function of the Fe content. Broken lines show linear approximations for pigeonite ($y = 0.00057x + 1.02$ where y is the band center in μm, x the Fe content) and diopside-hedenbergite ($y = 0.00054x + 0.93$). (c) band center as a function of the Mg content. Approximate lines for pigeonite ($y = -0.00059x + 1.05$) and diopside-hedenbergite ($y = -0.00058x + 0.98$) are shown. Note that sample 043 was omitted as an outlier from the linear approximation of diopside-hedenbergite in (b) and (c). (d) band center as a function of the Ca content.

The band centers of 2 μm absorption seem to follow an approximation line on the space of Fs and Wo contents (Figure 6a). The band centers of 2 μm absorption for pigeonite and augite move to longer wavelengths with an increase of Wo content. The average center wavelengths of 2 μm absorptions for diopside and hedenbergite remain almost constant at around 2.3 μm, but they scatter significantly around the average value, reflecting the asymmetry shape of 2 μm absorptions for some of the type-A spectra. The band centers of 2 μm absorptions of pigeonite move longer wavelengths with increasing Fs content. Overall, the band center of 2 μm absorption is separated between low- Ca clinopyroxene (pigeonite) and high-Ca pyroxene (augite, diopside, and hedenbergite) with a gap from 2.20-2.25 μm. Klima et al. [56] interpreted the gap to be a transition zone of mineral structure between P21/c and C2/c. Our results generally agree with their interpretation with one exception, i.e., sample 056 is assumed to be pigeonite in Klima et al. [56].

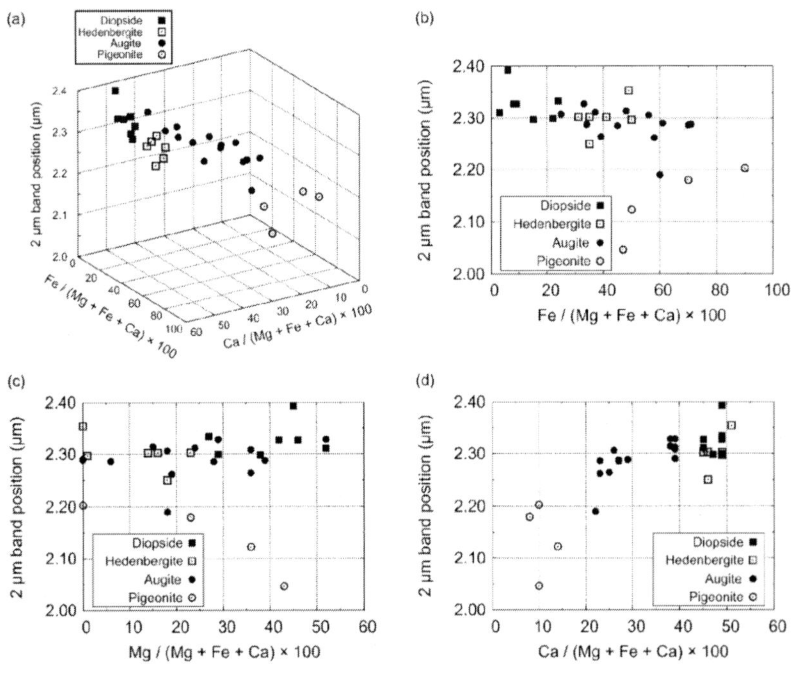

FIGURE 6: Center wavelengths of 2 μm band of Cpx samples as a function of the Mg, Fe and Ca contents. (a) band center as a function of the Fe and Ca contents. (b) band center as a function of the Fe content. (c) band center as a function of the Mg content. (d) band center as a function of the Ca content.

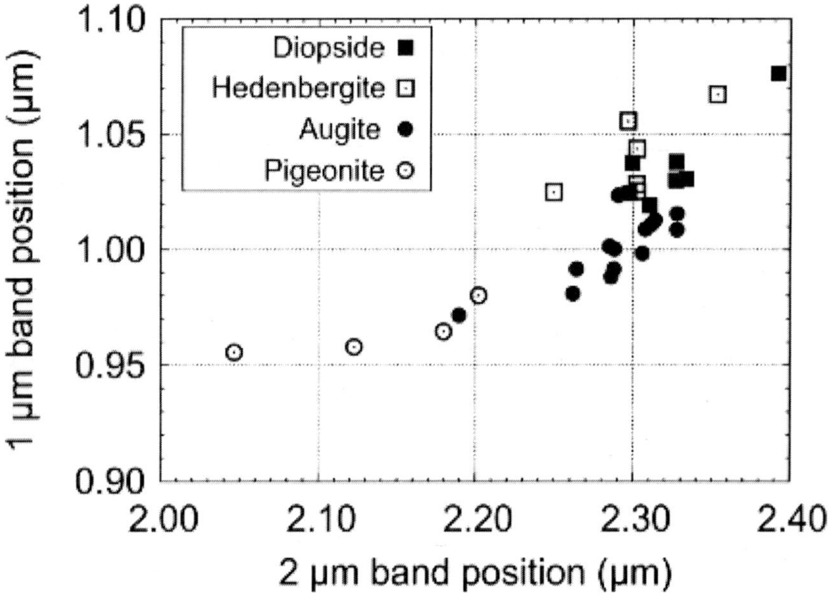

FIGURE 7: 1 μm band and 2 μm band diagram of synthetic Cpx samples calculated using the exchange Monte Carlo method.

The diagram between 1 μm and 2 μm band positions is shown in Figure 7. Generally low-Ca pyroxene locates in a shorter wavelength region, while high-Ca pyroxene locates in a longer wavelength region [19,41,52,56]. Our results are consistent with previous studies based on MGM analysis [41,56].

DISCUSSION

In order to avoid the local minimum problem, a Bayesian spectral deconvolution method with the exchange Monte Carlo algorithm has been applied to visible to near infrared reflectance spectra of synthetic Cpx with wide ranging Mg, Fe, and Ca contents. The results obtained in this study generally agree well with conventional MGM analyses. Here, we discuss some potential interpretations for the deconvolution results, as well as the discrepancies with previous results obtained by MGM analyses.

Fitting type-A spectra

Type-A spectra generally appear in high-Ca pyroxene such as diopside and hedenbergite, although previous studies suggest that there is no simple relationship between chemical composition of natural pyroxenes and type-A spectra [20,52,53]. The spectral data of samples 076, 037, 083, and 036 are categorized in type-A, which lack distinctive 2 μm bands. Since the absorption near 2 μm is mainly caused by spinallowed crystal field transition of Fe2+ in the M2 site, the absence of 2 μm band from type-A spectra is interpreted as a result of the replacement of Fe2+ in M2 site by other cations such as Ca2+ and Fe3+ [52,67]. On the other hand, spin-allowed crystal field transition due to the M1 site, with the absorption center locating near 1 and 1.2 μm, appear strong, and the two bands are combined to yield single broad band around 1 μm. It is difficult to fit type-A spectra with manually provided initial parameters (e.g., center, strength of Gaussian) using a conventional MGM algorithm, since the final solutions directly depend on the initial parameters, especially for the broad 1 μm band due to its non-Gaussian shape [56]. Although it is still difficult for the exchange Monte Carlo method to delineate the optimal number of Gaussians, we find that the center wavelengths of Gaussian functions for the broad 1 μm band do not move significantly with varying K. For example, the broad 1 μm band of sample 083 was fitted with two major Gaussian functions which could correspond to crystal field absorptions due to the M1 site whose center wavelengths of absorptions locate near 1 and 1.2 μm. Although the estimated errors are large, the overall result is consistent with previous deconvolution analyses of type-A spectra [53,56], suggesting that even with manually provided initial parameters, previous analyses obtained statistically optimal results. Since manually fitting type-A spectra is more difficult than type-B spectra, the exchange Monte Carlo method can be a useful tool for future deconvolution analysis for type-A spectra, of which are likely to appear in high-Ca pyroxene [52,53]. In addition, natural Cpx incorporate many minor compositions such as Al, Ti, Mn and Cr [49,50]. Such minor elements in natural Cpx yield more complex spectra than synthetic Cpx, thus the scheme presented in this paper can be useful to deconvolve such spectra.

1 μm band position vs. Ca-Fe-Mg content

Figure 5a indicates that the Cpx seem to follow a linear relationship among the 1 μm band position, with Fs and Wo contents. As shown as broken lines in Figures 5b and 5d, linear relationships are observed for pigeonite and diopside-hedenbergite with their Fs or En content. It should be noted that for diopside-hedenbergite, linear approximation was performed for all of the samples except sample 043. Sample 043 was omitted as an outlier because the spectrum shows higher reflectance and weaker absorptions compared with other samples, generally suggestive of the effect of glass [16,56,68]. Although it has been well documented for low-Ca pyroxene that the linear dependence of 1 μm band centers on Fs or En content [19,41,52,54,56], the relationship between the 1 μm band center and Fs or En content for high-Ca pyroxene has been poorly constrained by MGM analyses. Clénet et al. [41] analyzed only two high-Ca natural pyroxene samples in the diopside-hedenbergite region, thus no clear relationship has been derived. Although Klima et al. [56] analyzed 13 high-Ca pyroxene samples in the diopsidehedenbergite region; they observed no clear relationship between the 1 μm band center and Fs or En content. Our analyses show that the 1 μm band center of high-Ca pyroxene (i.e., diopside and hedenbergite) depends on Fs or En content with the band center moving longer wavelengths with increasing Fs content. This tendency is similar to the relationship between the 1 μm band center and the Fe/Mg ratio of Cafree orthopyroxene [64]. For Ca-saturated synthetic Cpx in which most of the M2 site is dominated by Ca2+, varying Fe/Mg ratio would affect only Fe/Mg in the M1 site because the M2 site is dominated with larger Ca2+ cations. Thus, the change of Fe/Mg would appear only in the 1 μm band center caused by a crystal field transition in the M1 site, but not in the 2 μm band center caused by the absorption due to the M2 site. We note, however, that large errors are included in the modeling results for type-A spectra (Table 1), thus additional data is necessary to confirm our interpretation.

Band center in three-dimensional spaces

As mentioned above, both the 1 μm and 2 μm bands seem to plot on approximate lines in the three-dimensional (3D) space of Fs-Wo content (Figures 5a and 6a). Low-Ca pyroxene samples (pigeonite) used in this study locates in the high-Fe and low-Ca regions with both the band centers being at shorter wavelengths. On the other hand, high-Ca pyroxene samples (diopside and hedenbergite) used in this study locates in the low-Fe and high-Ca regions with both the band centers being at longer wavelengths. Augite with intermediate-Wo contents distribute a in a linear trend between low-Ca and high-Ca pyroxene. Each mineral group displays three distinct clusters on these 3D spaces. The linear trend seen in both 1 μm and 2 μm bands could be understood from a crystallographic point of view. Whereas high-Fe and low-Ca Cpx samples contain more Fe2+ cations in both the M1 and M2 sites, leading to a decrease of the bond lengths due to the smaller size of Fe2+, which results in an increase in the crystal field splitting [34,69,70], low-Fe and high-Ca Cpx samples have larger Ca2+ cations, which dominate the M1 and M2 sites, leading to an increase in the bond lengths and a decrease in the crystal field splitting.

Application to future remote reflectance spectroscopy

Despite requiring meticulous parameter adjustment and prior information of mineralogy, MGMs using gradient descent method have been applied to reflectance spectra not only of laboratory data but also remote sensing data of the Moon [71], Mars [4,6,72-75], and some asteroids [76,77]. By applying the exchange Monte Carlo method to spectra of synthetic Cpx samples, we have shown that it is able to yield results consistent with both conventional gradient descent methods and crystal field theory. This means that the exchange Monte Carlo method could be applicable to at least some of the previous remote sensing data which have been analyzed using conventional MGM methods. Because the application of MGM analyses has been limited due to meticulous parameter adjustment, the exchange Monte Carlo method may have a potential to expand the applicability of MGM to a variety of space/ground-based observations, especially when we cannot obtain prior information of the target body beforehand. Remote sensing data of terrestrial reflectance spectroscopy have been validated with various reference spectra observed by in situ sample analyses, increasing the confidence of interpretation of remote sensing data [12,17,78,79]. However, such an analysis is not always available due to various geological contexts. The situation is more serious for space/ ground-based observations of small bodies in the solar system, because reflectance spectra are the only available compositional information for most of the small bodies such as Phobos, Deimos [80,81] and asteroid 1999 JU3 which is the target of Japanese space mission Hayabusa 2 [82]. In fact, asteroids are clustered based solely on reflectance spectra, although their detailed mineral compositions are poorly constrained [83]. Considering the large number of small bodies in the solar system, it is unlikely to send probes to each small body to conduct in situ sample analysis or sample return. Given the limited data and resource, interpreting reflectance spectra without assuming a priori information is essential not only for planetary science but also for mission planning. The exchange Monte Carlo method could be significantly useful under such circumstances, where meaningful compositional information other than reflectance spectra is not available.

CONCLUSIONS

We applied a Bayesian spectral deconvolution method with the exchange Monte Carlo algorithm to visible/near infrared reflectance spectra of synthetic clinopyroxene of diverse compositional variation, in order to avoid the local minimum problem and to remove the arbitrariness originated from initial parameters, inherent in the previous Modified Gaussian model. Our results indicate that the exchange Montel Carlo method is able to yield the consistent results obtained by conventional Modified Gaussian model. Since our model does not rely on a preliminary knowledge of the reflectance spectrum of mineral, the results suggest that previous spectrum analyses have obtained statistically optimal results. The successful application of our model to reflectance spectra of minerals indicates that this model could be applied to an automatic deconvolution analysis for a large spectral database, especially useful for space missions when preliminary knowledge of mineralogy is not available. Given the recent and future advancement of space missions, the exchange Monte Carlo method could be a useful tool for analyzing a wide range of minerals and remote sensing data of rocky bodies in the solar system.

ACKNOWLEDGMENT

This research utilizes spectra acquired with the NASA RELAB facility at Brown University. This work is supported in part by JSPS grant-in-aid 25120006 and TeNQ/Tokyo-dome.

REFERENCES

1. Goetz AFH, Vane G, Solomon JE, Rock BN (1985) Imaging spectrometry for earth remote sensing. Science 228: 1147-1153.
2. McCord TB (1988) Reflectance Spectroscopy in Planetary Science: Review and Strategy for the Future. NASA SP-493.
3. Soderblom LA (1992) The composition and mineralogy of the martian surface from spectroscopic observations: 0.3 μm to 50 μm. In: Kieffer HH, Jakosky BM, Snyder CW, Matthews MS (eds.) Mars. The University of Arizona Press.
4. Mustard JF, Sunshine JM (1995) Seeing through the dust: Martian crustal heterogeneity and links to the SNC meteorites. Science 267: 1623-1626.
5. Bibring JP, Langevin Y, Gendrin A, Gondet B, Poulet F, et al. (2005) Mars surface diversity as revealed by the OMEGA/Mars Express Observations. Science 307: 1576-1581.
6. Mustard JF, Poulet F, Gendrin A, Bibring JP, Langevin Y, et al. (2005) Olivine and pyroxene diversity in the crust of Mars. Science 307: 1594-1597.

7. Poulet F, Gomez C, Bibring JP, Langevin Y, Gondet B, et al. (2007) Martian Surface mineralogy from the Observatory for Mineralogy, Water, Ice and the activity on board the Mars Express spacecraft (OMEGA/MEx): Global mineral maps. J Geophys Res 112: E08S02.
8. Hapke B (2012) Theory of Reflectance and Emittance Spectroscopy(2ndedn) Cambridge University Press.
9. Straub DW, Burns RG, Pratt SF (1991) Spectral signature of oxidized pyroxenes: Implications to remote sensing of terrestrial planets. J Geophys Res 96: 18819-18830.
10. Mustard JF, Hays JE (1997) Effects of hyperfine particles on reflectance spectra from 0.3 to 25 μm. Icarus 125: 145-163.
11. Pelkey SM, Mustard JF, Murchie S, Clancy RT, Wolff M (2007) CRISM multispectral summary products: parameterizing mineral diversity on Mars from reflectance. J Geophys Res 112: E08S14.
12. Piatek JL, Hardgrove C, Moersch JE, Drake DM, Wyatt MB (2007) Surface and subsurface composition of the Life in the Atacama field sites from rover data and orbital image analysis. JGeophys Res 112: G04S04.
13. Cloutis EA, Gaffey MJ, Jackowski TL, Reed KL (1986) Calibrations of phase abundance, composition, and particle size distribution for olivine-orthopyroxene mixtures from reflectance spectra. J Geophys Res91: 11641-11653.
14. Poulet F, Erard S (2004) Nonlinear spectral mixing: Quantitative analysis of laboratory mineral mixtures. J Geophys Res 109: E02009.
15. Cheek LC, Pieters CM (2014) Reflectance spectroscopy of plagioclase-dominated mineral mixtures: Implications for characterizing lunar anorthosites remotely. Am Mineral 99: 1871-1892.
16. Horgan BHN, Cloutis EA, Mann P, Bell JF (2014) Near-infrared spectra of ferrous mineral mixtures and methods for their identification in planetary surface spectra. Icarus 234: 132-154.
17. Mustard JF, Sunshine JM (1999) Spectral analysis for Earth science: Investigations using remote sensing data. In: Rencz, AN (Ed.), Remote Sensing for the Earth Sciences: Manual of Remote Sensing (3rdedn.) John Wiley & Sons, Inc.
18. Murchie S, Arvidson R,Bedini P, Beisser K, Bibring JP (2007) Compact Reconnaissance Imaging Spectrometer for Mars (CRISM) on Mars Reconnaissance Orbiter (MRO). J Geophys Res 112: E05S03.
19. Adams JB (1974) Visible and Near-Infrared Diffuse Reflectance Spectra of Pyroxenes as Applied to Remote Sensing of Solid Objects in the solar system. J Geophys Res 79: 4829-4836.
20. Adams JB (1975) Interpretation of visible and near-infrared diffuse reflectance spectra of pyroxenes and other rock-forming minerals. In: Karr C (ed.) Infrared and Raman Spectroscopy of Lunar and Terrestrial Minerals. Academic Press, New York.
21. Gaffey MJ, Cloutis EA, Kelly MS, Reed KL (2002) Mineralogy of Asteroids. In: BottkeJr WF, Cellino A, Paolicchi P, Binzel RP (eds.), Asteroids III, The University of Arizona Press.
22. Burbine TH, Rivkin AS, Noble SK, Mothe-Diniz T, Bottke WF (2008) Oxygen and Asteroids. Rev Mineral Geochem 68: 273-343.
23. Gietzen KM, Lacy CHS, Ostrowski DR, Sears DW (2012) IRTF observations of S complex and other asteroids: Implications for surface compositions, the presence of clinopyroxenes, and their relationship to meteorites. Meteorit Planet Sci 47:1789-1808.
24. Reddy V, Nathues A, Le Corre L, Sierks H, Li JY (2012) Color and albedo heterogeneity of Vesta from Dawn. Science 336: 700-704.
25. Zambon F, De Sanctis MC, Schroder S, Tosi F, Longobardo A (2014) Spectral analysis of the bright materials on the asteroid Vesta. Icarus 240: 73-85.
26. Vilas F, Jarvis KS, Gaffey MJ (1994) Iron alteration minerals in the visible and near-infrared spectra of low-albedo asteroids. Icarus 109: 274-283.

27. Pieters CM, McFadden LA (1994) Meteorite and Asteroid Reflectance Spectroscopy: Clues to Early Solar System Processes. Annu Rev Earth PlanetSci 22: 457-497.
28. Hiroi T, Zolensky ME, Pieters CM, Lipschutz ME (1996) Thermal metamorphism of the C, G, B, and F asteroids seen from the 0.7 µm, 3 µm, and UV absorption strengths in comparison with carbonaceous chondrites. Meteorit Planet Sci 31: 321-327.
29. Hiroi T, Abe M, Kitazato K, Abe S, Clark BE, et al. (2006) Developing space weathering on the asteroid 25143 Itokawa. Nature 443: 56-58.
30. Bell JF, McSween HY, Crisp JA, Morris RV, Murchie SL, et al. (2000) Mineralogic and compositional properties of Martian soil and dust: Results from Mars Pathfinder. J Geophys Res 105: 1721-1755.
31. Hunt GR, Salisbury JW (1970) Visible and near-infrared spectra of minerals and rocks: I silicate minerals. Modern Geology 1: 283-300.
32. Hunt GR (1977) Spectral signatures of particulate minerals in the visible and near infrared. Geophysics 42: 501-513.
33. Clark RN, King TVV, Klejwa M, Swayze GA (1990) High spectral resolution reflectance spectroscopy of minerals. J Geophys Res 95: 12653-12680.
34. Burns RG (1993) Mineralogical applications of crystal field theory. Cambridge University Press, Cambridge.
35. Mayne RG, Sunshine JM, McSween HY, McCoy TJ, Corrigan CM, et al. (2010) Petrologic insights from the spectra of the unbrecciatedeucrites: Implications for Vesta and basaltic asteroids. Meteorit Planet Sci 45: 1074-1092.
36. Sunshine JM, Pieters CM, Pratt SF (1990) Deconvolution of Mineral Absorption Bands: An Improved Approach. J Geophys Res 95: 6955-6966.
37. Sunshine JM, Pieters CM (1993) Estimating modal abundances from the spectra of natural and laboratory pyroxene mixtures using the Modified Gaussian Model. JGeophys Res 98: 9075-9087.
38. Sunshine JM, Pieters CM, Pratt SF, McNaron-Brown KS (1999) Absorption Band Modeling in Reflectance Spectra: Availability of the Modified Gaussian Model. The 30th Lunar and Planetary Science Conference Abstract, #1306.
39. Tarantola A, Valette B (1982) Generalized nonlinear inverse problems solved using the least squares criterion. Rev Geophys Space Phys 20: 219-232.
40. Kanner LC, Mustard JF, Gendrin A (2007) Assessing the limits of the Modified Gaussian Model for remote spectroscopic studies of pyroxenes on Mars. Icarus 187: 442-456.
41. Clénet H, Pinet P, Daydou Y, Heuripeau F, Rosemberg C, et al. (2011) A new systematic approach using the Modified Gaussian Model: Insight for the characterization of chemical composition of olivines, pyroxenes and olivine-pyroxene mixtures. Icarus 213: 404-422.
42. Parente M, Makarewicz HD, Bishop JL (2011) Decomposition of mineral absorption bands using nonlinear least squares curve fitting: Application to Martian meteorites and CRISM data. Planet Space Sci 59: 423-442.
43. Makarewicz HD, Parente M, Bishop JL (2009) Deconvolution of VNIR spectra using modified Gaussian modeling (MGM) with automatic parameter initialization (API) applied to CRISM. Hyperspectral Image and Signal Processing: Evolution in Remote Sensing, WHISPERS '09 First Workshop pp: 1-5.
44. Swendsen RH, Wang JS (1986) Replica Monte Carlo Simulation of Spin-Glasses. Phys Rev Lett 57: 2607-2609.
45. Earl DJ, Deem MW (2005) Parallel tempering: Theory, applications, and new perspectives. PhysChemChemPhys 7: 3910-3916.
46. Nagata K, Sugita S, Okada M (2012) Bayesian spectral deconvolution with the exchange Monte Carlo method. Neural Networks 28: 82-89.
47. Hukushima K, Nemoto K (1996) Exchange Monte Carlo method and application to spin glass simulations. JPhysSocJpn 65: 1604-1608.
48. Kirkpatrick S (1984) Optimization by simulated annealing: quantitative studies. J Stat Phys 34: 975-986.

49. Cloutis EA (2002) Pyroxene reflectance spectra: Minor absorption bands and effects of elemental substitutions. J Geophys Res 107: 5039.
50. Morimoto N, Fabries J, Ferguson AK, Ginzburg IV, Ross M, et al. (1988) Nomenclature of pyroxenes. Mineral Mag 52: 535-550.
51. Lindsley DH (1983) Pyroxene thermometry. Am Mineral 68: 477-493.
52. Cloutis EA, Gaffey MJ (1991) Pyroxene Spectroscopy Revisited: Spectral-Compositional Correlations and Relationship to Geothermometry. J Geophys Res 96: 22809-22826.
53. Schade U, Wäsch R, Moroz L (2004) Near-infrared reflectance spectroscopy of Ca-rich clinopyroxenes and prospects for remote spectral characterization of planetary surfaces. Icarus 168: 80-92.
54. Hazen RM, Bell PM, Mao HK (1978) Effects of compositional variation on absorption spectra of lunar pyroxenes. Proceedings of the Lunar and Planetary Science Conference 9: 2919-2934.
55. Denevi BW, Lucey PG, Hochberg EJ, Steutel D (2007) Near-infrared optical constants of pyroxene as a function of iron and calcium content. J Geophys Res 112: E05009.
56. Klima RL, Dyar MD, Pieters CM (2011) Near-infrared spectra of clinopyroxenes: Effects of calcium content and crystal structure. Meteorit Planet Sci46: 379-395.
57. Akaike H (1980) Likelihood and the Bayes procedure. In: Bernardo JM, DeGroot MH, Lindley DV, Smith AFM (eds.), Bayesian statistics. University Press, Valencia31: 143-166.
58. Schwarz GE (1978) Estimating the dimension of a model. The Annals of Statistics 6: 461-464.
59. Pieters CM (1983) Strength of mineral absorption features in the transmitted component of near-infrared reflected light: first results from RELAB. J Geophys Res 88: 9534-9544.
60. Pieters CM, Hiroi T (2004) RELAB (Reflectance Experiment Laboratory): A NASA multiuser spectroscopy facility. Lunar and Planetary Science Conference, XXXV: #1720.
61. Turnock AC, Lindsley DH, Grover JE (1973) Synthesis and Unit Cell Parameters of Ca-Mg-Fe Pyroxenes. Am Mineral 58: 50-59.
62. Dowty E, Lindsley DH (1973) Mössbauer spectra of synthetic hedenbergite-ferrosilite pyroxenes. Am Mineral 58: 850-868.
63. Lindsley DH, Burnham CW (1970) Pyroxferroite: Stability and X-ray Crystallography of Synthetic $Ca_{0.15}Fe_{0.85}SiO_3$Pyroxenoid. Science 168: 364-367.
64. Klima RL, Pieters CM, Dyar MD (2007) Spectroscopy of synthetic Mg-Fe pyroxenes I: Spin-allowed and spin-forbidden crystal field bands in the visible and near-infrared. Meteorit PlanetSci 42: 235-253.
65. Goldman DS, Rossman GR (1977) The spectra of iron in orthopyroxene revisited: the splitting of the ground state. Am Mineral 62: 151-157.
66. Klima RL, Pieters CM, Dyar MD (2008) Characterization of the 1.2 μm M1 pyroxene band: Extracting cooling history from near-IR spectra of pyroxenes and pyroxene-dominated rocks. Meteorit Planet Sci 43: 1591-1604.
67. Rossman GR (1980) Pyroxene spectroscopy. Rev Mineral 7: 93-115.
68. Cloutis EA, Gaffey MJ, Smith DGW, Lambert RSTJ (1990) Reflectance spectra of glass-bearing mafic silicate mixtures and spectral deconvolution procedures. Icarus 86: 383-401.
69. Viswanathan K (1966) Unit cell dimensions and ionic substitutions in common clinopyroxenes. The American Mineralogist 51: 429-442.
70. Ohashi Y, Burnham CW, Finger LW (1975) The Effect of Ca-Fe Substitution on the Clinopyroxene Crystal Structure. Am Mineral 60: 423-434.
71. Hiroi T, Pieters C (1998) Modified Gaussian deconvolution of reflectance spectra of lunar soils. Lunar and Planetary Science conference XXIX: #1253.
72. Mustard JF, Murchie S, Erard S, Sunshine J (1997) In situ compositions of Martian volcanics: implications for the mantle. J Geophys Res 102: 25605-25615.

73. Gendrin A, Bibring JP, Mustard J, Kanner L, Mangold N, et al. (2006) Strong pyroxene absorption bands on Mars identified by OMEGA: geological counterpart. Lunar and Planetary Science conference XXXVII: #1858.
74. Baratoux D, Pinet P, Gendrin A, Kanner L, Mustard J, et al. (2007) Impact craters mineralogy from OMEGA data: implications on alteration history, ejecta emplacement, and subsurface composition. Proceedings of the Seventh International Mars Conference 3183.
75. Pinet PC, Heuripeau F, Clenet H, Chevrel S, Daydou Y, et al. (2007) Mafic mineralogy variations across Syrtis Major Shield and surroundings as inferred from visible-near-infrared spectroscopy by OMEGA Mars Express. Proceedings of the Seventh International Mars Conference 3146.
76. Binzel RP, Rivkin AS, Bus SJ, Sunshine JM, Burbine TH (2001) MUSES-C target asteroid (25143) 1998 SF36: A reddened ordinary chondrite. Meteorit Planet Sci 36: 1167-1172.
77. Hiroi T, Vilas F, Sunshine JM (1996) Discovery and analysis of minor absorption bands in S-asteroid visible reflectance spectra. Icarus 119: 202-208.
78. Kruse FA, Lefkoff AB, Boardman JW, Heidebrecht KB, Shapiro AT, et al. (1993) The Spectral Image Processing System (SIPS) – Interactive Visualization and Analysis of Imaging Spectrometer Data. Remote Sens Environ 44: 145-163.
79. van der Meer FD, van der Werff HMA, van Ruitenbeek FJA, Hecker CA, Bakker WH, et al. (2012) Multi- and hyperspectral geologic remote sensing: A review. Int J Appl Earth Obs 14: 112-128.
80. Fraeman AA, Murchie SL, Arvidson RE, Clark RN, Morris RV, et al. (2014) Spectral absorptions on Phobos and Deimos in the visible/near infrared wavelengths and their compositional constraints. Icarus 229: 196-205.
81. Pieters CM, Murchie S, Thosmas N, Britt D (2014) Composition of Surface Materials on the Moons of Mars. Planet Space Sci 102: 144-151.
82. Tsuda Y, Yoshikawa M, Abe M, Minamino H, Nakazawa S (2013) System design of the Hayabusa 2-Asteroid sample return mission to 1999 JU3. ActaAstronaut 91: 356-362.
83. DeMeo FE, Binzel RP, Slivan SM, Bus SJ (2009) An extension of the Bus asteroid taxonomy into the near-infrared. Icarus 202: 160-180.

CHAPTER 16

The Reconstruction of Paleo-Environment Albo-Aptian Sediments of the Massive El Hmaima North Tebessa North-Eastern Algeria

Diab Hamida*

Engineer status in sedimentary geology, National office of Geological and Mining Research (ORGM) Tébessa-Algeria

ABSTRACT

The paper is concerned with the studying of Albo-Aptian across North Africa to focused on sedimentological, tectonic and paleontological study of Albo-Aptian formations massive El Hmaima-North of Algeria. A systematic study was then initiated consisting of the sampling in a washing furniture samples of some Albo-Aptian rock units , and a careful sorting of microfossils and minerals under the lens of binocular microscope. Albo-Aptian is characterized by carbonate deposits rich in **ORGANIC MATTER** levels. This interval is marked by a large marine transgression typically connected to a eustatic rise in second order by tectonic events related to the opening of the western part of the **MEDITERRANEAN SEA** and/where the rifting of the central Atlantic and conditions paleogeographic. Mineralogically, the studied area are contain the important industries minerals such as; Calcium carbonate, Gypsum, Ferruginous nodules, pyrite, Quartz and clay minerals.

Keywords: Gypsum; Dolomites; Tectonic events; Microfossils; Quartz

INTRODUCTION

The reconstruction of paleo-environment of the studied area common Boulhef Dyr, using sedimentological approach. Sedimentology of facieses which is based on determination of the set of criteria as lithology, fossils continuous and sedimentary structures, allowed us to know the sedimentological and paleogeographic evolution of the Albo- Aptian sediments in the studied area.

The massive El Hmaima North Tébessa was located on the territory of the municipality of Boulhef Dyr, Daira of Morsott or twenty miles in the plain extends to the foot of Djebels Dyr, Chemla and Zitouna belonging to the southern area Mountains Mellegue which belongs to the Saharan Atlas mountains of the Eastern border Algerian - Tunisians. The massive El Hmaima North is located in the center of the sheet Morsott No. 178 (ech: 1/50 000) NNW of the city of Tebessa and there is a distance of about 15 km from it (Figure 1) [1].

★ The study area

Figure 1: Geographical location of the studied area from the topographical mapTébessa1/25.000.

The massive El Hmaima is composed of two multi-hectometric small clumps (El Hmaima North and El Hmaima South) emerging at the center of a large marl half buttonhole. Cenomanian is framed by escarpment Dj. Boulhef southeast, Dj. Chemla North-East and North Dj. Zaytuna opening to the west of the ditch Tébessa - Morsoft.

The present study are focused on one of these little clumps as El Hmaima North, which also called in some geological contributions as El Hmaima Dahara (Figure 2). El Hmaima North lies on 1600 m long and 600-700 m width, crossed by a single named Oued El Hmaima, NE-SW. Seasonal rains are mainly in spring and the climate is semi-arid to temporary flow. The vegetation is almost virtually absent developing relatively on northern slopes in the form of isolated trees [2].

Methodology

Sedimentology analysis is based on recognition of the terrain (unit delineation, measurement direction, dip, facies description, color, establishing a geologic section, and the systamatic sampling). In a laboratory test range is necessary to interpret such size, sedimentometry morphoscopic analysis. We exploit the results of the various tests in order to highlight the sedimentological parameters, which determine the mode and depositional environment. Petrographic study is to determine the type of facies and depositional environment. For cons, the tectonic study allows us to trace the tectonic phases in the regional context. The material used is existing in geology lab. Tébessa University, Algeria.

Figure 2: Photograph showing some exposure of Massive EL Hmaima North at the studied area.

The sediments studied were the subject of some chemical-weight analysis methods exposed by Boucart J and Berthois L. The principle is to weigh a few grams of the full sample and attack in a beaker (previously weighed) with HCL diluted 1/10 in unlimited quantity. Was heated a few minutes so as to promote the destruction attack and carbonates including dolomites, then washed with distilled water. Is allowed to settle and after drying in the oven, the beaker was weighed with the dry residue. After deducting the weight of carbonates.

Geologic Setting and Stratigraphy

This field study is limited to the north by latitude 35° 30' and south by the parallel 35° 23'. The western and eastern boundaries of the studied area are indicated by the meridian 8° 5' East and 8° 16' West (Figure 3).

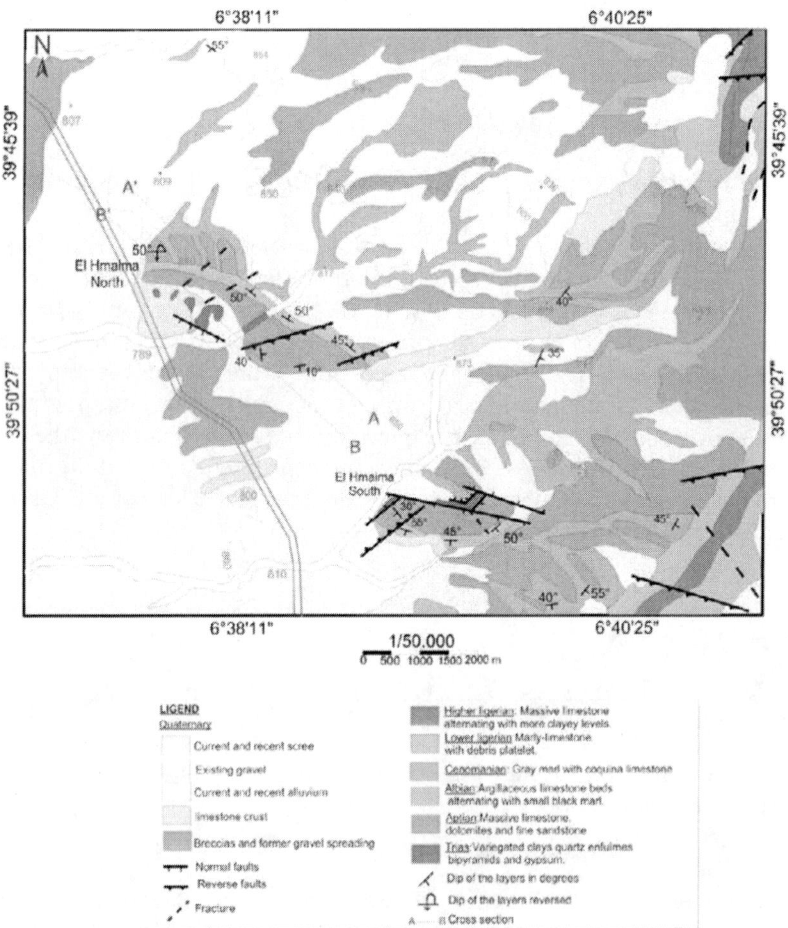

Figure 3: Geological formations of the area of intense study the geological map of Morsott (Algeria) to 1/50.000.

The field work is conducted for two geological sections, the first cut was made on the massive El Hmaima North SW-NE direction. The second cut was made on the massive Boulhaf-Dyr of NW-SE direction (Figure 4) encompassing all the Cretaceous. Samples were taken from the limestone banks of the Aptian, Albian, Vraconian, Turonian,Companien and Maastrichtian in order are fabricating thin blades. Associated with this work, samples were taken from the marly levels to perform washes marl [3,4].

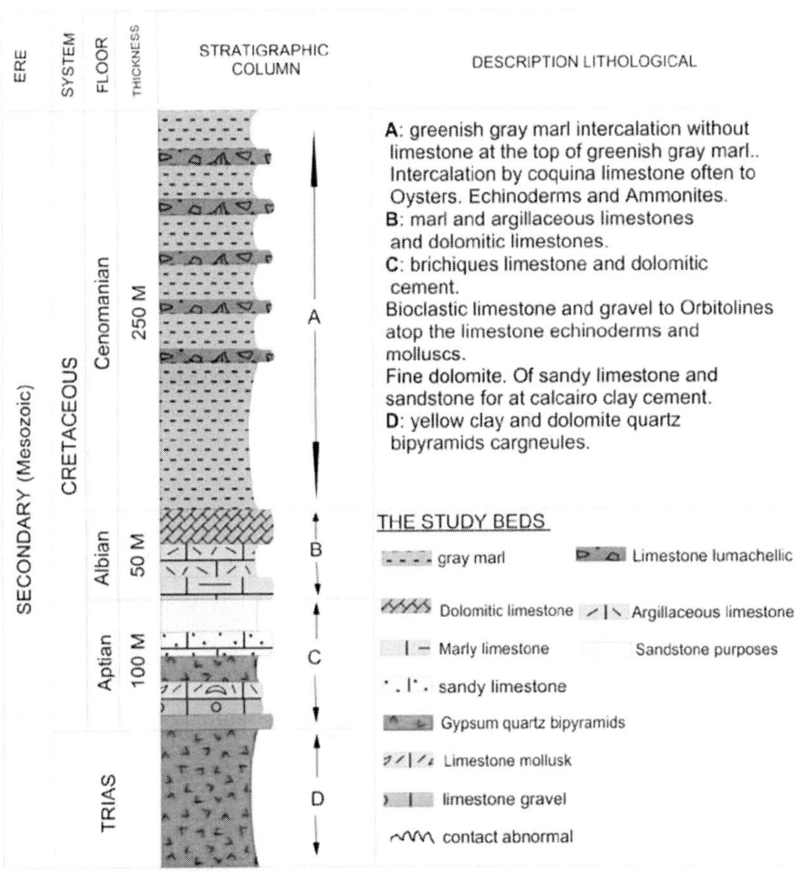

Figure 4: Stratigraphic log of the massive El Hmaima north at the studied area.

Triassic

Triassic appears on four cuts at the massive El Hmaima North, which outcrops tower along the Southwest side of the massif, especially in the western part where fills up the slope a few meters from the peak [5].

In the central part, Triassic appears at the lower part and the Eastern part of the outcrop. It is limited to a few some benches dolomites breccia intercalated with clay and sandstone levels softer rich baryto-celestite (Sr, Ba) SO4 to neoformation bipyramid smoky quartz and garnet crystals and dolomite in small shoals taped black breccia (Figures 5 and 6) [6].

Figure 5: Photograph showing crystals of garnet at the studied area.

In El Hmaima south,Triassic outcroped on the north flank. It is found in the Eastern part in the albo-alptians limestone with a sharp contact [7].

The largest outcrop of the Triassic in the region is located in the North and Northeast Mzouzia. It seems to fit the heart of El Hmaima diapir-Belkfif and it also represented the northern part of the diapir Mzouzia Boukhadra [8].

Figure 6: Photograph showing bipyramids crystals of quartz at the studied area.

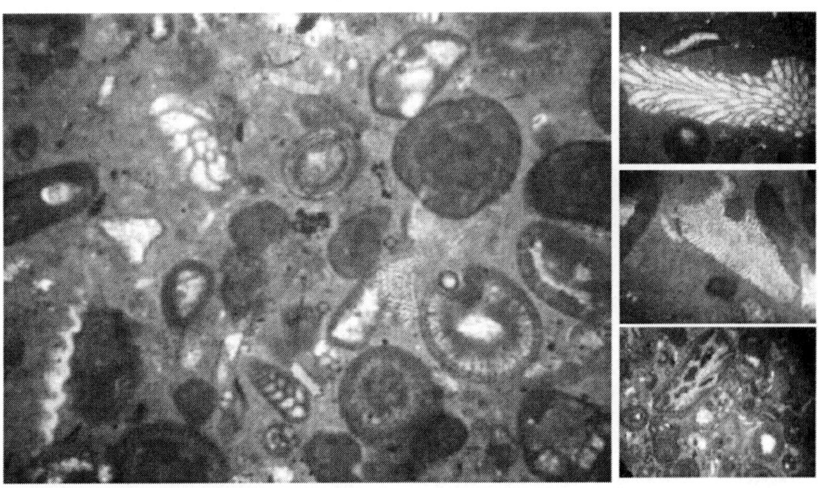

Figure 7: Photograph showing thin section of limestone with microfossil magnification X10000

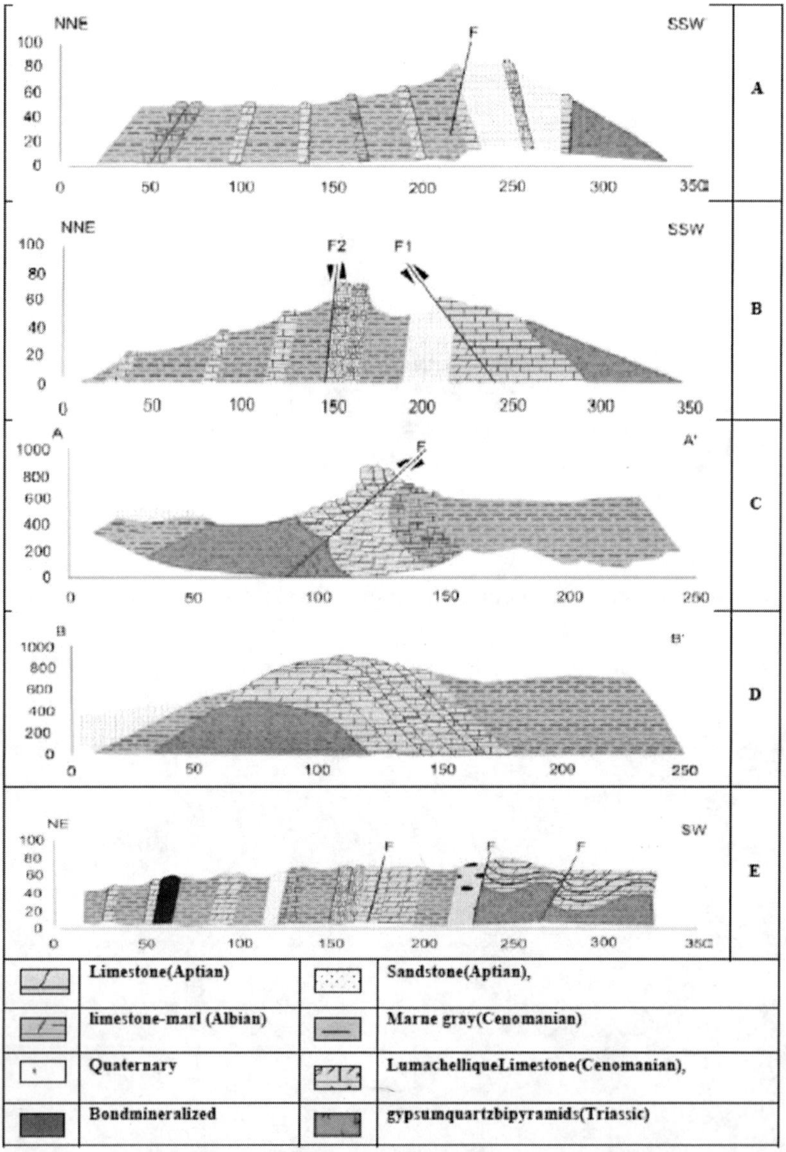

FIGURE 8: Cross-sections showing the sequence of different rocks units at the the study area. Cross-sections A, B, C, and D are signed through massive EL Hmaima North, while cross-sections E is signed through Wadi El Hmaim.

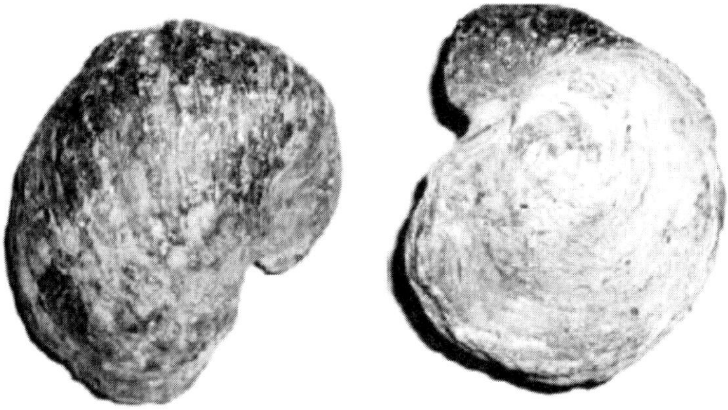

Figure 9: Photograph showing some fossils as Oyster and Exogera at the studied area.

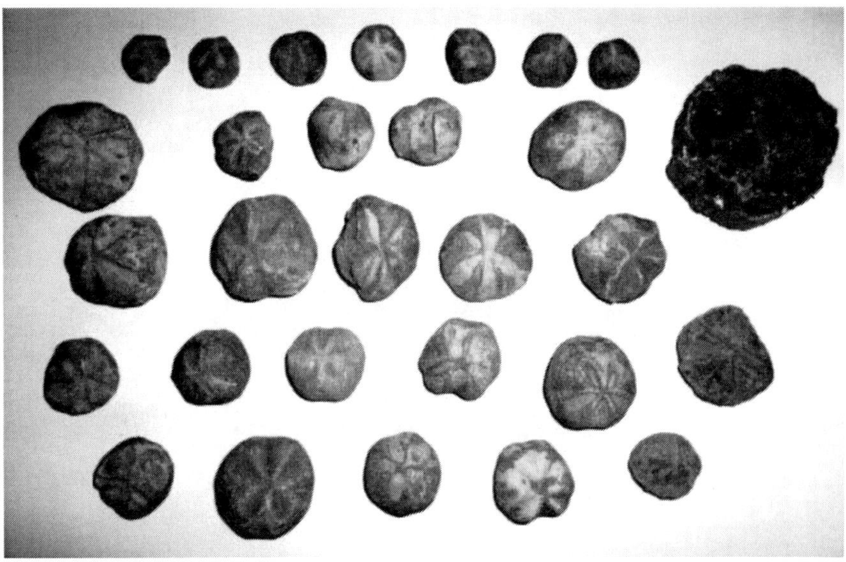

Figure10: Photograph showing Echinodermata fossils at the studied area.

Figure 11: Photograph showing Bivalves fossils at the studied area.

Figure 12: Photograph showing Trigonhdae and Myophorelle fossils at the studied area.

Figure 13: Photograph showing another type from bivalve fossils at the studied area.

Figure 14: Photograph showing the beaks of Belemnites at the studied area

Aptian

Aptian also well exposed in El Hmaima North represented by the sandstone levels with an intercalation metric limestone bank with micofossils (Figure 7) on a thickness of about 50 meters up the majority of the sequence (Figure 8A-D). It was consists of three types as;

* Limestone polyps and debris of organismes at the base.

* Rudist limestone (Figure 9)

* Limestone debris of various organisms such as echinoderms, bivalves, Polyps, algae (Figures 10-13) [9].

Albian

Albian of mountains Mellègue present as a marly facies and more generally very thick in Boukhadra area. The lower Albian has a powerful sequence of about 245 m [10] and follows the clansyasien with yellow marl facies intercalated sandy limestone (Figure 8A-D). North El Hmaima Albian is different from other sectors, where its smaller in thickness than neritic limestone facies [11].

Vraconian

Vraconian outcrops were at the north-east of North El Hmaima. it was overturned in tectonic contact with the Triassic to the east with the Albo-Aptian sandstone and carbonate. In the Eastern part, the Vraconian seems to be normal contact with the Albo-Aptian limestones and form a kind of incomplete closure periclinal (Figure 8A-D) [12].

In El Hmaima south Vraconian flush only on the south side, and show normal contact with the Aptian in the most part, where visible only to the southwestern corner of the massif. At the South-East end contact appears to have replayed and is masked by brecciation and the beaks of Belemnites (Figure 14). Vraconian has about 50 m thick, and the same lithological nature as El Hmaima North. At Djebel Belekfif, the Vraconian as small limited as Albo-Aptian sequence [13].

Cenomanian

Cenomanian surrounding El Hmaima North and El Hmaima South. It is limited by the Turonian limestones forming the cuesta of Djebels Boulhaf, Chemla, Zaytuna. The outcrop continuous in the northern flank of Djebel Boulhaf south of El Hmaima south. With about 250 m slightly greenish gray marl intercalation without limestone [14].

Turonian

Turonian is very remarkable in the region by large limestone beds that form ledges surrounding mountains among which Djebel Boulhaf, Djebel Chemla, Djebel Zitouna, Djebel Belkfif, Djebel Melloug. Turonian is almost similar in Djebel Belkfif and Djebel Boulhaf with the following units (Muriel Ruault-Djerrab, Bruno Ferré, & Fatiha Kechidbenkherouf)

- Limestone platelet (20 m).

- Very marly limestone and marl covered scree (about 220 m).

- Sub-lithographic gray massive limestone pink (60 m).

The thickness of Turonian is about 300 m. It is the ridge that forms the top of Djebel Melloug south of Djebel Mzouzia, which continuated to the Northwest cornice perched syncline Haoud Es sghir northwest of Djebel Boukhadra [10].

Senonian

Senonian is represented by a thick marl series in rare interbedded limestones and calcareous clay. Belkfif it reaches 1700 m thick where it flow into low-lying areas to the west of the massif [15]. The microscopic slides study of the thin section, from sedimentary formations limeston in the era of Aptian, showed the presence of a significant number of micro-fossil and these varied between foraminifère and calpionella in the picture.

Tectonics and Structure

Small clumps of El Hmaima North and El Hmaima South belong to a large anticlinal structure trended NE-SW. In the North East periclinal it is formed by Djebel Zitouna, Djebel Chemla and Djebel Boulhaf, while in the South-

West periclinal it is formed by Djebel Troubia and Djebel Es senane. The diapiric Triassic outcrops in El Hmaima North and El Hmaima south at south Belkfif.

El Hmaima North is a small elongated massive trended NW-SE with 1700 m long and 600 m width. It Can be divided into Eastern part and Western part [16].

Figure 15: Representation of tectonic in massive EL Hmaima North.

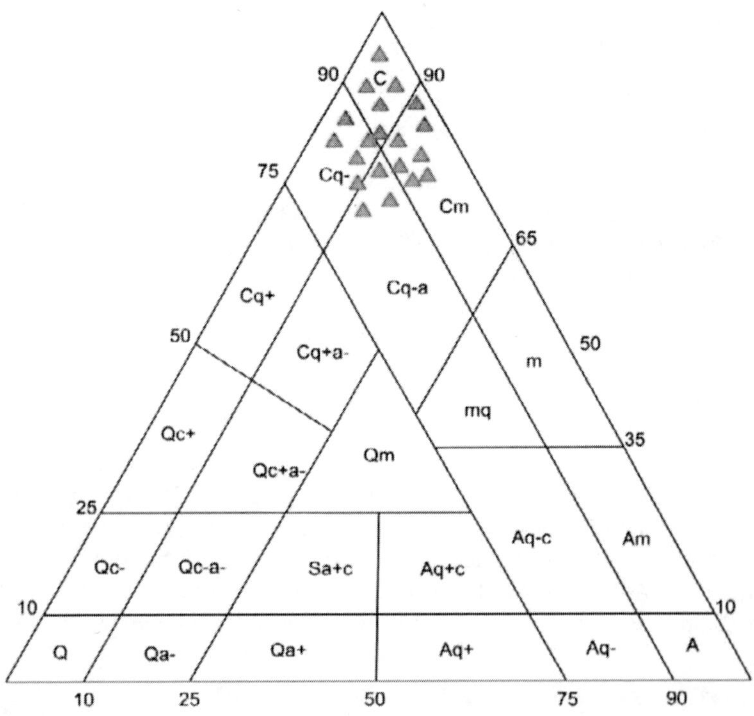

FIGURE 16: Triangular classification of sedimentary rocks three essential components: Quartz, Carbonate , Clay(A.Vatan, 1967.). Q: Quartz ; A: Clay; C: Carbonate; m: Marne ; +: many ; -: little.

The Eastern part has moderatly inclinal dip and stretches to the south-eastern end of the structure up to about 250 m northwest of Oued El Hmaima in contact with a fault oriented N160°. On the southwest side, it appears as a half-anticline. The layers of Albo-Aptian are normal stratigraphic contacted with brecciated Triassic dolomites and having a sweet dipping 10° to 20° NE and they become horizontaly to the South East, while the dip becomes stronger toward the top and the northeast side, reached 25° to 40° NE. The contact between the Albo-Aptian and Vraconian form a kind of sleeves complicated by small faults. It reappears on the Southwest side, where it unconformably overlies the Triassic dolomites form of small banks limestones directions (N5°- 60°) (Figure 15) [17].

The Western part has vertical to reverse dip and break picking N160° to the northwest end of the structure. It is characterized by vertical dip then overthrown by layers of Triassic to Vraconian. Going to the North West, the Albo-Aptian limestone disappears, and there is a contact between silicified limestone and Vraconian. The contact between Triassic and Vraconian also formed a kind of periclinal closure, and disjointed in southernwest flank which occupied by Triassic dolomites [18].

Mineralogical and Geochemical Aspects

Calcium carbonate (CaCO3)

The average value of 42% obtained for the samples 'furniture' reflects the membership there of to the petrographic class marl. However, extreme values allow us to distinguish, petrographic sense, clays (20% of the samples indeed contain less than 15% CaCO3) and marly limestones (about 20% have a rate greater than 65% CaCO3). The logical increase in carbonate limestone benches approach reflects a decrease in the water column, and conversely, an increasing proportion of clay is generally explained by a deepening of the depositional environment. Assuming normal conditions of oxygenation during deposition is therefore expected to find the most carbonate samples greater proportion of benthic organisms.`

Gypsum

'Gypsum marl': this term applies well to the Cenomanian marly levels. Nearly 10% of the samples indeed contain gypsum (sometimes clearly identified on the ground, sometimes only through the lens of the magnifying glass).

Ferruginous nodules and pyrite

Very common at all levels (present in about 30% of samples), these nodules are formed under reducing conditions, also necessary for the formation of pyrite. A certain effect levels, microorganisms (such as ostracods and foraminifera) are widely pyritized. Such an environment also allows the preservation of organic matter, as evidenced by the often dark color samples.

Quartz and clay minerals

Variations in quartz content can be explained logically according to these authors a greater or lesser distance from the continent-source.

Changes in relations between the various clay minerals is also explained by variations in paleoenvironmental conditions, including eustatic level.

Petrographic and Geochemical

Study methods

First time, we conducted an analysis based on macroscopic observations chromofaciologiques variations and identification of hardness, structure and texture of the rock. The thin refusal rocks are studied under a microscope to estimate roughly the presence or absence of light and heavy minerals that are often useful markers of the origins of rocks.

The sediments studied were the subject of some chemical-weight analysis methods exposed by Boucart J and Berthois L. The principle is to weigh a few grams of the full sample and attack in a beaker (previously weighed) with HCL diluted 1/10 in unlimited quantity. Was heated a few minutes so as to promote the destruction attack and carbonates including dolomites, then washed with distilled water, is allowed to settle and after drying in the oven, the beaker was weighed with the dry residue. After deducting the weight of carbonates.

The systematic position of the studied samples is shown in the triangular diagram A. Vatan which proposes to establish a descriptive classification to put a name on a rock. Other descriptive classifications have been adopted.

Chemical-weight analysis

Sediments studied two profiles are brittle variable colors (a color chart rocks is published by the National Research Council, Washington) changing between reddened, beigeâtre and rarely whitish. Reddening tells us about the climate conditions and demonstrate the presence of an oxidizing environment, continental. Sample processing using the method previously mentioned, gave the following results: According to the systematic A.Vatane (Figure 16).

Figure 17 is Showing the percent content of component of the studied rock units covered the studied area.

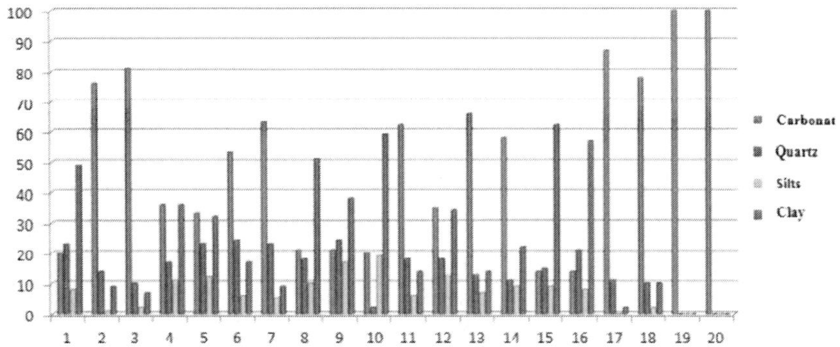

FIGURE 17: The percent content of component of the studied rock units covered the studied area.

RESULTS

In North Africa, the Aptian-Albian is characterized by carbonate deposits rich in organic matter levels. This interval is marked by a large marine transgression typically connected to a eustatic rise in second order by tectonic events related to the opening of the western part of the Mediterranean sea and / where the rifting of the central Atlantic and conditions paleogeographic. Thus, among the parameters that control sedimentation in this region, some are global such as eustatic variations and climatic parameters, while the others are local like tectonic events are local [19].

To isolate local factors, a sedimentary and paleontological analysis relied on a high-resolution biostratigraphy derived from fossils and microfossils (Figure 7), on sections located in different regions.

The Aptian sea in the studied area

Region Tébessa is almost completely covered by water during the Aptian, this fact is most typical installation reefs elongated wrinkles [20]. The fauna collected includes oysters, which living at the lower 200 or 250 funds. The early Aptian level was ended by 2 m thick dolomitic limestones intercalated with limestone gravel with a more or less thick layer of limestone molluscs (Figures 9,11-13) and the middle level becomes hot and restless [21,22].

The depth is increased and the environment become calm and allow to the sedimentation of calcareous sandstone, this time breaking the sea sedimentation allows for degrees [23].

Albian sea in the studied area

The Albian transition is made within a marly limestone layer indicates that the deep sea calm and cold. Transgression of the sea allows filed a considerable amount of sediments. Summit of the Albian is represented by dolomitic limestones, and argillaceous limestones at the middle level [24].

REFERENCES

1. Furon R (1950): Géologie de l'Afrique. Payot.
2. Vila JM (1980) La chaîne alpine d'Algérie orientale et des confins algéro-tunisiens. Thèse de doctorat, Université Paris 3: 665.
3. Odin GS, Odin C (1990) Echelle numérique des temps géologiques. Géochronique 35: 12-21.
4. DavidJL (1956)Etude géologique de la haute medjarda. Service de la carte géologique de l'Algérie 11: 304.
5. Herkat M (1999) La sédimentation de haut niveau marin du Crétacé supérieur de l'Atlas Saharien oriental et des Aurès. Stratigraphie séquentielle, analyse quantitative. Thèse de doctorat, Usthb, Alger.
6. Ouamer-Ali K (2009) Contribution à l'étude des sols à gypseux d'Algérie. Valorisation d'une banque de données et détermination des profils références. MémMag INA (El Harrach).
7. Djerrab MR (2008)Biostratigraphie et Paleoenvironnement du Crétacé moyen des Hameimats à partir de l'étude de la faune et de la microfaune - Coupe du Djebel Chemla, Morsott, NE Algérie.
8. BouzeOueA. (2001): Problématique de la structure des corps triasiques de lafrontièrealgéro-tunisienne. Conséquences métallogéniques. Communication orale au 1er séminaire sur la géologie du grand constantinois Univ. Constantine.
9. Busson G (1972) Principes, méthodes et résultats d'une étude stratigraphique du Mésozoïque saharien. Mémoire du Mus. Nat Hist Nat Paris 26: 441.

10. Dubourdieu G (1956)Etude géologique de la région de l'Ouenza (confins algérotunisiens). Publications du service de la carte géologique de l'Algérie, Alger, N.S 10: 659.
11. Durand DM,Fontbote JM (1960) Le problème de l'âge des nappes alpujarrides d'Andalousie, in Revue de géographie physique et de géologie dynamique.
12. Gradstein FM, Ogg JG, Smith AG, Bleeker W,Lourens LJ (2004) A new geologic time scale, with special reference to the Precambrian and Neogene.Episodes27: 83-100.
13. Beghoul M (1974) Etude géologique de la région des Hamimat (S.E. Constantinois). Application à la recherche pétrolière des confins algéro-tunisiens. Thèse Doct. Ing., Paris, pp. 127.
14. Cornet A, GouscovN (1952) Les eaux du Crétacé inférieur continental dans le Sahara alge´rien.
15. Chiki-Aouimeur F (1998) Les Rudistes du Crétacé supérieur de l'Algérie. Etude paléontologique, données paléoécologiques, biostratigraphiques et paléogéographiques. Thèse de doctorat, Usthb, Alger, pp. 198.
16. Chadi M (1990) Evolution tertiaire de la plate-forme crétacée externe de la chaînealpine d'Algérie orientale.
17. Bel F,Demargne F (1966) Etude géologique du Continental Terminal; DEC, ANRH, Alger, Algérie, 24 planches, pp. 22.
18. Dubourdieu G (1959) Esquisse géologique du Djebel Mesloula. Publications du service de la carte géologique de l'Algérie, Alger, N.S 21: 162.
19. Gignoux M,Moret L (1935) Géologie stratigraphique, éd. Masson.
20. Masse JP (1988) L'étagement bionomique des milieux benthiques néritiques actuels : signification bathymétrique et implications paléobathymétriques. Géol Médit., XV, 1, 91102.
21. Dunham RJ (1962) Classification of carbonate rocks according to depositional texture. In: HAMWE (ed.), Classification of carbonate rocks: American Association of Petroleum Geologists Memoir.
22. Folk RL (1962)Spectral subdivision of limestone types, In HAM WE (ed.), Classification of Carbonate Rocks-A Symposium: American Association of Petroleum Geologists Memoir 1: 62-84.
23. Fischer AG, Arthur MA (1977) Secular variations in the pelagic realm. Spec. Publ. Soc. econ. Paleont Miner 25: 19-50.
24. Sigal J(1952) Aperçu stratigraphique sur la Micropaléontologie du Crétacé. XIXème congr. Géol. Intern., Monogr. Région. 1ère série, Algérie, N°26, Alger.

CHAPTER 17

Rock Mass Characterization and Support Design for Underground Additional Surge Pool Cavern—A Case Study, India

Ajay Kumar Naithani*, Laishram Gopeshwor Singh, Prasnna Jain

National Institute of Rock Mechanics, Bengaluru, India

ABSTRACT

For better rock mass characterization and support design, 3D engineering geological mapping was carried for the heading portion of the under construction 200.00 m long, 68.75 m high and 20.20 m wide underground additional surge pool cavern of a Pranahitha-Chevella Sujala Sravanthi lift irrigation scheme package 8, India. To study cavern behavior, 3D geologic mapping of heading portion is very important for large cavern for predicting geologic conditions in benching down up to invert level, planning support system, selecting inclination for best location of supplemental rock bolt and choosing strategic locations for various types of instrumentation. The assessment of Tunnel Quality Index "Q" and Geomechanics classification for the granitic rock mass was done based on the information available of the rock joints and their nature and 3D geological logging. Hoek-Brown parameters were also determined by the statistical analysis of the results of a set of triaxial tests on core samples. On basis of geological characteristics and NMT Q-system chart, support system is recommended which includes rock bolt, steel fibre reinforced shotcrete and grouting. To evaluate the efficacy of the proposed support system, the capacity of support system is determined.

Keywords: Engineering Geology, Underground Cavern, Support System, Rock Bolt, Shotcrete

1. INTRODUCTION

Analytical, observational, and empirical are the main design approach for excavations in rock. In this paper, empirical approach for support design of additional surge pool cavern of a Pranahitha-Chevella Sujala Sravanthi lift irrigation scheme package 8 (PCSSLIS-P8) is discussed. Rock mass classifications as practiced in civil and mining engineering form an integral part of the empirical design methods, which is the most predominant design approach [1] . The main objectives of the rock mass classifications are to identify the most significant parameters influencing the behavior of a rock mass, divide area into rock mass classes of varying quality and provide quantitative data for engineering design purpose. Rock mass classifications have played an important role in estimating the strength and deformability of rock masses and in assessing the stability of rock slopes. They were also shown to have special uses for serving as an index to rock rippability, dredgeability, excavatability, cuttability, and cavability. For underground excavation, stable empirical approaches are developed based on the evaluation of a large number of case studies.

The major components of the PCSSLIS-P8 are: 4.133 km long and 10.00 m finished diameter "D" shaped twin tunnels, old surge pool (350 m long × 20 m width × 54 m height), 58 m long five numbers of draft tube tunnels, one pump house (215 m long × 25 m width × 54 m height) and five numbers, 50 m long horizontal and 150 m vertical shaft having 5.0 m finished diameter pressure mains, 80 m long delivery cistern and 5.85 km long gravity canal from delivery cistern to join flood flow canal. Lift height is about 126 m and five numbers of pump will be installed in the pump house cavity having 130 MW capacities each. The reengineering of the project was done and because of this additional surge pool is being constructed for increased discharge from 419 to 624 cumecs. Summary of input data of additional surge pool cavern used for support design as provided by sponsoring agency are given in Table 1. Sufficient lateral rock cover is available, and the vertical cover is more than 1D i.e. >70 m above the surge pool.

Table 1. Summary of input data.

1	Length of surge pool with approach for ventilation	200 + 25 m
2	Excavated width of cavern (*B*)	20.20 m
3	Clear width of cavern	20.00 m
4	Crown level	250.25 m
5	Spring level	240.00 m
6	Surge pit level	181.50 m
7	Height of surge pit wall	58.50 m
8	Height of overburden above crown (*H*)	70.75 (average)
9	Ground levels maximum and minimum above crown	321 m and 319 m
10	Rise of arc	10.25 m
11	Unit weight of rock (γ)	2.60 t/m^3
12	Average spacing of joints	0.750 m
13	Maximum upsurge level	239.9 m
14	Minimum downsurge level	214.8 m
15	Thickness of concrete lined bottom portion	300 mm
16	Rock ledge between old and new surge pool	100 m

For the underground cavern rock mass characterization was done based on 3D geologic mapping and laboratory test results. On basis of geological characteristics and NMT Q-system chart, support system is recommended and its efficacy is evaluated.

2. 3D GEOLOGICAL MAPPING

3D engineering geological mapping was done in 1:100 scale so that closely spaced geological discontinuities can be mapped (Figure 1). Geologic logging provides a permanent record of all geologic defects exposed on the walls and crown of an underground excavation. Rock type mapped was pink granite belongs to the Peninsular Gneissic Complex of Archaean age [2] [3] . Granite was coarse grained, hard and jointed in nature. The granite was generally fresh in nature. It was interpreted that same rock will be present during the benching of additional surge pool up to its invert level.

The details of the joint characteristics are given in Table 2. Joints are generally continuous and persistent, smooth-planar with unaltered to slightly altered joint walls. Staining has been recorded along the joint surfaces where the joints are tight and where opening is up to 3.0 mm, clay filling has been recorded. In general, the rock mass is characterized by dry condition or minor inflow i.e. <5.0 l/min.

3. LABORATORY TESTING

Selected rock core samples were tested for their physico-mechanical properties and test results as provided by MEIL are summarized in Table 3. The compressive strength of core specimens is ranging from 132 to 238 MPa and density varies between 2645 to 2695 kg/m^3. According to strength classification criterion for rock substance, the rocks are of very high strength [4] and density of material is high.

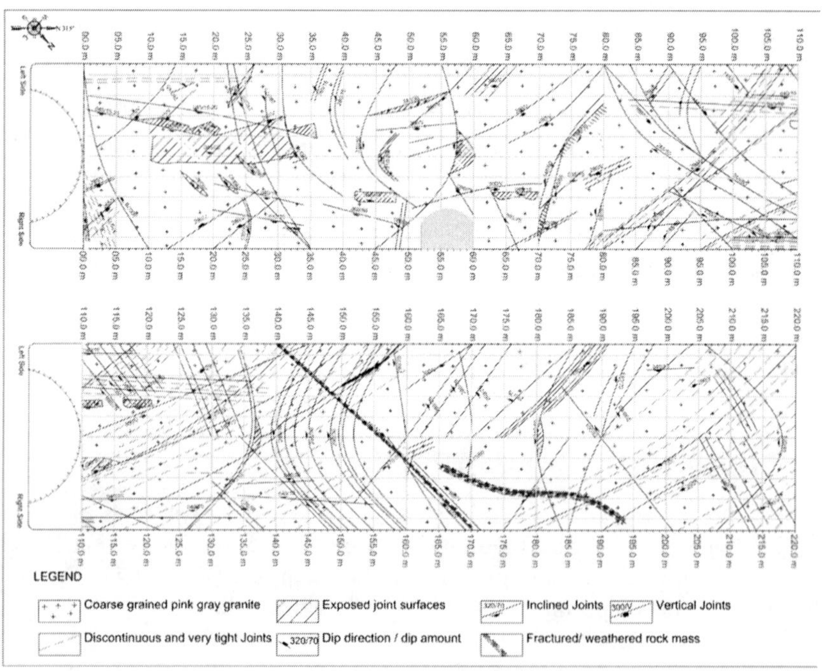

Figure 1. 3D Geological map of the heading portion.

Table 2. Joint sets recorded in coarse grained pink granite.

Joint sets	Azimuth/Dip Amount	Spacing (cm)	Strike length (m)	Roughness	Aperture (mm)	Infilling	GW
J1	280 - 300/V	30 - 150	>20	Smooth, planar	Tight	Fresh/clay coated	Dry
J2	035 - 045/10 - 25	30 - 100	>20	Smooth, planar	Tight	Fresh/clay coated	Dry
J3	280 - 310/65 - 75	60 - 200	>20	Smooth, planar	Tight	Fresh/clay coated	Dry
J4	280 - 300/30 - 50	75 - 200	>20	Smooth, planar	Tight	Fresh/clay coated (2 - 4 mm)	Dry
J5	130 - 145/50 - 70	>100	>20	Smooth, planar	Tight to 3 mm	Fresh to 3 - 5 mm clay filling	Dry
J6	240 - 260/40 - 50	>100	>15	Smooth, planar /undulating	Tight to 3 mm	Fresh to 3 - 5 mm slightly alter	Dry
J7	080 - 100/70 - 80	>100	>10	Smooth, planar	Tight	Fresh	Dry
J8	070 - 080/50	>100	>10	Smooth, planar	Tight	Fresh	Dry
J9	170 - 185/50 - 70	>100	>10	Smooth, planar	Tight	Fresh	Dry
JR1	330 - 345/30 - 50	>100	<10	Smooth, planar	Tight	Fresh/clay coated	Dry
JR2	300 - 310/20 - 30	>100	<10	Smooth, planar	Tight	Fresh	Dry
JR3	160 - 180/V	>100	>10	Smooth, planar	Tight	Slightly altered joint walls	Dry
JR4	240/15	>100	>10	Smooth, planar	Tight	Fresh	Dry
JR5	060/80	>100	>10	Smooth, planar	Tight	Fresh	Dry
JR6	340/75	>100	>10	Smooth, planar	Tight	Fresh	Dry

Notes: GW—Groundwater, JR—Random joint, V—Vertical.

4. ROCK MASS CLASSIFICATION

4.1. Tunnelling Quality Index (Q)

The Q-system was developed at NGI between 1971 and 1974 on the basis of approximately 200 case histories of tunnels and caverns [5] . They presented a useful correlation between the amount and type of permanent support and the Q with respect to tunnel stability. There has been a significant advance within support philosophy and technology in underground excavations since the introduction of the Q-system in 1974. After its introduction in 1974, two revisions of the support chart have been carried out. On the basis of 1050 examples mainly from Norwegian underground excavations an extensive updating was done in 1993 [6] . Based on more than 900 new examples from underground excavations in Norway, Switzerland and India, an updating was made in 2002. This update also included analytical research with respect to the thickness, spacing and reinforcement of reinforced ribs of sprayed concrete as a function of the load and the mass quality [7] .

Table 3. Results of lab tests to rock samples.

Rock Type	Elevation (m)	Density (kg/m³)	Uniaxial Compressive Strength-Dry (MPa)	Tensile Strength (MPa)	Modulus of Elasticity (GPa)	Poisson's Ratio	Cohesion (PMa)	Friction Angle
Pink granite	249.50	2695	212	11.50	74.30	-	47.81	46.09
Pink granite	255.50	-	-	11.10 - 12.90	-	-	-	-
Pink granite	260.00	-	-	9.50	-	-	-	-
Pink granite	261.50	2645	180	-	105.77	-	44.22	50.03
Pink granite	266.00	2690	-	10.80	-	-	45.38	49.24
Pink granite	273.50	2653	238	12.70	64.70	-	47.75	54.49
Pink granite	274.00	2695	191	11.00	71.36	0.25	38.14	47.37
Pink granite	285.00	2682	138	-	79.79	-	-	-

The Q-value gives a description of the rock mass stability of an underground opening in jointed rock masses. High Q-values indicates good stability and low values means poor stability. The numerical value of the index Q varies on a logarithmic scale from 0.001 to a maximum of 1000 and is defined by six parameters (Equation (1)). Q-value 0.001 is generally for exceptionally poor quality squeezing ground, while 1000 is for exceptionally good quality rock which is practically unjointed [5] .

$$Q = \frac{\mathrm{RQD}}{\mathrm{J_n}} \times \frac{J_r}{J_a} \times \frac{J_w}{\mathrm{SRF}} \tag{1}$$

where RQD is Rock Quality Designation (degree of jointing), J_n is Number of joint sets, J_r is Joint roughness number, J_a is Joint alteration number, J_w is Joint water reduction factor and SRF is Stress Reduction Factor

For the heading portion of additional surge pool the individual parameters were determined during geological mapping using tables that give numerical values to be assigned to a described situation. For the calculation of Q-values all the discontinuities per 5 m length and circumference were taken into consideration. An average piece size or block size can be determined using the same data i.e. discontinuities per 5 m length and circumference. The assessment of Q-values for the granitic rock mass, based on the information available of the rock joints and their nature and 3D geological logging, is tabulated in Table 4. The grade of rock mass based on the rock joints characteristics has the Q-values varying from 4.17 to 16.33, and it comes under fair to good rock mass category.

Table 4. Q-values recorded from the heading portion of the additional surge pool.

Chainage (m)	Rock Type	RQD (%)	J_n	J_r	J_a	J_w	SRF	Q	
								Value	Class
0 - 20	Coarse grained pink granite	92 - 98	2 to 3	Smooth planar	Slightly altered to unaltered joint wall	Minor inflow to dry excavation	Medium stress	10.22 to 16.33	Good
20 - 30	Coarse grained pink granite	95	2 + R	Smooth planar	Slightly altered joint walls	Minor inflow	Medium stress	7.92	Fair
30 - 65	Coarse grained pink granite	88 - 99	2 to 3	Smooth planar	Slightly altered to unaltered joint wall	Dry excavation	Medium stress	10.89 - 15.33	Good
65 - 105	Coarse grained pink granite	82 - 95	2 + R to 3	Smooth planar	Slightly altered joint walls	Minor inflow to dry excavation	Medium stress	4.72 - 7.92	Fair
105 - 125	Coarse grained pink granite	75 - 95	2 + R to 3	Smooth planar	Unaltered joint wall	Minor inflow to dry excavation	Medium stress	10.56 - 14.17	Good
125 - 195	Coarse grained pink granite	75 - 95	2 + R to 3	Smooth planar	Slightly altered joint walls	Minor inflow to dry excavation	Medium stress	4.17 - 7.92	Fair
195 - 210	Coarse grained pink granite	95 - 98	2 + R	Smooth planar	Unaltered joint wall	Dry excavation	Medium stress	15.83 - 16.33	Good
210 - 220	Coarse grained pink granite	95	2 + R	Smooth planar	Slightly altered joint walls	Dry excavation	Medium stress	7.92	Fair

RQD = Rock Quality Designation, Jn = Joint Set Number, Jr = Joint Roughness Number, Ja = Joint Alteration Number, Jw = Joint Water Reduction Factor and SRF = Stress Reduction Factor.

RQD = Rock Quality Designation, Jn = Joint Set Number, Jr = Joint Roughness Number, Ja = Joint Alteration Number, Jw = Joint Water Reduction Factor and SRF = Stress Reduction Factor.

Total 59 percent of area comes under fair rock mass category while 41 percent under good rock mass category. The average Q-value calculated is 9.58. The low Q-values are because of intersection of more joint sets in the excavated span of 5 m and joints surface characteristics.

4.2. Geomechanics Classification

The Geomechanics Classification, also known as the Rock Mass Rating system, was developed by Bieniawski during 1972-1973 on the basis of 49 case histories [8] . It was modified over the years as more case histories become available and to conform with international standards and procedures [9] . In 1984, 62 coal mining case histories were added and a further 78 tunneling and mining case histories collected by 1987. Last time it was modified in 1989 by Bieniawski amounting to 351 case histories. Since then it is being used in tunnels, chambers, mines, slopes and foundations projects. Most of the applications have been in the field of tunneling. This classification is one of the most commonly used rock mass classification system. This is based on the collection of field data and strength parameter. The six parameters which are used to classify a rock mass using RMR system are: uniaxial compressive strength of rock material (UCS), rock quality designation (RQD), spacing of discontinuities (SD), condition of discontinuities (CD), groundwater conditions (GW) and orientation of discontinuities (OD) (Equation (2)).

$$\mathrm{RMR} = \mathrm{UCS} + \mathrm{RQD} + \mathrm{SD} + \mathrm{CD} + \mathrm{GW} - \mathrm{OD} \tag{2}$$

In order to apply the RMR classification, the rock mass has to be divided into a number of structural regions such that certain features are more or less uniform within each region. Rock Mass Rating technique has been found to be quite useful due to the ease with which it can be practiced and its effectiveness in interpreting stability and recommending control measures. The RMR classification parameters are easily obtained either from borehole data or underground/ surface mapping [10] [11] [12] . Average stand-up time for an arched roof, cohesion and angle of internal friction, modulus of deformation, allowable bearing pressure, shear strength of rock mass and estimation of support pressure of rock mass may be obtained using RMR. For the heading portion of additional surge pool, RMR values are determined at every 5 m interval (Table 5). The grade of rock mass based on the 3D geological mapping and strength characteristics, has the RMR values varying from 53 to 71, and it comes under fair to good rock category. The average RMR-value calculated is 62.

4.3. Hoek-Brown Parameters

In order to use the Hoek-Brown criterion for estimating the strength and deformability of jointed rock masses, the value of the Geological Strength Index (GSI) for the rock mass, the uniaxial compressive strength (σ_{ci}) of the intact rock pieces, and the value of Hoek-Brown constant (m_i) for these intact rock pieces have been estimated. Geological Strength Index (GSI) was introduced by Hoek and Brown (1997) to provide a system for estimating the reduction in the rock mass strength for different geological conditions. The GSI can be related to the rock mass rating (RMR) or the modified rock-mass quality index (Q'). Modified rock-mass quality index is defined as (Equation (3)):

$$Q' = \frac{\mathrm{RQD}}{J_n} \times \frac{J_r}{J_a} \tag{3}$$

where RQD is the rock quality designation, J_n is the joint set number, J_r is the joint roughness number, J_a is the joint alteration number.

Hoek and Brown [13] suggested that GSI can be related to Q' and RMR by following equations (Equation (4) and Equation (5)). Bieniawski's RMR classification should be used for estimating GSI values for better rock masses (GSI > 25) and should not be used for poor quality rock masses.

Table 5. RMR-values determined at different chainage.

Chainage (m)	UCS (MPa)	RQD %	Spacing (cm)	Condition of Discontinuity					Ground water	Adjustment		RMR	
				Persistence (m)	Aperture (mm)	Roughness	Infilling (mm)	Weathering grade		Orientation	Rating	Rating	Description
0 - 20	212	92 - 98	60 - 200	>20	<0.1 - 1.5	Smooth	Soft > 5	W-I	Damp - dry	Fair	−5	66 - 71	Good rock
20 - 30	191	95	60 - 200	>20	1 - 5	Smooth	Soft > 5	W-I	Damp	Fair	−5	62	Good rock
30 - 65	180	88 - 99	60 - 200	>20	<0.1 - 1.5	Smooth	Soft > 5	W-I	Dry	Fair	−5	67 - 71	Good rock
65 - 195	138 - 180	75 - 95	20 - 60 & 60 - 200	>20	<0.1 & 1 - 5	Smooth	Soft > 5	W-I-WII	Damp - dry	Fair	−5	53 - 65	Fair to Good rock
195 - 220	238	95 - 98	60 - 200	>20	<0.1 - 1.5	Smooth	Soft > 5	W-I	Dry	Fair	−5	67 - 71	Good rock

Table 6. Rock mass classification of granite.

Category	Q		RMR		GSI Calculated/Estimated					
					From Q'		From RMR		From Hoek-Brown Chart	
	Value	Class	Value	Class	Value	Class	Value	Class	Value	Class
Minimum	4.17	Fair	53	Fair	56.85	Good	48	Fair	45	Fair
Maximum	16.33	Good	71	Good	69.14	Good	66	Good	65	Good
Average	9.58	Fair	62	Good	64.34	Good	57	Good	55	Fair
Mean	11.92	Good	63	Good	66.30	Good	58	Good	56	Good

$$\text{GSI} = 9 \ln Q' + 44 \quad (4)$$

$$\text{GSI} = \text{RMR}_{89} - 5 \quad (5)$$

For the additional surge pool GSI is calculated from Q', RMR and Hoek and Brown [13] chart. Hoek and Brown chart is based on geological description of the rock mass i.e. on the basis of interlocking and joint alteration. Minimum, maximum, average and mean values of Q, RMR and GSI are given in Table 6.

The values of σ_{ci} and m_i were determined by the statistical analysis of the results of a set of triaxial tests on core samples. After obtaining the test results, they were analysed to determine the uniaxial compressive strength

(σ_{ci}) of the intact rock pieces, and the value of Hoek-Brown constant (m_i) as described by Hoek and Brown [14] . A spreadsheet for the analysis of triaxial test data is given in Table 7.

For each sample the uniaxial compressive strength (σ_{ci}), the constant (m_i) and coefficient of determination (r^2) are calculated from Equations (6)-(8) respectively and values are given in Table 8. The Hoek-Brown parameters that describe the rock mass strength characteristics can be derived from GSI (Equation (9)).

$$\sigma_{ci}^2 = \frac{\sum y}{n} - \left[\frac{\sum xy - (\sum x \sum y / n)}{\sum x^2 - ((\sum x)^2 / n)} \right] \frac{\sum x}{n} \tag{6}$$

$$m_i = \frac{1}{\sigma_{ci}} \left[\frac{\sum xy - (\sum x \sum y / n)}{\sum x^2 - ((\sum x)^2 / n)} \right] \tag{7}$$

$$r^2 = \frac{\left[\sum xy - \left(\sum x \sum \frac{y}{n} \right) \right]^2}{\left[\sum x^2 - \frac{(\sum x)^2}{n} \right] \left[\sum y^2 - \frac{(\sum y)^2}{n} \right]} \tag{8}$$

$$m_b = m_i e \left\{ \frac{(\mathrm{GSI} - 100)}{9} \right\} \tag{9}$$

where m_b is the value of the Hoek-Brown constant m for the rock mass and mi is the Hoek-Brown constant for the intact rock.

Hoek-Brown constants "s" and "a" are depend upon the rock mass characteristics. For GSI > 25, i.e. rock masses of good to reasonable quality, the original Hoek-Brown criterion is applied with (Equation (10) and Equation (11)):

Table 7. Spreadsheet for the calculation of σ_{ci} and mi from triaxial test data.

$X(\sigma_3)$	σ_1	$y(\sigma_1-\sigma_3)^2$	xy	x^2	y^2
10	297	82,369	823,690	100	6,784,652,161
20	363	117,649	2,352,980	400	13,841,287,201
30	423	154,449	4,633,470	900	23,854,493,601
40	482	195,364	7,814,560	1600	38,167,092,496
100	1565	549,831	15,624,700	3000	82,647,525,459
Sum$x$		Sum$y$	Sumxy	Sumx^2	Sumy^2
			Elevation 261.50 m		
10	313	91,809	918,090	100	8,428,892,481
20	402	145,924	2,918,480	400	21,293,813,776
30	473	196,249	5,887,470	900	38,513,670,001
40	541	251,001	10,040,040	1600	63,001,502,001
100	1729	684,983	19,764,080	3000	131,237,878,259
Sum$x$		Sum$y$	Sumxy	Sumx^2	Sumy^2
			Elevation 266.00 m		
10	310	90,000	900,000	100	8,100,000,000
20	396	141,376	2,827,520	400	19,987,173,376
30	467	190,969	5,729,070	900	36,469,158,961
40	528	238,144	9,525,760	1600	56,712,564,736
100	1701	660,489	18,982,350	3000	121,268,897,073
Sum$x$		Sum$y$	Sumxy	Sumx^2	Sumy^2
			Elevation 273.50 m		
10	388	142,884	1,428,840	100	20,415,837,456
20	501	231,361	4,627,220	400	53,527,912,321
30	599	323,761	9,712,830	900	104,821,185,121
40	680	409,600	16,384,000	1600	167,772,160,000
100	2168	1,107,606	32,152,890	3000	346,537,094,898
Sum$x$		Sum$y$	Sumxy	Sumx^2	Sumy^2
			Elevation 274.00 m		
10	259	62,001	620,010	100	3,844,124,001
20	329	95,481	1,909,620	400	9,116,621,361
30	393	131,769	3,953,070	900	17,363,069,361
40	457	173,889	6,955,560	1600	30,237,384,321
100	1438	463,140	13,438,260	3000	60,561,199,044
Sum$x$		Sum$y$	Sumxy	Sumx^2	Sumy^2

Table 8. Rock mass properties for granite.

Rock Type	Elevation (m)	Uniaxial Compressive Strength (σ_{ci})	Constant (m_i)	Coefficient of determination (r^2)	Constant (m_b)
Pink granite	249.50	208.59	18.02	0.9	3.64
Pink granite	261.50	198.17	26.64	0.9	5.38
Pink granite	266.00	204.00	24.22	0.9	4.89
Pink granite	273.50	231.87	38.49	0.9	7.70
Pink granite	274.00	150.99	24.63	0.9	4.97

$$s = e\left\{\frac{(\mathrm{GSI}-100)}{9}\right\} \tag{10}$$

and

$$a = 0.5 \tag{11}$$

The rock mass strength can be characterized by a GSI value of 55 (fair category), which was used to establish the parameters (m_b, s, a etc.) required for the Hoek-Brown failure criterion. The constants " s " and " a " calculated are 0.0067 and 0.5 respectively. For average/fair category rock masses Hoek and Brown [13] assumed that post failure deformation occurs at a constant stress level, defined by the compressive strength of the broken rock mass. The reduction of the rock mass strength from the in situ to the broken state corresponds to the strain softening behaviour. Martin and Maybee [15] assumed that the failed rock behaves as a cohesionless frictional material. These values can be used for modelling because in the rock masses there are a sufficient number of closely spaced discontinuities with almost similar surface characteristics.

5. ESTIMATION OF SUPPORT PRESSURE AND GROUND SQUEEZING CONDITION

The rock mass quality (Q) is related with the ultimate support pressure requirement. An empirical equation relating rock mass quality Q and permanent support pressure was given by Barton et al. [5] which based on case records (Equation (12)). In this equation importance is given to joint roughness number. Better qualities of rock mass have their improved Q values from the dilatent property of interlocked non-planar rock joints, while the poorer qualities are dominated by more or less non-dilatent clay filled joints [5] . An improved empirical fit (Equation (13)) by incorporating number of joint sets (J_n) in Equation (12) is further suggested by Barton et al. [5] . When rock mass is intersected by three joint sets ($J_n = 9$) Equation (12) and Equation (13) will give an identical estimate of roof support pressure. When there are less than three joint sets Equation (13) will give a lower estimate of support pressure than Equation (12), and a higher estimate when there are more than three joint sets. When the number of joint sets falls below three, the degree of freedom for block movement is greatly reduced since three joint sets or two plus random is the limiting case for three-dimensional rock blocks. In those equations size of opening does not figure in the support pressure prediction. Singh et al. [16] also studied the effect of tunnel size, span ranging from 2 to 22 m on support pressure and inferred that they are independent.

In this study roof support and wall support pressure was estimated as per Equations ((14) and (15)), which is applicable for the non-squeezing ground condition [16] [17] . Grimstad and Barton [6] also agreed on the overburden correction factor from Equation (13).

$$P_{\text{roof}} = \frac{2.0}{J_r} Q^{-1/3} \tag{12}$$

$$P_{\text{roof}} = \frac{2 J_n^{1/2} (Q)^{-1/3}}{3 J_r} \tag{13}$$

$$P_{\text{roof}} = \frac{2.0}{J_r} Q^{-1/3} x f \tag{14}$$

$$P_{\text{Wall}} = \frac{2.0}{J_r} Q_w^{-1/3} x f \tag{15}$$

where P_{roof} is permanent/ultimate roof support pressure in kg/cm², Where P_{wall} is ultimate wall support pressure in kg/cm², J_r is joint roughness number, Q is rock mass quality, Q_w is wall quality/factor equal to $5Q$ for better qualities rock mass ($Q > 10$) and $2.5Q$ for intermediate qualities ($0.1 < Q < 10$), J_n is joint set number and f is correction factor for overburden. Correction factor for overburden can be estimated from Equation (16).

$$\begin{aligned} f &= 1 + \frac{(H - 320)}{800} \geq 1 \\ &= 1 + \frac{(70 - 320)}{800} = 0.69 \end{aligned} \tag{16}$$

where H is the height of overburden above crown in metres

Singh et al. [16] suggested an empirical approach (Equation (17)) based on case histories and by collecting Barton et al. [5] "Q" data and overburden (H) for the estimation of non-squeezing ground condition. Minimum Q-value is used for the estimation of ground squeezing condition. Above additional surge pool cavern maximum cover is 70 m hence ground condition is non-squeezing. The required support pressure for crown is be varying from 7.89 t/m² to 12.43 t/m² and for wall 4.61 t/m² to 9.16 t/m² (Table 9).

$$\begin{aligned} &H < 350 Q^{1/3} \\ &70 < 350 \times 4.17^{1/3} = 563 \end{aligned} \tag{17}$$

6. DESIGN OF SUPPORTS

As per hydraulic design, the additional surge pool is having an excavated width of 20.20 m and length 200 m. The bottom level of surge pool is kept at EL 181.50 m and crown level is kept at EL 250.25 m. The maximum upsurge level of surge

Table 9. Support pressure for the roof and walls.

Sr. No.	Chainage (m)	Q-value for roof	Q-value for wall	Joint roughness number for crown & wall	Joint alteration number for crown & wall	Ultimate roof support pressure (kg/cm^2)	Ultimate wall support pressure (kg/cm^2)	J_r/J_a	Friction Angle $\varphi_j = \tan^{-1}(J_r/J_a)$
1	0 - 5	10.22	51.10	1.0	1.0	0.922	0.539	1.0	45
2	5 - 10, 200 - 205	16.33	81.65	1.0	1.0	0.789	0.461	1.0	45
3	10 - 20	12.25	61.25	1.0	2.0	0.868	0.508	0.5	27
4	20 - 30, 65 - 70, 75 - 80, 130 - 135, 160 - 165, 210 - 220	7.92	19.80	1.0	2.0	1.004	0.740	0.5	27
5	30 - 35, 50 - 55	15.33	76.65	1.0	1.0	0.805	0.471	1.0	45
6	35 - 40	14.67	73.35	1.0	1.0	0.817	0.478	1.0	45
7	40 - 45, 195 - 200, 205 - 210	15.83	79.15	1.0	1.0	0.797	0.466	1.0	45
8	45 - 50	11.88	59.40	1.0	2.0	0.877	0.513	0.5	27
9	55 - 60	10.89	54.45	1.0	1.0	0.903	0.528	1.0	45
10	60 - 65	12.38	61.90	1.0	2.0	0.865	0.506	0.5	27
11	70 - 75, 80 - 85	7.67	19.17	1.0	2.0	1.014	0.748	0.5	27
12	85 - 90	4.89	12.22	1.0	2.0	1.178	0.869	0.5	27
13	90 - 95, 155 - 160 165 - 175, 185 - 195	7.33	18.32	1.0	2.0	1.030	0.759	0.5	27
14	95 - 100	4.72	11.80	1.0	2.0	1.192	0.879	0.5	27
15	100 - 105, 125 - 130, 175 - 180	6.83	17.07	1.0	2.0	1.054	0.777	0.5	27
16	105 - 110	12.50	62.50	1.0	1.0	0.862	0.504	1.0	45
17	110 - 115	13.67	68.35	1.0	1.0	0.837	0.489	1.0	45
18	115 - 120	14.17	70.85	1.0	1.0	0.827	0.484	1.0	45
19	120 - 125	10.56	52.80	1.0	1.0	0.912	0.533	1.0	45
20	135 - 155	4.56	11.40	1.0	2.0	1.206	0.889	0.5	27
21	180 - 185	4.17	10.42	1.0	2.0	1.243	0.916	0.5	27

pool works out to EL 239.90 m and minimum downsurge level works out to EL 214.80 m. As per design 300 mm thick concrete lined is proposed at the invert level of surge pool. For structural stability of surge pool segment above concrete lined portion, rock support arrangements were recommended based on rock mass quality Q and site geological condition. The objective of reinforcement system was to minimize deformations induced by the dead weight of loosened rock mass, as well as those induced by stress redistribution in the rock surrounding an excavation [18] .

The rock mass quality Q was developed after making a consistent relationship between Q, the excavation dimension, and the support actually used. The permanent support estimate is based on the rock mass quality Q, the support pressure, and the equivalent dimension and purpose of the excavation. The Equivalent Dimension (De) is applied by dividing the span or height (m) by the Excavation Support Ratio (ESR). The ESR for surge pool cavity as given in the ESR updated classification standard of NMT Q-system is applied to 1.0 [19] .

Bolt lengths depend on the dimensions of excavations and the length of rock bolts can be estimated from the excavation span (B) or height (H) and the excavation support ratio (ESR) [5] [20] . Lengths used in the roof arch are usually related to the span (Equation (18)), while lengths used in the walls are usually related to the height of excavations (Equation (19)).

$$L_{\text{roof}} = 2 + \frac{0.15B}{\text{ESR}} \tag{18}$$

$$L_{\text{walls}} = 2 + \frac{0.15H}{\text{ESR}} \tag{19}$$

where, $L_{\text{roof/walls}}$ are bolt length in metres for roof and walls, B is span in metres, H is excavation height in metres and ESR is the excavation support ratio.

By applying the above formula, the length of rock bolt for the crown and walls is calculated to be 5.03 m and 10.78 m respectively. The value of NMT Q-system chart proposed is 5.0 - 6.0 m and 11.50 - 13.0 m for crown and surge pit walls respectively.

The Norwegian Institute for Rock Blasting Technique has proposed a formula to estimate the length of the bolts in the central section of the opening [18] . By applying this, the length of rock bolt for crown of pump house is calculated to be 5.12 m (Equation (20)).

$$L = 1.40 + 0.184B \tag{20}$$

where B is the span of the opening in metres

The thickness of steel fibre reinforced shotcrete can be estimated as per equation (Equation (21)) from the ultimate support pressure (P_{roof}) and size of opening (B) [21] [22] [23] . The thickness of SFRS for crown and surge pool walls is calculated from the average Q-value to be 104 mm and

222 mm respectively. The value of NMT Q-system chart proposed is 80 - 100 mm and 120 - 140 mm for crown and surge pit walls respectively.

$$t_{fsc} = \frac{P_{\text{roof}} \times B \times F_{fsc}}{2q_{fsc}} \tag{21}$$

where, t_{fsc} is thickness of SFRS lining, P_{roof} is ultimate roof/wall support pressure, B is size of opening, F_{fsc} is mobilization factor for shotcrete (0.6 ± 0.05) and q_{fsc} is shear strength of fibre reinforced shotcrete (550 t/m^2)

The rock support arrangement includes steel fibre reinforced shotcrete, rock bolt, grouting and drainage holes provisions (Figure 2, Table 10). On the basis of geological mapping of the heading portion additional rock bolts of 6 m length is recommended at the centre of each grid between Ch 125 m and Ch 180 m (3 m on either side of centre line) and at Ch. 193 m (3 m on either side of centre line).

7. ESTIMATION OF SUPPORT SYSTEM CAPACITY

The capacity of support system consisting of SFRS, rock bolt and grouted arch/ rock column for surge pool cavern is determine using the integrated approach given by Singh et al. [21] , Singh and Goel [22] and IS: 15026 [23] .

The total support pressure ($u + p_{\text{roof/wall}}$) will be equal to the sum of capacities of support system (Equation (22)).

$$u + p_{\text{roof/wall}} = p_{sc} + p_{\text{bolt}} + p_{gt} \tag{22}$$

where,

u = seepage water pressure = 0.0 t/m^2.

p_{roof} = roof support pressure (varying from 7.89 to 12.43 t/m^2).

p_{wall} = wall support pressure (varying from 4.61 to 9.16 t/m^2).

p_{sc} = capacity of SFRS (t/m^2).

p_{bolt} = capacity of rock bolts (t/m^2).

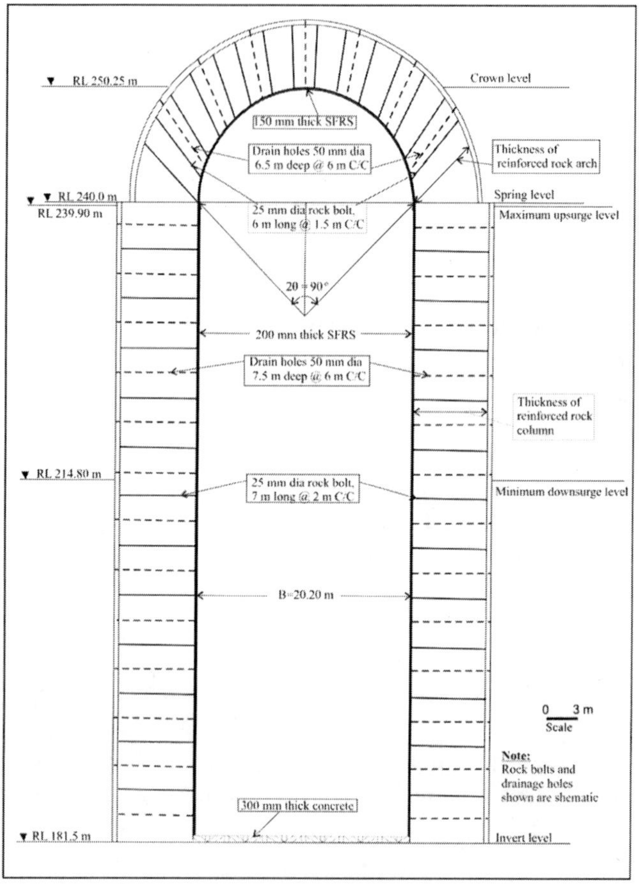

Figure 2. Support system of the surge pool cavern.

Table 10. Details of rock support arrangement.

Surge pool EL (m)	Required Support			Grouting	Drainage Arrangement
	Rock Bolts	Rock bolt spacing	Shotcrete		
Crown	6 m long, 25 mm diameter resin end anchored cement grouted rock bolts (Fe415)	1500 mm c/c (staggered)	150 mm thick steel fibre reinforced shotcrete	Up to 6.5 m and spacing should be decided on the trial basis	6.5 m long 50 mm diameter drain hole @ 6000 mm c/c
Side walls	7 m long, 25 mm diameter resin end anchored cement grouted rock bolts (Fe415)	2000 mm c/c (staggered)	200 mm thick steel fibre reinforced shotcrete	Up to 7.5 m and spacing should be decided on the trial basis	7.5 m long 50 mm diameter drain hole @ 6000 mm c/c up to maximum surge level

Note: Where additional support capacity is required to support local areas of weaker rock, bolts placed at the centre of each grid square will suffice.

p_{gt} = capacity of grouted arch/rock column (t/m^2).

It is assumed that the fibre reinforced shotcrete is intimately in contact with the rock mass and having the tendency to fail by shearing. Before putting shotcrete, the exposed surface should be properly cleaned and scaled because the strong bond between shotcrete and rock mass is the key to success in stabilizing a cavern The capacity of SFRS as estimated (Equation (23)) for roof and walls is 13.61 t/m^2 and 6.27 t/m^2 respectively.

$$p_{sc} = \frac{2q_{fsc} \times t_{fsc}}{BF_{fsc}} \tag{23}$$

where,

p_{sc} = capacity of SFRS lining (t/m^2).

q_{fsc} = shear strength of SFRS (550 t/m^2).

t_{fsc} = thickness of SFRS (0.150 m for roof; 0.200 m for walls).

B = size of opening (20.20 m for roof; 58.50 m for pump pit wall).

F_{fsc} = mobilization factor for shotcrete (0.6 ± 0.05 for higher for cavern).

The capacity of rock bolt is estimated (Equation (24)) and the minimum capacity for roof and surge pit walls calculated is 1.577 t/m^2, and 0.349 t/m^2 respectively.

$$p_{bolt} = \frac{2q_{crm} \times l' \sin\theta}{BF_s} \tag{24}$$

where,

p_{bolt} = capacity of rock bolt (t/m^2)

q_{crm} = UCS of reinforced rock mass (18.38 and 41.09 t/m^2 for roof and 10.34 and 23.11 t/m^2 for walls) (Equation 25)

l' = thickness of reinforced rock arch/rock column (5.125 m for roof and 4.00 m for walls) (Equations ((26) and (27)))

$$\theta = \theta^{\circ};\ \mathrm{Sin}\theta = 0.707$$

B = size of opening (20.20 m-roof; 58.50 m-pump pit wall)

F_s = mobilization factor for rock bolts

Singh et al. [21] proposed mobilization factors after back analysis of Barton et al. [5] support systems case studies. From 120 case histories, Thakur [24] confirmed these design criteria. For rock bolt mobilization factors (F_s) are calculated from Equations 28 and 29 for roof and walls respectively. For roof F_s values are varying from 3.996 to 4.181 while for walls values are ranging between 3.787 and 4.056.

$$q_{crm} = \left[\frac{P_{\text{bolt}}}{S_{\text{bolt}}^2} - u\right] \times \left[\frac{(1+\sin\varphi_j)}{1-\sin\varphi_j}\right]$$

$$\tan\varphi_j = \frac{J_r}{J_a} \tag{25}$$

$$l'_{\text{arch}} = l - \frac{\text{FAL}}{2} - \frac{S_{\text{bolt}}}{4} + S_{\text{rock}} \tag{26}$$

$$l'_{\text{column}} = l - \frac{\text{FAL}}{2} - \frac{S_{\text{bolt}}}{4} + S_{\text{rock}} - d \tag{27}$$

$$F_s = 3.25 \times p_{\text{roof}}^{0.1} \tag{28}$$

$$F_s = 3.25 \times p_{\text{Wall}}^{0.1} \tag{29}$$

where,

l = length of bolt (6 m for roof and 7 m for walls).

FAL = fixed anchor length (2.5 m).

S_{bolt} = spacing of bolt (1.5 m for roof and 2 m for walls).

S_{rock} = average spacing of joints (0.750 m).

d = depth of damaged rock due to blasting in walls (av. 2.0 m).

u = seepage pressure in the rock mass (0.00 t/m²).

J_r = joint roughness number.

J_a = joint alteration number.

p_{roof} = roof support pressure (varying from 7.89 to 12.43 t/m²).

p_{wall} = wall support pressure (varying from 4.61 to 9.16 t/m²).

The capacity of grouted rock arch/rock column is calculated by the Equation 30. The minimum grouted arch/rock column capacity for roof and surge pit walls calculated is 2.650 t/m² and 0.492 t/m² respectively.

$$p_{gt} = \frac{2q_{gt} \times l_{gt}}{BF_{gt}} \tag{30}$$

where,

p_{gt} = capacity of grouted arch/rock column (t/m²).

q_{gt} = UCS of grouted rock mass (18.38 and 41.09 t/m² for roof and 10.34 and 23.11 t/m² for walls).

l_{gt} = thickness of grouted arch/rock column (6.5 m for roof and 7.5 m for walls).

B = size of opening (20.20 m-roof; 58.50 m-pump pit wall).

F_s = mobilization factor for grouted arch/rock column.

For grouted arch/rock column mobilization factors (F_{gt}) are calculated from Equations 31 and 32 for roof and walls respectively. For roof F_{gt} values are varying from 3.932 to 4.610 while for walls values are ranging between 4.376 and 5.564. Total capacity of support system for roof and walls calculated at different Chainage is given in Table 11.

Table 11. Capacity of support system for the roof and walls.

Sr. No./ Chainage (m)	Ultimate roof support pressure (t/m²)	Ultimate wall support pressure (t/m²)	Capacity of SFRS(t/m²)		Capacity of rock bolt (t/m²)		Capacity of grouting		Total support capacity of support system (t/m²)	
			For roof	For walls	For roof	For Walls	For roof	For Walls	For roof	For walls
1	9.22	5.39	13.61	6.27	3.632	0.822	6.057	1.125	23.299	8.217
2	7.89	4.61	13.61	6.27	3.689	0.834	5.736	1.065	23.035	8.169
3	8.68	5.08	13.61	6.27	1.634	0.370	2.653	0.493	17.897	7.133
4	10.04	7.40	13.61	6.27	1.611	0.356	2.792	0.562	18.013	7.188
5	8.05	4.71	13.61	6.27	3.681	0.833	5.776	1.073	23.067	8.176
6	8.17	4.78	13.61	6.27	3.677	0.832	5.807	1.078	23.094	8.18
7	7.97	4.66	13.61	6.27	3.686	0.834	5.756	1.069	23.052	8.173
8	8.77	5.13	13.61	6.27	1.633	0.369	2.662	0.495	17.905	7.134
9	9.03	5.28	13.61	6.27	3.640	0.823	6.013	1.117	23.263	8.21
10	8.65	5.06	13.61	6.27	1.635	0.370	2.650	0.492	17.895	7.132
11	10.14	7.48	13.61	6.27	1.609	0.356	2.801	0.564	18.020	7.19
12	11.78	8.69	13.61	6.27	1.585	0.350	2.952	0.595	18.147	7.215
13	10.30	7.59	13.61	6.27	1.607	0.355	2.817	0.567	18.034	7.192
14	11.92	8.79	13.61	6.27	1.583	0.350	2.964	0.597	18.157	7.217
15	10.54	7.77	13.61	6.27	1.603	0.354	2.839	0.572	18.052	7.196
16	8.62	5.04	13.61	6.27	3.657	0.827	5.916	1.099	23.183	8.196
17	8.37	4.89	13.61	6.27	3.668	0.830	5.855	1.087	23.133	8.187
18	8.27	4.84	13.61	6.27	3.672	0.830	5.831	1.084	23.113	8.184
19	9.12	5.33	13.61	6.27	3.636	0.822	6.035	1.121	23.281	8.213
20	12.06	8.89	13.61	6.27	1.582	0.350	2.976	0.599	18.168	7.219
21	12.43	9.16	13.61	6.27	1.577	0.349	3.008	0.606	18.195	7.225

$$F_{gt} = 9.50 \times p_{\text{roof}}^{-0.35} \quad (31)$$

$$F_{gt} = 9.50 \times p_{\text{wall}}^{-0.35} \quad (32)$$

8. CONCLUSION

3D geologic mapping of heading portion using pilot and side slashing is very important for large cavern for predicting geologic conditions in benching down up to invert level. Geologic logging data were used for rock mass characterization and for support pressure estimation. Logging data were also used in planning tunnel support system and selecting best location and inclination of supplemental rock bolt. Support design empirical approaches are used. Empirical approaches are the best way for support design which is backed by a systematic approach to rock mass classification and providing a quantitative assessment of rock mass conditions. For structural stability, the rock support arrangement includes steel fibre reinforced shotcrete (SFRS), rock bolt, grouting and drainage hole provisions. Geologic logging data will also be very useful for choosing strategic locations for various types of instrumentation to study tunnel behavior. This cavern will be one of the biggest caverns in the world, so it is recommended that the support requirements may be re-evaluated in the light of the rock mass conditions revealed during the benching down of the cavern and the instrumentation data.

ACKNOWLEDGEMENTS

This paper is a part of sponsored project by M/s MEIL, so we sincerely thank the management of MEIL for the same. Authors are thankful to Director NIRM for the permission to send the manuscript for publication, encouragement and technical guidance.

CITE THIS PAPER

Naithani, A.K., Singh, L.G. and Jain, P. (2017) Rock Mass Characterization and Support Design for Underground Additional Surge Pool Cavern—A Case Study, India. Geomaterials, 7, 64-82. https://doi.org/10.4236/gm.2017.72006

REFERENCES

1. Bieniawski, Z.T. (1989) Engineering Rock Mass Classification. John Willey and Sons, New York, 251 p.
2. Ramam, P.K. and Murty, V.N. (2012) Geology of Andhra Pradesh. Geological Society of India, Bengaluru, 244 p.
3. Prakash, D. and Sharma, I.N. (2011) Metamorphic Evolution of Karimnagar granulite Terrane, Eastern Dharwar Craton, South India. Geological Magazine, 48, 112-132.
4. ISRM (1981) Suggested Methods for the Rock Characterization, Testing and Monitoring. ISRM Commission on Testing Methods, Pergamon Press, Oxford.
5. Barton, N., Lien, R. and Lunde, J. (1974) Engineering Classification of Rock Masses for the Design of Tunnel Support. Rock Mechanics, 6, 189-236.
6. Grimstad, E. and Barton, N. (1993) Updating of the Q-System for NMT. Proceedings of the International Symposium on Sprayed Concrete, Fagernes, 22-26 October 1993, 46-66.
7. Grimstad, E., Kankes, K., Bhasin, R., Magnussen, A. and Kaynia, A. (2002) Rock Mass Quality Q Used in Designing Reinforced Ribs of Sprayed Concrete and Energy

Absorption. Proceedings of International Symposium on Sprayed Concrete, Davos, 22-26 September 2002, 134-142.

8. Bieniawski, Z.T. (1973) Engineering Classification of Jointed Rock Masses. Transaction of the South African Institution of Civil Engineers, 15, 335-344.
9. Bieniawski, Z.T. (1979) The Geomechanics Classification in Rock Engineering Application. Proceedings 4th International Congress on Rock Mechanics, Montreux, 2-8 September 1979, Vol. 2, 41-48.
10. Gonzalez de Vallejo, L.I. (1983) A New Rock Classification System for Underground Assessment Using Surface Data. Proceedings International Symposium on Engineering Geology and Underground Construction, Lisbon, 12 September 1983, 85-94.
11. Nakao, K., Iihoshi, S. and Koyama, S. (1983) Statistical Reconsiderations on the Parameters for Geomechanics Classification. Proceeding 5th International Congress on Rock Mechanics, Melbourne, 10-15 April 1983, Vol. 1, B13-B16.
12. Naithani, A.K. (2007) RMR—A System for Characterizing Rock Mass Classification: A Case Study from Garhwal Himalaya, Uttarakhand. Journal Geological Society of India, 70, 627-640.
13. Hoek, E. and Brown, E.T. (1997) Practical Estimates of Rock Mass Strength. International Journal Rock Mechanics Mining Science, 34, 1165-1186.
14. Hoek, E. and Brown, E.T. (1980) Underground Excavations in Rock. London Institution of Mining and Metallurgy, London, 527 p.
15. Maybee, W.G. and Martin, C.D. (2000) The Strength of Hard-Rock Pillars. International Journal of Rock Mechanics and Mining, 37, 1239-1246.
16. Singh, B., Jethwa, J.L., Dube, A.K. and Singh, B. (1992) Correlation between Observed Support Pressure and Rock Mass Quality. Tunnelling and Underground Space Technology, 7, 59-74.
17. IS 13365 (1992) Quantitative Classification Systems of Rock Mass—Guidelines, Part-2, Rock Mass Quality for Prediction of Support Pressure in Underground Openings. Bureau of Indian Standards, New Delhi, 8 p.
18. Stillborg, B. (1994) Professional Users Handbook for Rock Bolting. Trans Tech Publication, Aedermannsdorf, 164 p.
19. NGI (2013) Using the Q-System—Rock Mass Classification and Support Design. NGI Publication, Oslo, 54 p.
20. Barton, N., Loset, F., Lien, R. and Lunde, J. (1980) Application of the Q-System in Design Decisions Concerning Dimensions and Appropriate Support for Underground Installations. International Conference on Sub-Surface Space, Rock Store, Stockholm, 1980, Vol. 2, 553-561.
21. Singh, B., Viladkar, M.N. and Samadhiya, N.K. (1995) A Semi-Empirical Method of the Design of Support Systems in Underground Openings. Tunnelling and Underground Space Technology, 3, 375-383.
22. Singh, B. and Goel, R.K. (2011) Engineering Rock Mass Classification. Elsevier Inc. Publication, Amsterdam, 364 p.
23. IS 15026 (2002) Tunnelling Methods in Rock Masses—Guidelines. Bureau of Indian Standards, New Delhi, 26 p.
24. Thakur, B. (1995) Semi-Empirical Method for Design of Supports in Underground Excavations. ME Thesis, IIT Roorkee, Roorkee.

Index